Sighting the Month

2020

Sighting the Month

Laoucet Ayari

First Edition

2020

Boulder, Colorado, USA

First Publication 2020

Produced in Boulder, Colorado, USA

ISBN 978-1-7331614-0-4

Preface

All cultures have relied on the phases of the moon to keep track of time, even though its cyclic rhythm is out of step with the daily and seasonal cycles. That is, a lunar month does not amount to a complete number of days, and several lunar months do not fit the cycle of seasons. This lack of synchronization made predicting recurring religious and social events plagued with discord and controversies for thousands of years. Also, all civilizations struggled with predicting the first day of the lunar month when the first crescent light becomes visible. The Babylonians, the Egyptians, and independently, the Mayans were meticulous in carrying lengthy observations of the heavens to understand and identify the moon cycle. Ultimately, they could predict eclipses, come up with the right criterion for new crescent moon visibility, and identify the 19-year Metonic period. However, predicting the day of the first new moonlight remained unsolved because the very first light appears at different locations from month to month.

Today, nearly all religions still follow the cycle of the lunar month in performing their sacred activities. Some religious branches have accepted the astronomical definition of the New Moon, which refers to the alignment of the moon with the sun as an alternative datum to the start of the month. They argue that modern societies need to plan ahead of time, and no one has ever predicted the first light from the moon. But many faithful have problems with such mathematical definition; they insist on the first light from the moon to be the start of the month per their holy references and not the dark moon of modern astronomy.

As the information age settled in, distances have shrunk with people around the world, being in direct contact, wanted to celebrate their holidays concurrently. However, festivals and religious observances remained in discord across human-made geopolitical boundaries. Events such as the Chinese New Year, the day of Easter, Ramadan, or the beginning of any lunar month, as defined by the first light from the moon, is not unique from one location to another. Published moon phases, while accurate, serve no purpose in the identification of the beginning of the traditional lunar month. For example, in 2019, Muslims around the world could not agree on when to celebrate the end of the fasting month of Ramadan. In another example, the Jewish People of Yahweh, for whom appointed times cannot be substituted by astronomical calendar dates, have given up on predicting first visibility. They argue that *"This has been tried over the centuries, but the moon cycle is complex for predicting a visible crescent moon. It will always be a 29 or 30-day month, but which one it is will occasionally only be confirmed by a sighting"*.

Throughout history, officials in charge of announcing the timeline for religious holidays and observances have been under pressure due to errors in determining the first day of the month. One can imagine the dissatisfaction of the population seeing a giant crescent in the sky lasting hours following sunset while being told today is the first day of the month. The first crescent is expected to be very thin and should last less than an hour once seen near the horizon after sunset. Another indicator would be the full moon, which is likely to occur around the 15th day of the month. The moon looking full on day 13 makes these officials certainly look bad. To diffuse tension and avoid blame, officials from multiple religious branches asked the population to look up the western horizon and report if they could sight the new light from the moon. As a result, holidays varied from country to country, usually over three days, and confusion settled in societies, especially in those where religion is essential in daily lives.

It should be noted that these controversies are nearly inexistent in the western and far-eastern countries, which ended reliance on the lunar and lunisolar calendars and adopted rule-based festivals. However, following the end of colonialism, officials in countries in the Middle East, Africa, and parts of Asia found themselves unprepared to make the correct decisions on when holidays ought to start. Although these countries have embraced western methods of running economies and societies, which rely on planning ahead of time, they maintained their traditional ways of life, which are often dictated by religious observances. In many cases, such observances cannot begin until a sighting of the new moon is confirmed, thus creating confusion and uncertainty, which affect preplanning the workdays, travel, and negatively impacting economies.

In general, politicians stay clear from controversies associated with the announcement of the new lunar month. They prefer to leave this matter under the jurisdiction of religious establishments. Religious organizations often face the pressure to reconcile between sacred texts, which are about eye-witness account of a sighting, and today's modern astronomical data. Because they have no stake in the matter, politicians consistently starved religious establishments from funding to have the subject studied in detail by competent academics. Often, the concerned religious establishments ended up consulting with amateur astronomers in an ad-hoc fashion but ultimately went back to religious texts to decide on how to proceed with a decision. As a result, errors kept occurring, as there are no sound criteria in place to determine when visibility of the new crescent moon is possible in their region.

The new moon has long been a matter of religion. It was the Pontiff who announced the number of days in a lunar month on the first day of a lunation, referred to as the *kalends* by the Romans -the origin of the word calendar. It was also the tradition of the Sanhedrin, the Rabbinical Supreme Court in Judaism, to declare the start of the month, known as *Chodesh*. For Muslims, it is the Mufti, the highest religious authority, who announces the beginning of the month, or *Baraa*, especially the start of the fasting month of Ramadan.

Most scientists tend to stay away from controversial religious matters, and the topic of the new moonlight is not an exception. The scientists who got involved with the subject were either jailed or ignored by the establishment. Ironically, on the one hand, religious people have never discussed the matter in cross-faith settings, even though it is the same moon, and all religions ordained the faithful to keep track of the new moon. On the other hand, scientists know well that there is an outstanding technical problem to be solved. The obvious question that no one from either side is eager to pose: Could a verifiable visibility law be an alternative to the social/religious requirement of sighting a new moon?

Because the first light from the moon is a global, single, and predictable event that happens at a precise universal time over a particular geographic location, celebrations and observances should be revised so that everyone around the world celebrates and observes at the correct time. Our purpose is to add a new lunar phase to the traditional four (*new moon, first quarter, full moon, and last quarter*) and have a precise time for *first moonlight*. And because it is one moon, there should be concordance between celebrations in different faiths.

The US-based non-profit and non-governmental Witness Organization, which promotes astronomy and space exploration for peaceful purposes, decided to tackle the question of first crescent moon visibility. The organization includes engineers and scientists from several countries, including the USA, Saudi Arabia, Canada, Tunisia, Algeria, Argentina, Libya, India, Pakistan, France, Belgium, Korea, Singapore, and Germany. The Organization sponsored three multidisciplinary meetings between 2008 and 2014 to discuss the topic of crescent first-visibility. Several meetings with decision-makers in several governments around the world have led to the conclusion that there is a need to address the problem from a technical perspective. In the year 2008, the organization produced Al-Shaahed, a unique system designed specifically to carry observation around the world and to document the new crescent at the earliest time following conjunction. The system is conceived to self-orient and locate the position of the moon in the sky, then point and track its descends towards the horizon while collecting as much light as possible during twilight conditions. The system documents the image of the new moon and transmits it to a control center for review. These functions require the assembly of a combination of technologies in a single astronomical device. Later, during the third multidisciplinary meeting on new crescent sighting, which occurred on June 16, 2014, at the Sheraton Hotel in Tunis, the photometric concept for crescent visibility was proposed by the organization. Later, the organization produced specifications for a competition to design a control center and an observatory on donated land. The design of the facilities emphasizes on public use and outreach.

This book collects results from the effort made to solve the question of first crescent moon visibility. We have used some mathematical developments in establishing the relevant parameters involved in new moonlight visibility. Even though some topics may seem advanced for some readers, explanations are provided to ensure access. Efforts are made so that the readers are not prevented from grasping the key concepts which are communicated in plain words, pictorially and mathematically. In the meantime, we apologize to our advanced readers who are well-versed in the topic; they may find some of the material not concise. The intention here is to communicate with the broader readership.

Contents

Abbreviations

AD	*Anni Domini Nostri Jesu Christi:* means *in the year of Our Lord Jesus Christ*. Non-Christians may not be entirely comfortable with this designation. Instead, the neutral abbreviation of "CE" and "BCE" for Common Era and Before Common Era, respectively, were proposed as an alternative.
AFMC	Air Force Materiel Command (USA)
AFOSR	Air Force Office of Scientific Research (USA)
AFRL	Air Force Research Laboratory
AH	"Anno Hegirae," year of flight, or separation
Alt-Az	Altitude-Azimuth coordinate system or telescope mount
BC	see AD
BCE	see AD
BIPM	Bureau International des Poids et Mesures
ca.	*Circa* (Latin): meaning around or about.
CCD	Charge-coupled Device: a silicon surface of light-sensitive elements. Photons impacting this surface produce electric charges which channel through an etched circuit on the surface.
CE	see AD
CIE	Commission Internationale de l'eclairage
CMOS	Complementary Metal–Oxide–Semiconductor: a type of integrated circuits used in microprocessors, including an image sensor
DORIS	Doppler Orbitography and Radiopositioning Integrated by Satellite
Eid	The day celebrated by Muslims after the month of Ramadan
ET	Ephemeris Time
GIS	Geographic Information System
GPS	Global Positioning System
GR	Gregorian calendar date
GST	Greenwich sidereal time
Hadith	Words of the Prophet Mohammed
Hijra	Separation, migration or flight (sometimes written as Hegirae or Hijri)
Hilal	Crescent moon
HMNAO	Her Majesty's Nautical Almanac Office
IAU	International Astronomical Union
ICE	Interactive Computer Ephemeris, U.S. Naval Observatory Almanac Office
IDL	Interactive Data Language -software for data representation and display
IERS	International Earth Rotation and Reference Systems Service
JPL	Jet Propulsion Laboratory, NASA

JW	Jewish calendar date
LLRO	Lunar Laser Ranging Observatories
LST	Local Sidereal Time
MPSAS	Magnitude per arc-second
MSISE	Mass spectrometer and incoherent scatter radar model of the atmosphere through the exosphere.
NASA	National Aeronautics and Space Administration (USA)
NOAA	National Oceanic and Atmospheric Administration
NSF	National Science Foundation
ONR	Office of Naval Research
Qur'an	Word of Allah in Islam
SAAO	South African Astronomical Observatory
SI	The International System of Units (from French, Système International d'unités) is the metric system.
SLR	Satellite Laser Ranging
SRS	Satellite Remote Sensing
TAI	International Atomic Time
TDB	Barycentric Dynamic Time
TDT	Terrestrial Dynamic Time
USNO	U.S. Naval Observatory
UT	Universal Time
UTC	Coordinated Universal Time
VLBI	Very Long Baseline Interferometry
WGS84	The World Geodetic System of 1984
WGS84	The World Geodetic System of 1984. Later revision in 2004

Chapter 1 Introduction

You talk when you cease being at peace with your own thoughts.

Khalil Gibran

Ancient cultures have extensive references to the moon in historical records, which include works of art, literature, social traditions, religious observances, and war rituals. As old cultures grew in sophistication, they relied more and more on the moon cycles for planning socio-economic and religious activities. Research in anthropology has revealed that planning is one of the critical attributes that differentiate humankind from the rest of the earth's co-inhabitants. Moreover, it is this very issue of planning that prompted the need for measuring elapsing time.

Today, in the drive to understand our evolution and make sense of our dynamic environment, we notice that the rhythm in the cyclic light from the moon has shaped cultures in many ways. To understand the influence of the moon on our evolutionary path, we imagine a world without it. Dark and dangerous nights would have either forced our primitive ancestors to follow a dominant defensive evolutionary posture or assume an offensive stand to survive the perils of the night. In either case, it is conceivable that we would have occupied other habitats, lived on some other diets, and developed more specialized sensing capabilities to deal with the darkness of the night. Even if we had taken the same evolutionary track as we did, we would not have been able to have a reference for short and midterm periods, such as those made available by the phases of the lunar cycle. The phases of the moon provided humanity with a natural calendar to keep track of time passage, plan, synchronize, interact, and travel further away from home. The invention of artificial light presents a parallel to the role of the moonlight in extending our capacity to interact with the environment. Our capacity to survive natural perils of the night during our early history is due to the availability of moonlight, and our evolution would have been altered or at least delayed in a moonless world.

Humanity has evolved over many thousands of years, yet the faithful moon remained constant watching over their progress until the arrival of the first civilizations. These civilizations cherished the moon and gave it an essential role in their interpretation of their observable world. Physical records show details of religious festivals going back at least 4000 years to the time of the Babylonians. We also know that all cultures have assigned rituals through the days of the lunar cycle. However, despite the astronomical advances made in identifying the lunar cycle, the prediction of the day of first-light has remained a mystery through the ages and could not be established with certainty ahead of time. This uncertainty has led the highest religious authority in each culture to oversee which day the cycle begins, thus prescribing when to initiate sacred rituals over the days of the lunar month. The situation remained the same until these days and age.

Very few scientists and astronomers ventured to tackle the subject of when the new moonlight is first-visible despite the pressing need for such information in several religions. Is this lack of interest due to the fear of violating the doctrine of separation of science and religion? We are referring here to the widely understood notion that there is a clear separation line drawn between scientific works and faith-based studies. This separation was put in place by scientists from multiple disciplines at the core definition of what constitutes scientific works. In opposition to other intellectual efforts, science must involve (a) carrying observations of physical manifestations, (b) performing experiments, and (c) deriving laws governing the physical and natural world. In the western world, many scientists fought infringements by some religious camps eager to teach creation as a science, which has resulted in passionate debates at times. This disconnect[1] between science and religion made it hard for many scholars to step-in and address topics that may bring doubts about their intentions. There may be a few exceptions where science and religion have interacted again during the past few decades, however. An example of such interaction and relevance to our subject happened when events from a few centuries ago did not fit current astronomical computations. Researchers in the field of Paleoastronomy have established that the rotation of the earth has been slowing down due to the tug from the moon and other sources[2]. By comparing eclipses and other events documented in religious texts, and other historical records with astronomical predictions, a historical adjustment in the rate of rotation of the earth was established.

The question of why the problem of predicting the new moonlight has been outstanding is a matter of opinion. However, it is fair to say that this is not a straightforward problem due to its multidisciplinary nature and the requirement for field measurements. Despite the availability of relevant data from the field of atmospheric optics and modern computational capabilities in astronomy, there are no developments worth mentioning outside a couple of

---

[1] Interest in science was initiated and promoted by the Christian church with its work during the Age of Translation (from Arabic texts) 1000-1200 CE

[2] Friction from earth's geophysical layers, the atmosphere and other weather-related effects such as El-Niño.

efforts. Obviously, such work requires financial support. However, most governments are either ill-advised or unable to realize the benefits of solving the problem.

1.1 Persistent lunar month: an age-old problem

Despite the significant role the moon played in the evolution of humankind, there are two problems with the use of the lunar calendar. The first one, which is the subject of the book, is the longstanding issue with the utilization of the first light as the start of the month. The second problem is related to its cycle. It is well-known that the lunar calendar is out of sync with the seasons and is therefore useless for livelihood activities such as farming, fishing, health management, education, food industries, tourism, and entertainment. We are referring here to the yearly eleven-day shift of the lunar cycle, which generates a translation in consecutive seasons in about a decade. This has led time-reckoning priests in antiquities to introduce an intercalary month every 2.7 years to bring the lunar calendar in-phase with the seasons. This well-documented historical topic is addressed in the following chapter when we discuss the history of calendars.

The lunar month begins with the first visual sighting of the new moon crescent, with no consideration given to the fact that first visibility may not be constrained to the local geographic location. It was not until the digital era that such determination has become common knowledge. However, throughout history, announcements were made to declare the start of the month. The Rabbinic institution lit fires atop of the mountains, while Muslims made announcements from the minarets of their mosques, and Christians rang church bells to welcome the new month. The start of a new month was considered universal and was therefore communicated over hundreds of miles. The new moon was the silver light first seen over the western horizon after sunset. Later, and with the emergence of science, astronomers took over the job of priests to keep track of the lunar cycle, producing almanacs which replaced direct sighting. These almanacs have the new moon defined to occur at the time of conjunction. That is when the sun and the moon have the same geocentric ecliptic longitude. The conjunction does not necessarily occur when the sun and the moon are closest when seen from earth or when the right ascensions are equal (see Section 8.3.5 for the definitions of ecliptic longitude and right ascensions coordinates). The moon and the sun can be up to several degrees apart when the month begins according to such definition. While the definition of conjunction has the advantage of precisely specifying a starting point in time for the lunar month, there is no light reflected off the moon that can be seen on earth at that time, and therefore conjunction has no relation whatsoever with the traditional crescent moon.

Besides, the definition of conjunction is mathematical and does not relate to a photometric manifestation of the moonlight, as referred to in old religious texts. There is no doubt that conjunction is an acceptable definition for the start of the month, but it cannot be the basis for a fundamental law for visibility due to its inherent reference frame dependency. For a physical law to be valid, it must be frame-invariant. Finally, the use of conjunction to initiate a cycle has led to a situation in which lunar months are different from one another. One can compute the angular separation between the moon and the sun at published times of conjunction to realize the lack of uniformity.

Today, despite our clear understanding of the astronomical elements of the moon, we have yet to find an accurate prediction for the start of its photometric cycle, which consists of the first visible light reflected off its surface. Such information remains unavailable despite the pressing need for it in cultures and religions. We emphasize here the need for the start (origin) of the photometric cycle as it has been exploited for a long time and according to instructions from religious texts and social customs.

No one contests the correctness of the astronomical lunar calendar offered by specialists, and everyone accepts the astronomic predictions of the moon as a fundamental in Planetary Science. The photometry of the moon is also well understood and is often utilized in the calibration of space instruments. However, as people are set to initiate their rituals and celebrations according to their almanacs and per official announcements, they look up the horizon sky but cannot find the new crescent. The new silver light from the moon is sometimes seen up to two days past the astronomical new moon time. When asked about the discrepancies, the entrusted specialists correctly state that the predictions are accurate, and the new light is first-visible only from other geographic locations or is not visible because the moon is very low in the horizon sky. People wanted direction on when to start the new month but received no satisfying answers. The start of the month referred to in the canons of their religions and traditions is not the first of the month as defined in Astronomical Almanacs; it is the first visible light from the moon instead.

There are currently two schools of thought in the matter of lunar month initiation: Some countries and religious branches rely on astronomy while others still rely on direct sighting in fulfillment of the literal word of their religious texts. In recent years, there is a tendency for those who rely on direct sighting to use what was coined as "guidance" from astronomical calculations. The calculations refer to specific empirical criteria for sighting that state that the first light from the moon for a month is visible when one or a combination of physical parameters, meets a certain threshold. The parameters used in these criteria include the duration between the sun and moon settings, the thickness of the crescent (or the surface illuminated in percent, the elongation of the moon (from the sun), and the elevation of the moon at sunset. While such criteria did not lift controversies surrounding the new month, they managed to technically anchor the topic, bringing it forward for discussion, and preventing errors.

1.2 Objectives and Motivations

There are many opinions on when the first light from the moon is visible, but only very few speculate on the physics of the matter. To remain impartial and provide a purely technical treatment, we have proposed two approaches. The first is instrument-based while the second is analytical. The first approach consists of a system named Alshaahed, meaning the Witness in Arabic, which documents sightings around the world and therefore brings an authentication for the beginning of the month. The second approach is to model the physics of the problem of the first light and establish a verifiable fundamental physical law that predicts when and where the new light from the moon is first perceived on the face of the earth. Both solutions have merit and could be exploited to support one another.

The ultimate objective is to produce the foundation for a new photometric calendar which would lift the controversy surrounding the beginning of the lunar month. To this end, efforts in the first approach have led to the development of a unique optomechanical system to track and document the earliest visible lunar crescent following conjunction from any location. The system, coined The Witness, has capabilities to

- estimate the location and time of first possible new moon visibility at any geographic location based on the different proposed criteria for first-visibility,
- locate the position of the moon in the sky, point to and track the new crescent as it descends towards the horizon,
- collect as much moonlight photons as possible in twilight conditions, and
- document the image of the new moon and transmit data to a remote location for review and validation.

These functions required a combination of technologies put together in a unique portable device. Components of the system were produced and tested in many locations with success.

In the second approach, we have developed an analytical solution to the problem of first visibility. The solution has two flavors: one is radiometric, and the other is photometric. Both solutions are verifiable through measurements and observation. The solutions evaluate the threshold contrast between the crescent moon and the transient twilight background sky brightness. The radiometric solution is first derived for a monochromatic light and subsequently integrated over the visible spectrum, weighted by the standard human eye efficiency in detecting wavelength over the visible spectrum. The photometric solution uses published data on the human threshold for light detectability and transient background sky measurement data from unique instrumentation built by The Witness team.

The objective of this book is not to change the conjunction-based lunar calendar, also known as the astronomical lunar calendar. Instead, it proposes an extension that states when and where the new moon is first-visible under a clear sky and perfect conditions. We want to clarify further that this book does not address merits, correctness or controversies, and is not about religious topics, historical facts, or calendars. We are merely revisiting a topic that has been historically handled by religious institutions. Also, we are proposing a path to a solution derived from pure science without infringing on interpretations, opinions, or discussing controversies. Finally, the results presented in Chapter 11 of this book are not of professional calendarists grade, even though we took great care in the compilation of the results. This work ought to lay the foundation for an official professional version of what we propose herein. Such a professional version would account for some details ignored herein for simplicity, such as the effect of atmospheric refraction in specific analyses and other atmospheric details of the line of sight, for which first visibility occurs. Derivations often use fundamental knowledge taken from several disciplines, including mathematics, astronomy, atmospheric optics, physics, and computer science. At times, the reader is required to have a minimum university-level knowledge of certain materials, in which case we provide pointers for pre-requisite information. Specific topics are also addressed independently so that the reader may temporarily skip sections and go all the way to the end of the book, where we present some significant results.

1.3 The Witness System

Named after the Witness Organization, The Witness System is primarily intended for the documentation and verification of new moon sighting. The system operates a network of mobile and fixed observatories designed for capturing the young crescent moon. Digital images taken from different sites are computer-processed in a standardized procedure and then transmitted to a coordinating control center for authentication. The control center analyzes the images together with their associated data and subsequently issues authentication for when and where the new moon is first observed.

For someone who does not rely on the new light from the moon in determining when a holiday might be, the Witness system might seem to be an over-the-top solution. However, given the disruptive effect of the uncertainty on economies and societies, one can quickly realize that such a system is worth the effort. For example, entire Muslim nations are put on hold waiting for a decision to be made on whether someone could have seen the new moon. The Hindu faith relies on the moon for most rituals. For example, it is ideal to fast on a new moon and offer food to ancestors, and newly married couples go outside to witness the new moonlight. One can easily verify that the uncertainty in the timing of the new month varies over three days from one point to another around the world. Timely initiation of prayers and sacred rituals is of paramount importance to the faithful of all religions. It is stressful for devoted people to realize that their efforts may have been compromised by errors, despite the availability of the necessary tools to solve the problem.

The Witness system is proposed to go beyond moon sighting, impacting scientific advancement in host countries, especially in regions where environmental monitoring is critical. Besides, the system has the potential to jump-start a new scientific culture in optical engineering, atmospheric science, computer engineering, and telecommunication, which is especially beneficial for places in need for such projects. The use of experimental physics and hands-on characteristics of the system fit the interest of young and future generations in astronomy, and telecommunication. Should many mobile stations be deployed and made accessible to students, the system will inevitably keep young people engaged with applied science and valuable discoveries. Engineers and high-tech business owners in technologically advanced parts of the world grew up involved with science, instrument usage, and technologies. They are part of a trans-generation technical culture whose origin is rooted in specific needs that manifested at some point in time. The Witness is designed to turn the old need for a uniform calendar into hands-on and practical scientific programs capable of impacting a substantial number of engineers and scientists across multiple disciplines. The modularity of the system provides a platform for incremental teaching and development for users who are interested in research and discovery.

Once the observation stations are mass-produced, the exchange of information and experience among users can become a standard practice due to the sheer number of related topics and details. We expect that the system generates on-line science discussion groups where users share their latest findings and improvements to the hardware. This interaction will inevitably culminate in regular international meetings that focus on specific topics and subsystems.

The baseline design of the Witness system is structured to have multiple options, and because of its modular nature, the choices of what options to retain can either be selected from the start or added to meet the user's experience and growth. The top-level requirements of The Witness system take into consideration several details which we cover next.

From the essential purpose to operation and function, one immediately recognizes the need to rethink the way we want to observe, document, and verify the light from a new moon. We shall address the fundamental characteristics of classical telescopes and give a vision for an observation system that responds to these alternative needs in Chapter 1. The devised mobile observation stations can run either in manual or auto-mode. When in auto-mode, a mobile station is capable of self-locating using a global positioning system, self-aligning to accurately point and track the moon, perform autonomous detection and perform auto-dialing to communicate with the control center when the window of opportunity for viewing comes to a close. The Manual Mode provides access to all capabilities of the observatory to adjust critical parameters that are not in the default settings. The Manual Mode can be useful for investigating specific capabilities that are part of the system but are not in the default setting, deal with potential adverse conditions, and troubleshooting. Stationary observatories are designed to be used as a testbed for investigating the effects of polarization, IR observation, detector evaluation, and interferometry. They are also used for training and public outreach.

When viewing gets disrupted in classical astronomy, work may be continued the following evening, and therefore, there are nightly opportunities for observation. In contrast, new moon detection has a window of about half an hour a month to achieve positive detection and verification, providing the conditions for visibility are favorable. If it happens to be overcast, one will have to wait until the next month, which is not an option for decision making. Therefore, the system components must work reliably under all circumstances. Transportation is always associated with a potential for trouble and mishap, which require sturdy hardware and proper protection for delicate hardware. Finally, the weather conditions are critical in selecting which observation sites are adequate for detection. Timely information on weather conditions not only is an integral part of deciding which site to choose but also in determining what additional equipment needs to be made available with observation equipment. Our team members have dealt with extreme heat and cold, very windy conditions and wild animals bear, baboon, aggressive cows posing a threat. Sites are usually not well characterized due to the lack of time to perform the necessary research; we end up trespassing on private property on multiple occasions.

There are significant differences between classical optical telescopes, which are traditionally conceived to resolve faint celestial objects against a dark sky, and a lunar observation system dedicated to new moonlight extraction during twilight. When it comes to viewing performance, two aspects are essential for classical telescopes: First, the ability of the telescope to collect light and focus it on a small area, thereby increasing the brightness of the image so that it can be seen by the naked eye. The amount of light the telescope can collect depends mostly on its size; the larger the telescope, the more light it can collect. The second aspect pertains to the ability of the telescope to focus the light image on a small area without distorting it. A distorted image appears blurry to the viewer. Therefore, the optical elements must be manufactured to accurate shapes using quality materials. It is this requirement that is responsible for most of the price difference between low and high-end telescopes. The lunar observation system is designed to filter out scattered sunlight from the layers of the atmosphere by taking advantage of the differences between its spectral content and polarity and those of moonlight. We shall review the historical development of optical instruments and discuss the lunar observation tools designed to aid in extracting and resolving the new moon crescent from scattered sunlight during twilight and even daylight conditions. We will also address the detection of moonlight outside the human vision in other regions of the light spectrum.

Similar to other nighttime classical observation systems, the optical assembly of the Witness system needs to keep its target still in the field of view. The motion of the moon differs from other celestial objects. Its trajectory during the time of new light detection is nearly a straight line and could be handled using only a one-axis drive mechanism. In order to track any object for an extended period, one would need two independent drive mechanisms working in harmony because the trajectory of the object is usually nonlinear. Tracking devices used in classical astronomy differ in type and performance. There are two basic types of commercially available drives, although there are variations of each: the first type is called Altazimuth, also referred to as Alt-Az. As its name indicates, the drive has one mechanism which covers the component of the motion along the azimuth axis, while the other covers the component of the motion in the altitude direction. The other type of drive is called Equatorial. It also has two mechanisms. One covers the component of the motion along the plane of the Equator, while the other covers the component of the motion about an axis parallel to the earth's axis of rotation.

The Witness system needs to rotate at the angular speed of the earth's rotation, and therefore needs only one drive system. In order to keep the moon in the field of view, the optical system is oriented due west in a plane, which normal makes an angle equal to the latitude of the observer. The axis of rotation is made parallel to the earth's axis. It is, however, recommended to maintain a two-drive system to automatically locate the moon, as the technology is inexpensive and readily available. The performance of a drive is measured by the drift accumulated over time. Imaging faint objects can take hours, but in our case, the crescent moon is available for observation over less than an hour. Fortunately, there is enough light and concern about limitation in drift comes not from how faint the crescent is but rather the collection of the few photons, especially when the moon is near the horizon.

Unlike other observation systems, the information resulting from new moonlight observation needs to be immediately documented and communicated over a secure network. Therefore, data and voice links are made integral to the observation system. Also, because observations take place in remote areas varying from month to month, satellite communication is the preferable method of communication.

To ensure quality science, documenting the new moon through imaging requires the use of sensors with a higher signal to noise ratio and quality like those used for advanced nighttime astronomy. Operating during twilight and setting up an observation station while the sun is still above the horizon makes the detector signal noise due to heat, a handicap. Therefore, controlling the temperature or keeping it to a minimum in the design of the observation system is addressed.

There are many differences between classical night-time telescopes and instruments for new moon detection during twilight. Even during setup before sunset, given the new moon position is close to the sun, pointing an optical system towards the sun can cause instant permanent loss of vision should someone looks straight in the telescope. Some detectors are not made for twilight conditions; their design is, in fact, for nighttime astronomy where the intent is to gather dim light overtime against a dark background sky.

Because new moon detection takes place once a month during evening twilight, the system can be used the rest of the time as a global instrument for education, research, and communication. Once deployed, the system can be expanded to support other academic, government, and commercial uses. For example, the control center may include any or all of the following: 1) an observatory for astronomy, 2) receivers for secure satellite-based communication, and 3) fixed ground terminals for high-speed network communication. The different options open the door to hands-on technical work not only for scientists and engineers but also for the public at large to contribute.

Ultimately, the Witness system is expected to synchronize naked-eye sighting, instrument-aided detection, and even satellite imaging to eliminate confusion caused by erroneous sighting claims. Details of The Witness system are the subject of Chapter 3.

1.4 Key additional contributions

In addition to The Witness system, there are several contributions in this work, which may be of interest to geophysicists, paleo-astronomers, aerospace engineers, calendarists, religious leaders, historians, and government planners.

The computer *software Nuria* is developed to synthesize direct and indirect solar illuminations of the earth during daytime and twilight phases. Lunar illumination is added to the simulation to deliver a global view of the interaction of the solar light and the new moonlight over time. This global view, which consists of contour lines depicting the solar and lunar terminators together with twilight isoclines, gives a unique pictorial insight as to how the relative photometry of the moon and the sun interact as time elapses during the time of the sighting. The IDL software, which we wrote as a teaching aid, explains why a new moon may not be visible over parts of the globe and why people should not waste their time attempting to observe the new moon under certain conditions. The graphics display can be supplemented using layers made of a stack-up of sky cloud fractions obtained from prior historical satellite data around the time of the simulation in prior years. When present, the cloud cover west of the observer can be detrimental to a sighting. Clouds can stack up over hundreds of kilometers in a westerly horizon where the new moon is setting. The software platform is also used to go back in time and reconstruct the illumination scenes which occurred during historical events. We shall reconstruct the time crucifixion, the time of death of the prophet Mohammed, and other events during his life when the Arabic calendar was in use. As we will see later in 0, the Hijri calendar came into effect several years after the Prophet Mohammed passed away, and while they use the same names for the months, they do not refer to the same months due to the practice of intercalation in the Arabic calendar. We shall discuss the details of the differences and revise the dates of the historical events during the time of the prophet in Chapter 1

In what may be the most crucial development of this book, we establish an interpretation of the new moon visibility based on the characteristics of the scattered light from the sun and the moonlight during twilight. The formulation expresses that sighting of the crescent moon occurs when the contrast of its brightness against that of the background sky reaches the human eye threshold for perception. We built instrumentation to track the sky in the direction of the setting moon and measure its varying brightness as the sun drops gradually below the horizon. Separately, we extrapolate the brightness of the crescent moon from published data. After sunset, the sky brightness gradually decreases, causing an increase in contrast. However, at the same time, the new crescent moon brightness also decreases gradually as the air mass traveled by its photons increases. Does the sky brightness decrease fast enough to overcome the loss of moon brightness due to air mass increase, producing enough contrast before the moon sets below the horizon? We shall show that when the height of the moon above the horizon at the time of sunset is less than a critical value, the contrast decreases monotonically, and therefore, it is impossible to sight a new moon. When the moon is above such threshold, the contrast starts increasing, as the atmospheric layers with the highest density become immersed in the earth's shadow, blocking sunlight scattering, which is the main obstacle for the earliest new moon sighting. We present this development in support of an alternative to the requirement of having someone physically sight the new moonlight. Could a verifiable crescent visibility law be a substitute for direct visibility, primarily when the horizon sky is overcast, or when the location of first visibility occurs at inaccessible locations, such as the middle of the ocean? The answer to this question is a definite yes.

In other contributions, we discuss specific photometric calendric data of interest to paleo astronomers, historians, and religious scholars. We shall evaluate critical historical events and records based on what we shall learn from the photometry of the moon in Chapter 1.

First, we shall evaluate specific Mayan lunar calendar dates as they would have occurred between the 4th and 8th century CE in Mesoamerica. The Mayan lunar day count, which happens to be in elapsed time from the old moon, rather than in current time from a new moon, is also shown to be prone to errors. The great Mayan astronomers tracked the moon and precisely identified the Metonic lunar cycle. We will show, however, that they were facing the same controversy that the old world has been dealing with over a few thousand years, even though they choose the old moon for their calendar for ease of sighting.

The second topic which received ample attention in the same Chapter was related to the Islamic calendar, which replaced the Arabic a few years after the death of the prophet Mohammed. As mentioned earlier, Muslims rely on the sighting of the new moon for the start of the holy month of Ramadan, the Hajj, and other practices. By relying on our findings, we established a centurial Ramadan Calendar, giving the start and the end of the photometric lunar month of Ramadan based on the time of the first sighting and the geographic location of where the new moonlight is first observed.
Christians around the world celebrate Good Friday in remembrance of the crucifixion event. Solar and lunar illuminations during the event happen to exhibit symmetric coincident terminators. This condition happens to occur only 160 times during the first three millennia of the Common Era. A partial lunar eclipse could be seen in Jerusalem right after sunset as what was expected to be a full moon was rising. The sighting of a lunar eclipse in Jerusalem during the event is discussed considering the earth's slowing rate of rotation due to tidal lunar pull and other factors. In addition, the timeline of the umbral eclipse is calculated and addressed in conjunction with the reported historical event. Finally, People of Jewish faith also rely on the moon for religious celebrations. The Jewish calendar is a rule-based lunisolar one, which means its lunar months are adjusted so that the calendar remains in-step with the seasons. The announcement of the new lunar month was not free of historical controversies and was not acceptable to many, such as the People of Yahweh, for whom appointed times cannot be substituted by astronomical calendar dates. Like many of the Muslim faith, they have also given up on predicting first visibility. The new crescent visibility based on contrast measures and the photometric calendar may be a substitute for direct sighting. To illustrate such a possibility, the dateline for the Jewish New Year, which is the Molad of the month of Tishrei, is computed based on the sighting law and tabulated for 33 years.
Finally, we address the lunisolar Arabic calendar, discuss its relation to the old Jewish calendar, and its transition to the Islamic purely lunar calendar, also known as the Hijri calendar, following the emergence of Islam in Arabia. The timeline before, during, and after the introduction of the Islamic calendar reveals some interesting historical details that may be of interest to religious scholars and historians.

1.5 New moon sighting parametrization

When a new moon is born, its coordinates are close to those of the sun. And given the observation of a new moon occurs during the early evening twilight in the direction of the western horizon sky, the altitude of a setting object diminishes in a near-straight line over a small angle. In addition, all setting objects, including the moon and the sun, have parallel trajectories. This leads to leads to major simplifications of pertinent astronomical parameters controlled by the relative moon-sun positions. In addition, equations from spherical geometry describing scattered sun rays and rays from the new moon during twilight conditions are simplified using approximation theory. Series expansions and linearization of the geometric description of scattered sunlight in the layers of the atmosphere lead to other important simplifications. For a setting sky near the horizon, additional approximations are introduced when the scattered angle of sunlight is found to be nearly constant during the observation time. The sunrays responsible for the background sky brightness as the earth shadow rises can be formulated in simple and manageable forms and compared to the brightness of light reflected off the new moon. The brightness of the new moon is derived from extrapolations of approximations of the moon brightness measured by multiple authors over the past few decades. The formulation of the contrast between the brightness of the background sky and the brightness of the new moon is then proposed as a basis for a novel photometric measure that controls new moonlight visibility. This measure is validated using field measurements using a calibrated sensor mounted on a tracking device set to move at a setting moon rate. The brightness measured by the sensor is related to the depression angle of the sun. The logarithmic formulation of the contrast relates both scattered sunlight and moonlight attenuations to incoming radiations outside the atmosphere through a standard attenuation procedure that uses atmospheric optical thickness. The atmospheric optical thickness along an arbitrary path is assumed to be proportional to air density. The MSISE empirical model of the atmosphere provided by the Community Coordinated Modeling Center at Goddard Space Flight Center was used for the lower altitudes while at higher elevations, the model was supplemented by data from the early Soviet program. The contrast is then expressed in a form having all unknown quantities as a bias, which can be dropped for comparative assessments. The main finding here is in the establishment of the limit angle of sun depression that allows for the contrast rate to first switch sign and subsequently reaches a critical value. This criterion is then recast to be interpreted in terms of astronomical parameters and provides the basis for a novel definition of the new moonlight quantification. It is this quantification that has led to the development of the photometric calendar. That is a calendar that emphasizes the light of the new moon and not the mathematical definitions of conjunction. The timetable for new moon for each lunar month is provided with additional parameters: the geographic coordinates (longitude and latitude of the birthplace of the new moon, including coordinates at time conjunction and the geographic location for visibility as time progresses.

1.6 Book organization

Because the matter of new moon sighting is of concern to people from all sorts of backgrounds, and because the topic itself is multidisciplinary in nature, there is a deliberate attempt to provide basic and introductory information about all subjects covered herein. It is our intention to make the different materials used in establishing the results accessible for the widest readership. In addition, the material of the book is sectioned so that one can skip through matters that are familiar without loss missing the important elements of the work.

The book is intended to provide a baseline for what has been accomplished so far by an organization of enthusiasts from different backgrounds who believed at one point in time that there is a genuine problem to solve. The members of the organization were enchanted with the "magic" of identifying the first dim light from the moon. The first moonlight manifestation is a recommended event for everyone to experience. It only requires the knowledge of the time and position of where to look on the western horizon.

Chapter 2 provides a review of the history of time measurement and calendric development over the ages. We shall highlight the role of the moon as a time-keeping celestial clock, the divine intervention in most religions and the controversies surrounding calendars and dating, and the parallel development of science in world cultures. From prehistory to these days and age, calendric development has been associated with flourishing cultures, and therefore, our topic brings a unique historical perspective to the understanding of how humans evolved to where we are today.

Chapter 3 presents details about the proposal of The Witness Organization to leverage technology and unify efforts to solve the problem of the new moon. We cover the details of documenting first visibility, the baseline instrumentation needed for the task, and possible extensions and options. We shall emphasize the economic and educational benefits of such a system, especially in remote areas, in need to jump-start their economies. Among the options or future development of the system, novel new moon sighting methods are proposed, including sightings from the International Space Station and even laser-pumping moonlight from the surface of the moon surface.

In chapter 4, we address the cyclic changes in the moonlight as perceived from the earth's surface. We provide a temporal and geographical representation of how the new moon illuminates the surface of the Earth in relation to the illumination from the sun during twilight and present a method to visualize the relative settings of the moon and the sun. There are opinions that call for tracking the new moonlight during the daytime. We discuss the merits and drawbacks of such a possibility and introduce the concept of the photometric lunar calendar, which has the new moon sighting as the origin of the lunar month.

In Chapter 5, the geometry of the light rays responsible for the evening twilight sky illuminance in the direction of the setting moon is derived function of a dipping sun below the horizon. The scattered sunlight and moonlight are characterized from the perspective of an observer placed at the surface of the earth. Subsequently, we derive a simple solution to the earth's shadow height and the optical path cover by a lunar photon in its journey through the layers of the atmosphere.

In Chapter 6, we discuss the moonlight brightness measurements from open literature. The new moon crescent geometry is derived and is shown to track with the phase angle for small sun-moon separation angles. The surface brightness of the moon is then characterized to account for the view angle of the moon. Except near a new moon and a full moon, the surface brightness seems to be nearly constant when accounting for the view factor. It must be mentioned that data and measurements at small separation angles are needed to characterize the brightness of the moon. Such a task is in development at the US Bureau of standards at the writing of this text.

In Chapter 7, we address the mechanisms for light absorption, attenuation, and scattering through the layers of the atmosphere. A geometric model describing the propagation of the new moonlight and the sunlight scatter, and its illumination of the background sky in the direction of a new moon is formulated. We subsequently derive a logarithmic measure of contrast relating both scattered sunlight and moonlight attenuation. By integrating over the visible light spectrum and making use of the standard human eye efficiency in perceiving different wavelengths, a bulk contrast measure of new moon brightness is derived. Measurements show that first-order scattering dominates the illumination of the sky during twilight when new moon sighting takes place. It is also shown that the contrast measure depends only on the scattering angle. We shall present a photometric methodology for new crescent moon sighting and a radiometric logarithmic measure. The latter is vetted through actual measurements of the sky brightness at the end of the chapter.

Chapter 8 presents topics from spherical astronomy that are relevant to our topic. The proximity of the moon to the sun in the hours following conjunction allows the use of approximation theory to constrain the problem of new moon visibility to the necessary astronomical parameters. The position of the moon is approximated and is given relative to the position of the sun.

In Chapter 9, we bring together the optical condition for new crescent moon visibility, which we derive in Chapter 7 and the astronomical description of setting celestial bodies from Chapter 8 to establish the time and location of optimal visibility and the region of the globe where first and subsequent visibility occurs. We shall provide a linear model, which is shown to be sufficient to explain first visibility. A quadratic model is also derived at the end of the chapter, which we close by showing that the condition for visibility of the crescent moon is in agreement with published data on crescent moon sighting.

Chapter 10 applies the results of Chapter 9 to observances in Judaism, Christianity, and Islam. The chapter also revisits important topics in history and paleoastronomy. We show, for example, that the Mayan were not able to predict the new moon either, despite their mastery of astronomy. The transition from the old Arabic calendar to the Islamic calendar is studied to retrieve the dates of some key events related to the emergence of Islam.

Chapter 11 provides tables of data reflecting the results of Chapter 10. The reader may find the Jewish New Year based on photometry for the years 2014-2045, The crescent first-visibility defining the start and end of Ramaḍan for a century and the dates of Jesus Crucifixion under the particular condition of orthogonal illumination which took place on April 1 of the year 33 CE.

Finally, Chapter 12 summarizes some of the key findings of this effort and the need to synchronize celebrations across time zones and geopolitical boundaries and reflects on what hampered the development of a photometric calendar by specialists.

It is s our hope to help decision-makers make informed decisions about the start of the month. I would like to remind leaders that the anxiety from the never-ending uncertainty on these dates is monumental for those who consider them sacred. Our intention is public education and outreach on a universal topic, one that is of interest to mankind, irrespective of background, religion, or ethnicity. This book attempts to explain when and where the new lunar cycle begins.

Chapter 2 The Calendar

Many good references address the topic of the calendar with accurate knowledge about its history and technical making. This is a large subject outside the scope of the unification of the dates of celebration and the detail of first moonlight visibility. We never intended to pay attention to the history of the calendar or to its technical making. The topic imposed itself as we ventured deeper in the details of new moonlight and how it was used over the centuries, and we could not resist the urge to summarize the topic from our subject's perspective. In this chapter, we present a limited account of the historical development of the calendar from a western narrative. The reader will notice that certain developments, such as those from early Asian civilizations, are not addressed. This does not mean that they lack importance; they simply either do not feed into the matter of first moon visibility, and the controversies associated with the calendar or there is a lack of access to documentation.

In the following sections of this chapter, we start by discussing the fundamental rhythmic nature of time-keeping, the use of cyclic rotations of the earth and the moon to keep track of time and define what a calendar is about. Very much like anything else that is shared, calendars have been surrounded with controversies. We will discuss these controversies in Section 2.4 and continue to highlight them throughout the presentation of the development of the calendar. For generations, all cultures related to the first light of the month, and to the number of days since the start of the lunar cycle. These cultures kept track of the days in a cycle using the calendar, considered to be the most important achievement of mankind by historians and even holy in major religions. We classify the calendar in this text as prehistoric, ancient, ensuing, western and secular. We shall go back in time and trace the evolution of the calendar. In Section 2.5, we address how the moon is perceived in a primitive context and discuss a number of prehistoric lunar calendars from markings found in archeological digs, weigh in some aspects of the calendric nature of the Stonehenge monument in England, discuss the role of the moon as a calendar in paganism and witchcraft, and address the role of the Horizon calendar as a natural mean for humans to notice the cyclic changes in the rise and set points of the sun and moon over the seasons.

In Section 2.6, we discuss ancient Egyptian, Babylonian, and Mayan Calendars. The invention of the calendar by the Egyptians and the astronomical observation made by the Babylonians resulted in the delivery of an accurate lunisolar calendar in-step with the seasons. The solar calendar was also a product of these first civilizations. The Babylonians measured time down to a unit of 4 minutes and mastered algebra, propelling humanity out of darkness into an era of science and civil society until foreign occupation brought their advances to a halt. The people of Mesoamerica have independently developed knowledge in science, culture, medicine, and agriculture. These descendants of Mongolians and other Asian groups who migrated to the Americas about 13,500 years were sky-watchers who developed advanced calendars using a completely different approach to time measurement.

Ensuing calendars, including the Hebrew, Arabic, East-African, Islamic, Chinese, South Asian, Persian and Hellenic calendars are all derived from the Babylonian heritage and are the subject of Section 2.7. Later, with the rise of Rome, the need for a calendar was immediate Western calendars are the subject of Section 2.8. We shall cover the Roman republic calendar, the Julian calendar, Computus (Latin for computation - a calculation that determines the date of Easter), the French Roman Emperor calendar and the Ecclesiastical calendars in a chronological way which parallels the development of science. Section 2.9 is the subject of the Eastern Orthodox calendar. The question of controversy is addressed again in Section 2.10, where the subject of calendar unification of the date of Easter was revisited. As science stepped out of the cradle of religion, Secular calendric efforts were proposed. We shall discuss the French Republican calendar, Kodak's International fixed calendar and the Henke-Henry permanent calendar in Section 2.11.

In Section 2.12, we discuss the astronomical lunar calendar and the effort of the Nautical Almanac Office at the US Naval Observatory to bring closure to the thorny mathematical problem of identifying the orbital characteristics of the moon. Unlike other celestial bodies, predicting the orbital behavior of the moon had been a long-standing difficult problem because of the effects of many perturbations. Accurate tables were finally produced using the Hill-Brown moon theory in 1919. The transition from the lunar (or Luni-solar) calendar to the solar calendar has taken place since and, in the process, traditional and functional needs of the lunar calendar were inadvertently dismissed. Renewed interest in first light visibility resurfaced in the 1980s by key members of the Nautical Almanac Office at the US Naval Observatory including the late Chief Le Roy Doggett. The office was alerted by Muslim astronomers who sought to produce a photometric lunar calendar to bring order to a state of chaos surrounding the observance of Ramadan and other celebrations. The office of Almanac's attention resulted in multiple important publications addressing semi-empirical criteria for first visibility involving parameters such as elongation, crescent thickness, etc. These contributions from members of the office, as well as others, did not address the fundamentals probably since no funding agencies were available as the need was not established. Nevertheless, the support of the Nautical Almanac Office was major from a scientific perspective. While aware of the general idea behind the contributions of the 1980s and beyond, we kept from studying them so that we possibly come up with alternative views or even independent confirmation of what has been done. And as mentioned earlier, even the historical account was produced at the end of the documentation of this work. It is up to the reader to decide if our approach is rewarded.

It is important to note that most calendars have evolved over time and their final versions are not necessarily identical to their original form. The lessons learned from studying the calendar is an important part of this effort, as we open a window into the past and explain historical matters of relevance to mankind. These conclusions are presented in Section 2.13 at the end of the Chapter.

2.1 Cyclic rhythm

For millions of years, life on earth adapted to the rhythm of the sun for daily livelihood matters in the form of food and other existential processes. Examples of such adaptation are abundant in nature. After World War II, while mapping the bottom of the oceans using sonar scattering, the US Navy discovered the largest migration of the living on earth: The Diel vertical migration. It happens that earth's biomass in oceans and lakes has a diurnal pattern of movement that follows solar daylight. Life has also adapted to all forms of seasonal variation. Some animals go into hibernation in winter to re-emerge when it is warmer. Even when the sun is not available at higher latitudes, life managed to pick the rhythm of moonlight. Diel vertical migration of zooplankton across the entire Arctic, including open seas, was found to be driven by lunar illumination during polar nights (Kim S. Last, 2016). This departure from the 24-hr solar day to a 24.8-hr lunar-day takes place in winter when the moon is above the horizon. In addition, a pattern of mass descent of zooplankton from surface waters to depths of about 50 m occurs regularly every 29.5 days in winter, coincident with the full moon period. Such a pattern is thought to occur to avoid moonlight-enabled predation. Many examples of reliance on the position of stars and other heavenly bodies for gathering food, migration, long-distance ground, and sea navigation can be observed in nature. In fact, anywhere a cyclic rhythm is made available, life seems to connect and establish a fundamental existential rapport. This connection is observed everywhere, including in the behavior of mankind.

In our planet, we distinguish three important cycles: The diurnal cycle, the lunar cycle, and the seasonal cycle. Despite seasonal variation in the length of the day, by as much as two to three folds in certain geographic locations, the cyclic nature of the phases of the day provides the main reference as time elapses. Dependency on the phases of the day, whether direct or indirect, is immediately observed in the behavior of the living and is due to the cyclic rotation of the planet around its axis. Since remote antiquities, daily activities of mankind relied on the illumination of the planet and its cyclic rhythm. Unlike modern 24-hour clock time, old cultures began the unit of the day with the phase of sunset and evening twilight, followed by the period of night rest, which ends at morning twilight. The same day continues with the subsequent morning following sunrise, midday -time at which the sun is highest in the sky- and ending with the afternoon period. The unit of the day did not start in the morning with sunrise or in the middle of the night as dictated by the 24-hour clock. It started with twilight because the day count was based on the light from the moon, which is first observed during twilight when a lunar cycle begins. In all cultures, daily social, economic and religious activities relate to the position of the sun in the sky, morning and evening twilight phases and to the position of the stars at night.

Early short-term activities involving several days may have relied on finger designations for a set of days, while near-term counting must have pointed to the phases of the moon, e.g., finger designation of a set of new and/or full moons. The dependency of the living on the phases of the moon has for origin a cyclic rotation as well: the rotation of the moon around the earth. The phases of the moon play an essential role in the food supply of many living entities due to tidal patterns. Certain water-born food is more accessible during certain moon phases, which results in many animals extending their diurnal activities into the night for hunting and gathering. In the case of humans, all cultures have depended on the moon phases for social and religious festivities for at least the past several thousand years. It is also well-known that humans, distinguished by their capacity to plan and/or their insistence to go to war, use the phases of the moon to initiate aggression around new moon nights to take advantage of lack of visibility. In fact, many historic wars were initiated around a new moon. As humans evolved, their reliance grew more and more sophisticated to include complex social and religious activities thanks to prolonged mild weather and stable governance.

The third cycle happens to be the manifestation of seasonal variation due to earth rotation around the sun. Life on earth also responded to seasonal variations in climatic conditions, exploiting the cyclic nature of the earth rotation the around the sun, e.g., animal migration, reproduction food gathering, storage, etc. Most creatures with off-springs that rely on a few weeks of nurturing tend to give birth in the spring, taking advantage of the seasonal warmth and giving a chance to an offspring to survive.

Biological cycles and rhythms run deep in all forms of life. They could be daily, monthly, annually, or periodic running over several years. They are at the root of all sorts of forms of life varying over time to generate a stunning huge diversity. Life possesses several essential attributes as discussed by Sasselov (Sasselov, 2012): (a) chemical, in essence, (b) system not in equilibrium, (c) adaptive and self-optimizing, (d) compartmentalized, and (d) uses molecules suited for water. The main mechanism behind the third attribute relies on rhythmic facilitation. That is, life adapts to the environment it occupies and seeks to optimize its exchange of resources through rhythmic adaptation. Anytime we encouter a natural cyclic rhythm, life tends to connect with its beat. Living entities tend to be in a perpetual search for equilibrium, finds a temporary shelter -unstable equilibrium- in the cyclic behavior it encounters. For now, we have only life on earth to observe. We hope to encounter more examples in our neighborhood within the solar system soon. Over the past decade, we have confirmed that our planetary solar system is not unique; this revelation comes from the cyclic nature of the motion of planets around their parent stars. In the hunt for extraterrestrial life, we need to think in terms of how life could exploit the cyclic behavior of planets and their moons, and detect signals as living systems transit between different states. Of relevance to these questions is the fact that everything from energy to light and matter is characterized by its frequency content (Eigen properties). The behavior of all manifestations is basically a combination of proportions of its modal content. In other words, cycles or clocks are everywhere and in everything; they constitute the fundamental basis for all processes in the world of matter. Even individual structures and entities possess natural modes associated with mechanical, kinetic, potential, and thermal energy exchange. Clocks exist at all dimensional levels: at the celestial, atomic and particle levels. These clocks are optimal in nature. They are the outcome of solving an optimization problem, i.e., any Eigen-value problem (mathematical framework for identifying them) can be written as an optimization statement, e.g. (Johnson, 1985).

In recent times, mankind went beyond celestial objects and started making own tools to keep track of time. We produced the hourglass, and then invented the mechanical clock running on gears to measure time without help from the heavens. Big Ben was "big deal," and the Swiss gained fame with their accurate development of precision gear systems. Soon after, the people of the new world identified atomic resonance and forged the Coordinated Universal Time using atomic clocks with errors so small that precision time seemed to be no longer an issue (today's precision is better than losing a beat every quintillion tick), until we found new applications requiring higher and higher precision from imaging the nanoscale to geo-positioning within a few millimeters from space. Timekeeping, which started with reliance on cyclic celestial phenomena evolved into mechanical cyclic systems and later went down to cyclic atomic resonance at cryogenic temperatures. Such feat came very fast and, in our journey, we have not finished one important matter pertaining to the use of our old companion for marking celebrations. We shall refer to the concept of the photometric lunar month as the missing achievement.

2.2 What is a calendar?

With the emergence of agriculture, mankind started organizing at a higher socio-economic level. And as humans grew more sophisticated in planning farther in the future, they realized that they had a need for a predictable and ordered system for time measurement. Such need led humanity to invent what historian hailed as the most important achievement: The Calendar.

The **Calendar** could be defined as ***an agreeable system for organizing and measuring time.*** This definition implies the requirement to adopt the system by active members of society. With the birth of strong civilized states in Eastern Africa's Egypt and separately in Mesopotamia, each civilization could see its calendar internally adopted to run its affairs in an organized manner. Priest-astronomers took charge of the development and maintenance of the Calendar. There was no record of high-level interaction between the powers of the time in the matter of calendar, despite the similarities in their constructs. There was no need then to establish an agreeable and unified calendar system across the two civilizations which were separated by a vast swath of land. As a matter of fact, the calendar was probably an area of competition and state secret, but the information must have filtered through between the two sides as the similarity is undeniable.

Perhaps the best illustration of the need for agreement across different societies showed up much later in pre-Islam Arabia between divided tribes in the resetting and unification of the Arabic calendar (see Section 2.7.3). Pilgrims to Mecca showed up at various times around the assigned pilgrimage dates, and even one month off whenever intercalation is in effect. This was bad for trade which happened in parallel with pagan worship. This discord has led to an agreement between the different tribes to reset and unify the Arabic Calendar. Later, when Islam settled in Arabia, Muslims had to start a unified calendar, for managing their religion and affairs. The old calendar of the city of Medina, which was long maintained by the local Jewish tribes, was discarded in favor of a revised version of the Arabic calendar that does not introduce intercalary months as dictated by the Qur'an. This subject will be discussed in detail in Sections 2.7.2- 2.7.4. The key point here pertains to the agreement that took place for a calendar to take hold.

On a global level, there is no better example to illustrate the agreement of calendric usage than that of the current Gregorian calendar. People of different faith including Jews, Muslims, Protestants, followers of the Orthodox Church and even from communist China resisted the adoption of Pope Gregory's new calendar because of religious concerns that the calendar, as corrected, was rooted in Catholicism even though everyone knew that its origin was the Julian calendar which in turn was derived from the Egyptian calendar. Today the entire globe uses the Gregorian calendar for several reasons. First, because of its correctness and scientific foundation, second because of the need to do business and communicate on a global scale. The gradual removal of religious nomenclature in the Calendar dating system (e.g. the old use of the year of the lord, A.D., and B.C. for "After Dominion" and "Before Christ" are now more and more replaced by C.E. which stands for "Common Era" and B.C.E for "Before Common Era ", respectively) helped with the dissemination during recent times. Even though the common adoption of the Gregorian calendar was imposed by a need to do business and communicate globally and not necessarily by a voluntary agreement, it happens that such adoption ultimately reflected the triumph of science as it introduced the right correction to its predecessor, the Julian calendar, finally fitting the motion of the earth with the heavens. This subject will be addressed in detail in Section 2.8.

The definition of the calendar refers to the effort of organizing and measuring time. All calendars use units for short term reference and other units for near term reference and measurement. For example, In the Arabic, Hebrew, Gregorian, and other calendars, the days are collected in sets of seven, and then four of these sets seem to fairly coincide with the moon cycle. And in turn, twelve sets of lunar months almost coincided with the solar year (within 11 days). The correction of the discrepancy has led to the lunisolar calendars. Because of our dependence on long term decisions and because of the seasonal variations of the lunar calendar, the lunar calendar and its subsequent (overhauled) lunisolar versions were dropped altogether for the more season-coherent solar calendar in administrative, business and legal matters, while the lunar calendar remained the essential reference for celebrations and observances.

2.3 The moon cycles

Romeo:
Lady, by yonder blessèd moon I vow,
That tips with silver all these fruit-tree tops—
JULIET:
O, swear not by the moon, th' inconstant moon,
That monthly changes in her circle orb,
Lest that thy love prove likewise variable.

Romeo and Juliette
by William Shakespeare

Little Juliette knew that the longtime companion of our earth is constant, faithful, and a source of nutrition for many the living. The inconstancy manifested by either the moon phases or the time of moonrise or moonset is from Juliet's perspective and has nothing to do with the behavior of the moon. The moon maintains a near-perfect cycle; its clock has been exploited by countless life forms, including humans.

There are five cycles associated with the motion of the moon from which five periods can be identified: The s*idereal cycle,* the *tropical cycle*, the *synodic cycle,* the *draconic cycle*, and the *anomalistic cycle*. The sidereal period is the time it takes the moon to return to the same position about background distant stars which are fixed in the celestial sphere. This period lasts about 27.32 days. Depending on the reference stars used, this period varies slightly. When the point of Airy (Equinoxes) is used, the cycle is called the tropical cycle. The Synodic period is the time it takes the moon to return to the same position relative to the sun when observed from the earth and lasts 29.53 days. The difference between the two periods is because earth is also moving in its journey around the sun. The synodic cycle is the one commonly used in lunar calendars, including the astronomical lunar calendar which will be discussed in Section 2.12. It is also the basis for the reflected light off the moon and the phases as perceived from the earth surface. The draconic period which lasts 27.21 days and consists in the moon returning to the same orbital node, one of the two points where the moon traverses earth's orbital plane around the sun. Finally, the anomalistic cycle corresponds to the time it takes the moon to come back to its closest point to earth, i.e., its perigee, and lasts 27.55 days. In the current work, we define a sixth cycle: the average photometric cycle which amounts to the time from a first visible crescent to the next one at the start of the subsequent month. The length of the photometric cycle is nearly the same as the length of the Synodic cycle (less than five minutes difference) with maximum absolute variations between cycles less than five hours.

2.4 Controversies with calendars

There are multiple controversies associated with calendars which could be rooted in one of several matters including their technical construct, imperfections, inaccuracies, or stemming from discord on proposed corrections to imperfections, or in certain cases man-made, driven by a thirst for power and control, economics, human ego, to name a few. In this section, we address a few of these controversies and attempt to understand something about our human nature.

As we start going through the history of the calendar in the following sections, prehistoric markings found in archeological digs amounting to what seems to be the days of the lunar month uncovered our first controversy. Evaluation of archeological finds led to the realization that it was difficult for early humans to witness the new moon, that is the first day of the month. In fact, this has led to the casting of doubt on the nature of the artifacts themselves. This topic ended up being a charged matter when such archeological finds were interpreted in a way to elevate a race or people with a certain background while disgarding what others might have achieved. Even though we have had examples of this in the past with the discovery of the Neanderthal, less than a half-century later, we fall in the same ego-trap of wanting to be more than what we are. Subsequently, as we discuss the important role of the moon in old pagan witchcraft which relies heavily on the moon phases, we find that witchcraft had no magical answer to our fundamental question of the start of the lunar month. In fact, spells planned for the new moon may be also controversial as they have no accurate account of when a new moon starts either.

From a technical perspective, fitting the calendar with the seasons has been a major ordeal for mankind. Historically, there have been two controversies with the invention of the calendar by ancient civilizations. The first controversy is associated with fitting the lunar calendar to the seasons, while the second pertains to identifying the start of the lunar month.

For the people of early civilization, it became immediately clear that there are issues with fitting predictions to the motion of the heavens. The lunar calendar is not in sync with the seasons due to the fact that the cycle of twelve-month lunar year is short of the twelve-month solar year by about eleven days. This difference accumulates over time to make a lunar month migrate from one season to a prior one in less than a decade. Muslims know about this firsthand. Because they maintain a purely lunar calendar for religious activities, they experience the fasting month of Ramadan drifting through a complete solar year cycle in a span of about 33 years. Almost all ancient civilization: Egyptians, Babylonians, Chinese, Mayans and later Jews, and Arab pagans (before Islam) had to address this drift for economic, social, religious and other weather-related reasons. They brought up the lunar year in sync with the solar one by introducing a system of intercalation to transform the lunar calendar into a lunisolar one. Ancient communities without a strong state to maintain a calendar and keep track of time were challenged by when the month starts and when intercalations need to occur to bring the calendar in-step with the seasons so that migration begins or seeding takes place. For example, in pagan Arabia, multiple calendars were used depending on the tribe. This has created a lack of synchronization between the tribes to travel for trade, rituals such as the Hajj to Kaaba prior to Islam. Muslims adopted a purely lunar calendar as instructed by the Qur'an, affirming the sanctity of four of the twelve months of pagan Arabia, and placing such designation to have occurred as early as the day of creation of the heavens and earth. The sacred months are designated for worship, accentuated spiritualties, charitable deeds, fasting, and abstinence from conflicts. Letting the lunar month drift made it easier for the devoted people as they did not have to develop and maintain observatories or institutions to manage observance of the faith. It suffices that any one person sees the new crescent moon to make everyone start fasting the month of Ramadan. However, this is not as easy as one might think, seeing the first new crescent moonlight can be a challenge. Confusion and controversies about when the month starts persisted until this day and age. Jews, known for their knowledge of the calendar from the time of Babylon, have faced this same issue long before Muslims. They have devised a system in which the rabbinic institution tracks the new moon and put in place a reward system for anyone who could sight the new moon and reports it. A huge fire is then started atop of the mountain to announce the beginning of the month so that rituals would take place over specific days of the month (see details in Section 2.7.2). The added power to the rabbinic institution was not well-received by the powerful and rich of the time who worked to fail the system by starting fires one day earlier, bringing confusion to those who live in remote areas, and who in-turn supposed to relay the information by lighting a fire, thus propagating the announcement throughout the region. The Roman governor ultimately destroyed the Jewish temple, as it was perceived as decent to have separate powers from those of the state, thus ending the traditional system of lunar month initiation and announcement.

From ancient times, when astronomers started using the moon for marking time, they have been faced with the problem of when the new crescent would first appear marking the start of the lunar month. Criteria for new crescent visibility have been proposed by ancient Babylonians all the way to the present day (Maurice, 1997). These efforts were not successful in the beginning as reliable orbital parameters were unavailable for the moon due to the multiple and complex influences exerted on its body. Once the proper formulae for the orbit were established, and computer equipment was made available, new visibility criteria started emerging. While these criteria had a degree of success in predicting visibility, they were not systematic in addressing first visibility and therefore did not get wide-acceptance by everyone. In modern days, western astronomers were quick to call the problem solved by defining the start of the lunar month as the time of conjunction and had ephemeris data to support their claims, but many religious communities did not care about conjunction; they kept looking for the new moon crescent.

The solar calendar was not free of controversies either. Drift over the years due to the solar year being 365.25 days, and not a complete number of days leads to a substantial shift in the seasons over the centuries. Early astronomers understood the problem with the imperfection but did not have a clear and uniform solution until the introduction of the reforms of the Gregorian calendar (see Section 2.8.6). Note that Orthodox Christians would not accept the Gregorian correction because of religious considerations that we shall address in Section2.9. Note that humanity used the lunar calendar for most of its history, and the reign of the Gregorian calendar has yet to be universal, as we will discuss the unification efforts of the date of Easter in Section 2.10. Nevertheless, the unification of celebrations is a need that all people from all backgrounds want to see achieved irrespective of their background, whether religious, race, economic, or education.

As a new moon is born, its position lies between the earth and the sun and is said to be in conjunction with the sun as seen from earth. At conjunction, there is no sunlight reflected off the moon. As time passes, and beyond a certain moon phase angle, the thin crescent can be observed during twilight after the sun drops sufficiently low below the horizon. The earth shadow rises high enough above the surface, preventing scattering of the light from the already-set sun by molecules and aerosols. There is no new moon crescent observation during night time because, when close to conjunction, both the sun and the moon have already set at night time. Therefore, sighting the new moon start being achieved only during twilight and that is possible over a window of minutes following sunset. One may argue that with proper instrumentation, we can observe a new lunar crescent during the daytime, through longer wave detection outside the visible spectrum. Our opinion is that this is not an option for three reasons. First, such radiation come from absorbed sunlight which then gets re-emitted by the moon in the Infra-Red (IR) regime, in a way similar to observing the glow of a hot object in the dark, which de-scopes the topic of the new month from its social, historical and religious context. Second, such an effort requires the use of detectors capable of sensing radiations outside the visible spectrum. Third, even if such capability is made available to all people, it does not make sense to invite the population to the danger of staring at the sun.

We may state that, with reference to the new lunar month, all religious and social events related to the first visible crescent of light reflected off the moon and not to the onset of conjunction. The commonly used lunar phases provide the lunar percent illumination, and the start of its cycle based on conjunction, which is a non-physical origin. Such definition has several drawbacks and is shown to even be the source of added controversies. Muslims, Jews, Christians, Chinese, Hindu, and others have a need for a clear definition of when the new moonlight is visible. However, the astronomical lunar calendar does not accurately relate to the new light from the moon. Add to the frustration of the non-cognizant person, who has no background in astronomy, the fact that the moon crescent itself may be impossible to see due to astronomical reasons and / or atmospheric conditions. For Muslims, the difference in opinion on how to determine the beginning of the month of Ramadan (i.e., by naked-eye, instrument-aided sighting or sun-moon conjunction) has hampered efforts to adopt a unified methodology that establishes when a lunar month should start. Occurrences of erroneous sightings around the world have only added to the discrepancies between countries where the new month is declared. The guidelines on sighting, known as the "Istanbul recommendations" were created in response to the problem and yielded some success: Everyone agreed that the beginning of the month must be based on the sighting of the new light and not on conjunction. However, the controversy surrounding the declaration of the new month remains, even with the new guidelines. The new crescent is first visible at a certain moment in time and at a certain point on earth. Such time and point of first visibility vary from month to month and is independent of the geography of the observer. It is unacceptable that people in many countries are asked by officials to go out and look for the new light from the moon in days where it is physically impossible to see it. In these days and ages, modern civil societies require plans to be drawn in advance –this is the very issue that distinguishes humans from other co-inhabitants of mother earth. Unfortunately, many of the decision-makers are either lacking the necessary skills to properly decide, or they are ill-advised and are prone to making mistakes in declaring the start of fasting and other religious events. After making erroneous decisions on what day constitutes the beginning of the Lunar month, criteria were wrongfully adjusted, and finger-pointing and blame took place in many parts of the world. Local science representatives blamed the religious scholars that they do not want to change, and in return, religious scholars do not accept the blame by referring to clauses from sacred texts. Religious scholars seem to be locked on the requirement of sighting using the human eye. And some astronomers have been stuck on the use of calculations and a state of discord has settled with no solution in sight. Mankind struggled to predict the date and time of a photometric new moon over thousands of years. Even with advances in the field of astronomy, predictions did not settle the issue. As a matter of fact, all scientific papers addressing the photometry of the moon, beginning from the 1930s stopped at large phase angles and did not venture into the early hours following conjunction since a systematic study requires an expensive and difficult global perspective beyond national geographies.

2.5 Prehistoric and primitive calendars

2.5.1 Archeological calendars

Archeologists studying vanished cultures tend to have a keen interest in how prehistoric and primitive humans utilize elements of the sky in their rituals. It is generally agreed that celestial objects were fundamental to early believe systems and social interactions. Among others, Hadingham (Hadingham, 1984) demonstrated the connections between the observations of early sky-watchers and sacred aspects of their lives through academic investigations. In a landmark paper published in 1970 and subsequent works, Marshack, a Fellow at the Harvard Peabody Museum published photos of archeological artifacts from the ice age era that seemed to have a connection to the moon - see details in (Marshack, 1991). His novel work on "cognitive archaeology" was recognized by anthropologists as an important contribution to the field. He investigated the early human cognitive capabilities through over 100 artifacts from French and Spanish caves. He concluded that the extraordinary achievements of the Ice Age must be viewed, not as the initial spark of human artwork and symbols, but instead as a fundamental episode of "evolving hominid and human-capacity, including the level of capacity apparently present in Homo erectus and the Neanderthals." His work was coined as revolutionary in some publishing circles. But not everyone agreed. It was argued that it was difficult to prove such connection between the markings observed in some artifacts and the moon (Hadingham, 1984). Hadingham stated that because with observations of this kind, the number of days in each month or part of a month would vary depending on how an observer would treat the invisible days around a new moon[3]. The larger number of possibilities makes the theory that the markings are attached to the lunar phase unfairly more probable and that Marshack's work is not provable beyond any doubts. Nevertheless, he states that the lunar theory is the most sensible explanation of certain Ice Age engravings and that "his most vital achievement was to show that the earliest attempts to record the passage of time began during the last ice age from about 35,000 to 9,000 BCE".

---

[3] This observation infers that the new moon controversy has plagued even the science community in their interpretation of records from early sky-watchers.

European early modern humans (also known as Cro-Magnon see Figure 2.5-1) appeared between 43,000 and 45,000 years ago in Italy, France, and Britain, reaching the arctic about 40,000 years ago. Through genetic analysis of their remains, it is established that they mixed with the Neanderthals who were more adapted to the cold of the ice age and whose presence dates to 400,000 years ago. It was one of the descendants of these early humans who produced the markings on the bone tool of Blanchard, France around 28,000 BCE, shown in Figure 2.5-2 (see also details in Figure 2.5-3). These markings which may be indicating moon phases (Crowley & Heyer, 1991) was a considerable find showing that someone was actively transferring information into a record

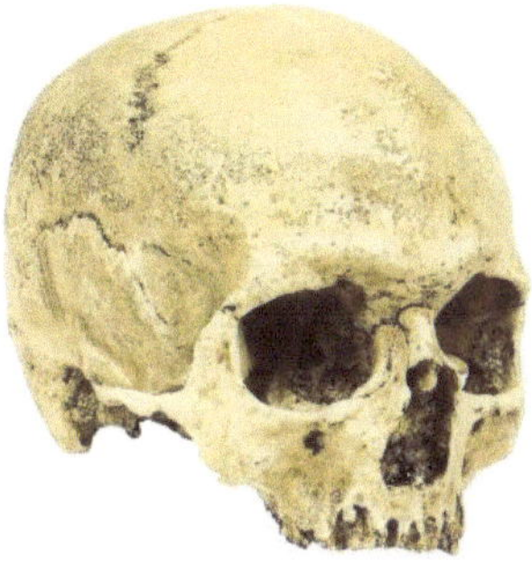

Figure 2.5-1 "Old man of Cro-Magnon," Courtesy of the Smithsonian National Museum

Figure 2.5-2 Engraved bone from the Blanchard shelter, c. 40,000-25,000 BCE -Musée d'Archéologie Nationale, France © RMN-Grand Palais/Art Resource

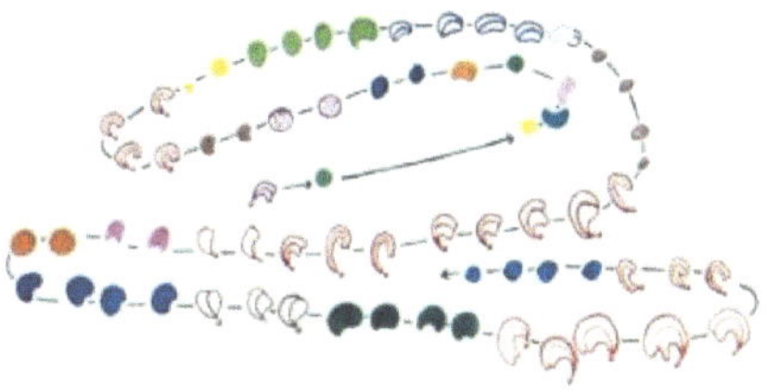

Figure 2.5-3 Schematic of marks on Blanchard bone (Harvard University, Peabody Museum)

It is well known that excavations in Africa revealed that our first ancestors roamed the east African grasslands some 2.8 Million years ago. Recent discoveries in South Africa have shown that beads, engravings, and sophisticated stone and bone tools were in-use in southern Africa 75,000 years ago. Organic artifacts from Border Cave, South Africa, shows that inhabitants of this cave have carved bones to use as media for marking and produced ornaments in the form of marine shell beads and ostrich eggshell beads about over 40,000 years ago (Francesco d'Erricoa L. B., 2012). Radiocarbon age estimate places the emergence of modern hunter-gatherer adaptation, to about 44,000 y ago.

Figure 2.5-4: Baboon fibula from 1BS Lower B-C presenting an incomplete sequence of 29 notches (scale 1 cm). Permission to reproduce photo granted by Professor d'Errico

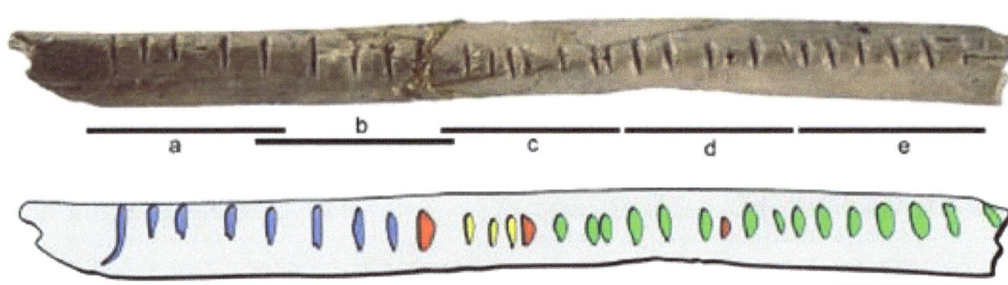

Figure 2.5-5. Photo and interpretive drawing of the notched baboon fibula with notches which were made by the same tool shown are shown in the same color. Scale = 1 cm). Permission to reproduce art granted by Professor d'Errico

The first skywatchers were probably from the motherland of Africa and did not have to cross to Europe to start watching the sky. The first calendar known today may be recorded in a Baboon bone dating between 41,000 and 42,000 BCE from the Lebombo Mountain archeological site in South Africa, shown in Figure 2.5-4 (Beaumont, 1973). The polished texture of the bone suggests extensive and long-term usage. There are also indications from the spacing that the notches were not carried at the same time.

Even if the engravings details shown in Figure 2.5-5 do not constitute counting, they represent the essential precursor step to developing such a system as it was correctly pointed out by Marshack. The drive to make sense of the natural rhythm is clearly manifested by such action.

Engraved bones were the subject of a well-documented and publicized find that again Marshack has promoted to the forefront of archeology in his investigation of the Leplacard Eagle bone from about 11,000 BCE. He presented an argument by providing comparisons with other engraved bones and artifacts from contemporary "primitive" cultures, such as the Yakuts of Siberia and the Nicobars of Malaysia with their calendar sticks, e.g. (Thorpe, 1995). Later, he argued that the bone plaque from the Grotte de Tais in France from about the same era was structured in sets of 29 notches. Other examples of lunar tallies in bones came from the former Soviet Union, where a 10,000 years old moon calendar was announced following a dig in the Kamchatka peninsula which took place in 1978.

There is a general tendency to underestimate the cognitive abilities of our early ancestors. Some go as far as adopting a narrow perspective on interpreting and extrapolating views from limited finds. It is always prudent to keep in mind that what has emerged from digs and old records is limited information which is probably "the tip of the iceberg." And finally, we have seen that controversy with the start of the lunar month came in a topic that is completely unrelated to the subject of celebration of religious and social of festivities. Some have already recognized the difficulty in establishing the earliest calendars we know so far due to the lack of visibility of the first light from the moon.

2.5.2 Wiccan calendar

When the moon is full, and witches are summoned
the poet's evening begins

Anonymous

The dialectic interaction between human genders has been a key driver of our historical development. Issues between genders remain outstanding in these days and age with diverse manifestations in cultures and societies. One thing for sure, the dominance of men over women is fast crumbling, and the long dominion of men is fast ending. Women are dominating the medical field in many countries today, as males are falling behind in the competition for the limited number of seats in medical schools. Proposals of quotas for women are entertained. In North Africa, the dean of one prominent veterinary school calls the 70-30% female-to-male rate of success a new phenomenon. Now that physical violence against women is deemed unacceptable in civil societies, some are arguing that we are witnessing the re-emergence of the female gender's ancient tradition of healing, care, and wisdom. The history of the female gender is often one of injustice and suffering in many parts of the world. This is a sad story that has yet to fully come to light, as research is needed to understand the multiple facets of what happened in the past, and what do we carry with us today from long endured practices of oppression, mutilation, slavery, abuse and homicide.

Long before biblical times, pagans around the world practiced forms of rituals linked to the phases of the moon. Wicca, also known as pagan witchcraft: a current movement which relies on old Witchcraft practices of magical knowledge to influence certain physical manifestations is full of examples which rely on the moon. This early form of religion claimed by its followers to date to more than 35,000 years ago and was led predominantly by women who oversaw the family and the communities of the hunter-gatherer modern man. As males ventured for hunting and fishing to support families, women organized through helping one another back in camps caring for the young and old and when needed, defend the premises. This horizontal form of organization by the female gender is said to be practical, effective and free from conflict, in opposition to male forms of organizations which are often run vertically through dominance, competition and threat of elimination. Capabilities in the hands of females must have brought immediate attention and competition from the male gender for pure economic advantages. We know that in later times, the Babylonian and Egyptian priests invented the concept of "what goes in the sky influences what happens on the ground" and made a living from the supernatural. Later, as biblical religions came to the scene, witchcraft had to be eliminated from the scene and end up subjected to prosecution for the past 12 centuries by religious leaders in male-dominant societies around the world. Seen as a threat by the early male priesthood, witches were subjected to the cruelest treatments. Witchcraft was intentionally or unintentionally equated to Satanism, something that they had nothing to do with. Ironically, Satanism is a product of the prosecutors. In England witchcraft, has been in hiding for the last six centuries and was practiced secretly until the repeal of the Witchcraft Act of 1951 with the enactment of the Fraudulent Mediums Act. The Witchcraft Act, passed in 1735 by the parliament, makes it a crime for someone to claim that a person has magical powers or is practicing witchcraft punishable by a maximum of one year of imprisonment. The law of 1735 came a little too late for Janet Horne, the last person to be executed in England under prior witchcraft laws enacted over two centuries, used as a basis for the cycle of violence against women.

Early male-dominated cultures under the rule of physical force could not realize the importance of the moon to the female gender for tracking whether they are going to bear a new life or reset in a natural renewal cycle. Women have long cherished the moon for its association to their personal bodies and their private matters. As healers and caretakers, they accumulated knowledge about plants and wild foods. It is in this context that suspicions, driven by pure ignorance, and in occasions pretexts for vengeance and hate-fueled crusade against the female gender in a dark episode of a witch-hunt, that took place until recent times in the name of religion. We limit this writing to western Wicca as it is addressed in open and free literature. Most if not all other cultures have had some parallel historical developments but lacked written material. Even western literature did not salvage the bulk of information on old witchcraft. Medieval laws on witchcraft had forfeiture of the properties of the convicted in addition to the death penalty. This has resulted in a complete loss of the necessary information for adequate scrutiny. Modern Wicca carries some of the old traditions, but many pockets have had several adoptions of other practices such as astrology and fortune-telling.

The moon is central to all ancient witchcraft rituals and keeping track of the lunar cycle is essential to the practice. It is argued that the moon has multiple effects on the living and its action on oceans and other large bodies of water is evidence for its effect on the rest of the earth inhabitants. It is also claimed that heightened emotions and increase of human violence are evidence of the powers of a full moon and, after all, aren't we 70 % water? Wiccans consider the full moon phase to be the most powerful of all phases, offering magical culmination while the dark moon is known to offer no magic. The waxing moon enables additional energies while the waning moon is a vehicle for banishing effects. Plants used for magic rituals are collected with the moon phase in mind to leverage its powers on plants. Herbal potency is associated with the full moon.

Celebrations and rituals are like what is found in today's religions. In fact, some witchcraft followers claim that most religions borrow their rituals from pagan holidays called '***Sabbats.'*** The name is from the same origin as Sabbath, which means sacred, and corresponds to the night and day of a full moon. The rituals carried on Sabbats are called ***Esbats.*** Wiccan covenants follow a pre-established calendar centered around 12 and occasionally 13 Sabbats in a solar year that begins on Halloween. Despite attempts to trace early western European pagan usage of a rudimentary calendar, there is no definite literature that supports a one used in ancient rituals. There is often a confusion that Wiccan calendars pertain to forms of the zodiac. These calendars cannot be the original ones used by Wicca in Europe and England. And the wide consensus today is that the usage of the 12 (or in certain years 13) moon calendar came to the western world only recently from the Greeks who inherited the concept from the Babylonians.

Most rituals depend on the correct phase of the moon. Very few rituals are conducted on a dark moon, which is a time for reflection and preparation for welcoming the crescent new moon. The new moon is associated with rituals and omens across cultures. Witches consider it as a time of renewal and action. When they sight the new crescent, some would precede the start of their spell by lighting a white candle saying:

Moon, moon, maiden moon
Welcome! Welcome to my home Welcome
to my heart Welcome! Welcome! Welcome!

In the West, some would flip a coin when they see a new moon, while people in the Middle East, see it as a confirmation for whatever they are doing. The waxing crescent, between a new moon and a full moon, is a phase of illumination and drawing of what is intended for accomplishment. The transition between the Dark and New moon has been difficult to define and hampers the commencement of rituals. The sacred full moon is the time of culmination and gathering. The phase of a Waning moon is exploited for banishing spells, severance, and termination of relations and leaving old and familiar settings.

Transitions between moon phases are not well understood from both temporal and geographic standpoint. Witchcraft does not have a magical answer to the question of when the new month starts. And because spells are about timing, then witchcraft also lacks accurate knowledge about the subject.

2.5.3 Stonehenge calendar

Circle of Stone,
you put a pedometer on the sun
and timed the moon.
Computer rocks,
Giant monoliths
made a calendar of daily span.

Carol Earle Chapin

Stonehenge is an ancient monument in the form of a circle of large stones weighing up to 37 tons each, which have been placed in a well-designed pattern. Some are vertical boulders standing alone and some stand in pairs, capped with a third boulder overhang in the form of an archway. This monument is in England and is certainly the most important archeological monument in Europe.

The Monument was first named Stonehenge in the twelfth century by Wace, an Anglo-Norman from the Jersey island -an island with a mix of French and English cultures located in the English Channel near the French coast. Stonehenge simply means hanging stones. One of the mysteries of this monument is in the fact that these large stones were brought from as far as 200 miles from the site and were artificially placed by men. Legends and speculations about the purpose of Stonehenge fill the relatively young history of the British people which is known for its abundance of mysteries. For about five centuries after Wace's writing, many suggestions were made about who might have erected these structures and during which era. There was no writing, no drawings left behind by the ancient people of the land. Only a set of out-of-place lonely boulders scattered in what seemed to be an act of some man from a distant past. One thing about Stonehenge was its uniqueness; no other ancient culture has erected a similar monument. But unfortunately, there are no other clues found from the builders to be found so far; no arts, no symbols, and no writing left by the builders to tell a tale. Over time, archeologists seem to have struggled for a long time to reconstruct the history of the site.

In 1771, John Smith, who was known for his work on smallpox, made the claim that the Stonehenge is an ancient form of a calendar that was used to track planetary motions. He related 30 stones in one circle to the 12 signs of the Zodiac to yield the ancient 360 days in a year. His claim was built on the work of William Stukeley of the society of antiquaries who in 1740 was first to realize that the monument's main axis is so oriented to align with the sunrise direction at midsummer which happens to be the longest day of the year. Despite several unfounded claims, Stukeley was the first to use a scientific approach in the assessment of how the original temple would have looked. Later in 1796, Henry Wansey noted that the three miles open view to the horizon indicates that indeed the temple had an observatory function that relates to the rise and setting of celestial objects.

Later in the 1800s, the restored monument erroneously referred to as the temple of the druids (after the Celtic priests) was subjected to vandalism from visitors, literally shipping large pieces of the stones for souvenirs with the help of the local farmers who were more than happy to rent the tools for the task. Therefore, many Boulders are missing today and were removed by transients. The other unfortunate restoration of the site was carried at different times over the past three centuries when archeology was in its infancy before the emerging of dating techniques, computer mapping and genetic identification of human remains. The monument was actually built and updated over centuries by different people. The Stones themselves were certainly moved several times by different people over the centuries. Despite these adversities, the modern work in archeology and astronomy has established several important widely accepted facts. Stonehenge was built over a long span of time. Excavations revealed several burial chambers in the site dating all the way to 4000 BCE. Many deformed skeletons were unearthed from the chambers suggest that the individuals are from different locations, including the Alpes and the Mediterranean coast, which may support the idea that the site is used for healing purposes. All sorts of theories were proposed about who may have built Stonehenge. What we know today from carbon dating is that it was not the druids who built the site as long speculated by historians before carbon dating emerged. In fact, the monument was erected over a millennium before the Celts moved in. There is a total agreement that multiple tribes contributed to the site. Artifacts and bones suggest that the original builders were the Neolithic agrarians, followed by multiple native tribes with advanced tools and means have brought the site to its most advanced state.

Even today, the complete history of Stonehenge has yet to be told. Hidden beneath the landscape surrounding Stonehenge a wealth of archeological finds were announced in 2014 by archaeologists from the University of Birmingham in the UK. Four years of detailed surveying of the landscape and underground mapping of the surrounding using advanced techniques yielded a treasure of 17 previously unknown buried monuments, many have circular shapes, making Stonehenge the centerpiece of an entire layout with promises of future findings. A particularly long burial mound dating between 3000 and 2000 BCE was discovered. The mound sits on top of a large timber building dating to 4000 BCE may have been used for processing bodies of the dead in rituals. Earthwork, in the form of linear walls with multiple Boulders, was also uncovered in the survey, pointing to an evolution of the site and reflects substantial activities at some later period.

In the 1960s, with the advent of computers, Gerald Hawkins (Hawkins & White, 1965), a British astronomer working at Boston University proposed that the megalithic stones of Stonehenge operated as an astronomical calendar and was used to predict solstices, equinoxes and even eclipses. Hawkins used an IBM 7090, a popular fully transistorized mainframe computer of the day with 32-kilobyte core memory, to calculate the directions of rising and setting of the sun and the moon over the seasons. He assumed the building year of the monument was 1500 BCE (site is now known to be much older) and carried the astronomical computations for the latitude of the site. Using a pair of surveys of the site monoliths, he determined the sun and moon rise and setting directions vis-a-vis the layout of the site at the extreme declinations of midsummer and midwinter (24 degrees North and South, respectively). The same was done for the time of extreme declinations of the moon. Nodal regressions of the moon cause the maximum declination of the moon to vary between 29° and 18.7° North and south in about nine-year period, the rising and setting corresponding to the North and South declinations were considered over monument pairs of objects. An apparent horizon of 0.6° was adopted together with 0.47° for atmospheric refraction and parallax angles of the sun of the moon of 0.0025° and 0.9508°, respectively. Hawkins concluded that observations carried by Stonehenge people were made from distant stones over mounds. He also confirmed that what is known as an "avenue" in the site marked the direction of midsummer sunrise, a long-known fact about the site.

Critics argued that the builders did not have the necessary knowledge of astronomy and that the dark skies would have prevented such observation in the first place. But even if we assume that the first monument did not have an intended astronomical function, these locations for rising and setting of the sun and the moon would have been noticed over centuries, resulting in improvements to the monument through addition or reposition of stones which was confirmed through archeological work. On the matter of weather, it should be noted that the weather in England was very different from that of the current era when the early builders embarked on their construction task. England has emerged from the warm Holocene Climate Optimum, about 7000-3000 BCE. Early construction of the site was during a period marked by an increase in temperature at higher latitudes and less sea ice present near the Arctic. The northwest of Europe experienced warming, while there was cooling in the South. Some research showed that winters might have been warmer by 3 to 9 °C while summer was warmer by 2 to 6 °C in northern Siberia. The tropical ocean surface was 1 °C warmer around 3000 BCE. Alaska had summer temperatures 2–3 °C warmer than the present. On the ability of the Stonehenge men to understand astronomy, the answer is without any doubt. First, there are several old cultures that erected massive stone structures, including some in Northern Europe, and therefore excluding the people of Stonehenge would not make sense. Stonehenge required a massive undertaken, even if it was built over several periods and especially if the large stones had to be brought to the site from far away. It is then safe to conclude that there is an important purpose behind going to such length and effort which is not directly linked to food, shelter, or reproduction. From the burial chambers, where the dead were laid in a uniform direction, the evidence that parts of the site were used for sacrifice and from what we know about other cultures that left intricate documentation, there is a high probability that the site had a religious purpose. All other believe systems had priests, or leading figures in charge of rituals, erection, and maintenance of monuments and other functions. The people of Stonehenge must have had such leadership and therefore were capable of an organization at a high level. The lack of drawings, scripting or written commemorations from the sites does not point to a lack of know-how. It was possibly imposed by the sanctity of the site. Other shrines are known to require strict behavioral codes. The question then is why such information was not recovered elsewhere outside Stonehenge. The answer is that the population of the region was scarce. After all, the site is on an island and contacts were not that frequent with the mainland. The last glacial maximum brought a remarkable depopulation of Britain and northern Europe around 23,000 years ago. It was about the time the last cousin of man has completely vanished. Humans made a comeback to the region about 15.000 years ago, as the Ice Age ended. What we know today is that most of the DNA of the people of the region comes from only several thousand hunters who roamed the region after the end of the ice age, at which time the sea started to rise gradually until about 7500 years ago when Britain became completely isolated from Europe. The isolation of several thousand-people trapped on the island with no easy supply of new visitors prevented the development of writing and other civil manifestations. But this isolation lasting 20 to 30 centuries had only caused a short-lived delay. In 2016, a nearly intact one-meter diameter wooden wheel from 800-1100 BCE was unearthed in Cambridgeshire with a nearby horse spine, exotic glass, pots, advanced textiles, and tools. The complete wheel with the hub as part of a horse-pulled chariot in a thriving community occupying circular dwellings. This finding which was recovered from about 2 meters below the surface took experts by surprise and pushed back on the idea that inhabitants of the island lacked in knowledge and in the mastery of the technology of the time. A smaller incomplete wheel measuring 0.8 meters in diameter was recovered in the 1990s and dated to 1300 BCE. The moist wet landscape of England tends to destroy archeological artifacts and poses a challenge to the writing of the history of the island. Earlier wheels recovered in In Europe date to the Copper Age, at least 2,500 BCE. Wheeled transportation apparatus appeared 4000-4500 BCE in the Sumerian culture and in Central Europe.

From the connection to the seasons and their interest in the moon, we conclude that the people of the Stonehenge monument displayed a unique ingenuity in their erection of a device capable of tracking the seasons. There is no doubt that the need to keep track of the seasons is for planning purposes. Finally, why no records in the form of drawings or symbols were left by the builders and planners of the Stonehenge? Certain believe systems prohibit (or discourages drawing entities such as animals and the alike). Could it be that the people of Stonehenge were prohibited from such activities? Given the size of the population of the island, the achievement of Stonehenge is certainly remarkable.

2.5.4 Horizon calendars

It is inevitable that early humans across the world with habitat free of obstruction to the West and/or to the East noticed that the sun sets at different points in the horizon as the seasons changed. This variation is cyclic, moving back and forth between two extreme landmarks in the horizon every year. These extreme landmarks correspond to summer and winter solstices, and therefore over a full year, the point of sunset (or sunrise) makes a round trip between the two extreme landmarks. Such observation was probably transmitted through generations, signaling important seasonal events such as the tribal movement for hunting, fishing, planting seeds and other activities. The points of sunset and sunrise provided mankind with a physical scale for time-keeping referred to as the Horizon Calendar which was discussed by American Ethnographers in the 19th century as discussed by Hubert (Hubert, 1998).

The Zuni culture of the American Southwest paid attention to the seasons through observation of the sun, and its points of rising and setting. Their ancient habitat was ideal in providing distinct features in the form of multiple peaks and depressions in the horizon to reference seasonal changes in the position where the sunsets. The sun priest tracks the sun through the seasons and calls for the pertinent ceremonies associated with the seasons.

Section 9.3.2 provides a formula for the azimuth angle of a setting celestial body in the western horizon. It will be shown that a setting azimuth angle depends on the declination angle of the celestial body and the geographic latitude of the observer. The declination of the sun is extreme at the solstices, and as it sets, its azimuth angle reaches an extreme angular distance relative to the west. Over the year, the setting point of the sun oscillates symmetrically about the western direction. The amplitude of such oscillation is equal to about 23.4 degrees at the equator and grows slowly as the latitude increases in either hemisphere. At the Equinoxes, the sun sets in the West, that is the Azimuth 270 deg. and reaches extreme positions away from the West at the summer and winter solstices, as shown in Figure 2.5-6. Designers of ancient calendars that we discuss next in Section 2.6 were fully aware of such fact.

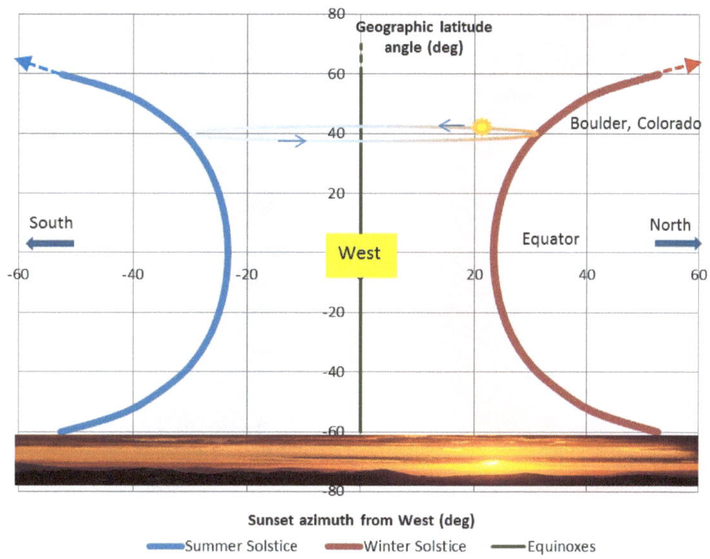

Figure 2.5-6 Sunset Azimuth varies with the season between two extreme angles at the solstices, The variation of such position depends on the observer's latitude

2.6 Ancient calendars

2.6.1 The Egyptian calendar

The invention of the Calendar was hailed as the most important achievement of the ancient Egyptians (White, 1970) and because it is widely accepted that scientific knowledge was closely associated with the Calendar in ancient civilizations, it may be safe to state that science was born in Egypt. Egyptian priests acquired new powers through the maintenance and exploitation of science, probably brushing aside the role of female healers who dominated the scene for centuries. Priests guarded the tools to measure time, considered as a state secret, starting with ancient Egyptians who monitored celestial bodies prior to the first dynasty.

As glaciers retreated at the beginning of the Holocene 11,700 years ago, and the warm interglacial period settled in, it took a couple of millennia for the Egyptians to start settling in Northeastern Africa. Warming continued until about 8,200 years ago when a lapse of 400 years made the Savanna turn into a desert and farming along the banks of the mighty Nile River begun. The sudden switch from a wet green environment to the largest desert on earth is not fully understood. It was a different time where the earth's tilt was 24.1 degrees versus today's 23.5 degrees. The desert grew to form a 700 Km stretch around the Nile over three sides while the Mediterranean Sea level kept increasing in the North. It took only 2000 years after the 8.2 KY event for the Egyptians to establish the first date in the history of mankind; the year was 4241 BCE. Within a few centuries, a state by today's many standards has risen with central governance, laws, arts, fine architecture, and technology over a unified Egypt. The desert and sea provided shelter from all foreign temptations for the next three thousand years. By 2500 BCE Egypt had three calendars: a lunar one, a solar civil calendar, and a lunisolar calendar, that is a lunar calendar in-step with the solar calendar. When the solar calendar was devised, the Egyptians still had their religious events based on the timing of the moon.

The Egyptians initially determined the length of the year to be 365 days that they partitioned into 12 months of 30 days and an addition of 5 days that were added to the year by the Egyptian God Thoth and which are the birthdays of Osiris, Isis, Horus, Nephthys and Set according to their mythology. Later, the Egyptians realized that Sirius, the brightest star in the sky, ascended in the sky aligned with the sun once a year coinciding with the Nile flood. That day became the Egyptians' New Year and the first day of the month of Thoth, a day of ceremonies and celebrations that started with the appearance of Sirius on top of the obelisks precisely aligned with the observation point. This has ultimately resulted in the discovery of the actual year length of 365 ¼ days length of the year. The Egyptians counted the years starting from the ascension of each King. The Babylonians also restart the year counting with each King until the year 311 BCE, with the ruling of Seleucus, they switched to a continuous dating system.

2.6.2 The Babylonian calendar

While it was the great Egyptians who pioneered the fields of astronomy and time measurement about 6,000 years ago, it was the Babylonians who were the ancient masters of astronomy.

Mesopotamia is in what is today Iraq. It was called Sumer, pronounced "Shomer," home of the earliest civilization. Together with the Indus Valley and Egyptian civilizations, Mesopotamia propelled humans to reign supreme over the planet. Settlements in the Ubaid period run all the way back to 6500 BCE in south Mesopotamia and to 5300 BCE in the North. From about 7500 to 6000 years ago, during the glacial retreat and melting of the Ice Age, the Persian Gulf or the Arabian Gulf (depending on which bank of the Gulf you stand) was growing further north, challenging the farming settlers. The moving boundaries of the Gulf must have forced a continuous restoration effort of irrigation networks under constant threat from the ever-expanding Gulf water. These settlers were the descendants of the ancient Sumerian people (Arthur Cotterell, 1980). The Ubaidian culture pioneered grain cultivation and built urban dwellings between 6000 and 7000 years ago with large villages built of mats and clay starting the civilization for mankind.

Ultimately, social and scientific development emerged, including proto writing (forms of communication using symbols, a precursor to writing), which was invented 4500 years ago. Astronomy records going back to more than 4300 years ago, may be some of the first information to be recorded. Reference to the twelve lunar months and the addition of an intercalary month to bring the lunar year in step with the seasons came later, going back to the period of the Sumerian city-state of Ur 2294 to 2187 BCE at which time the intercalary months had an 8-year cycle. Sumerians were non-Semite people who spoke a separate language that no language is derived from today but coexisted with the Akkadian Semitic speakers who conquered them around 2270 BCE. Sumerians rose to power again from 2100-2000 BCE and the Sumerian coastal city of Eridu became the world's first city, where multiple cultures fused leading to an unprecedented demographic growth with a healthy food supply. The city saw the partition of labor in a thriving economy. But ultimately, the Akkadians eclipsed the Sumerians.

The ancient Akkadian city of Babylon was a small cultural center in Mesopotamia when the Semitic Amorites emerged from western Mesopotamia (modern-day Syria) and settled in different cities including Babylon. They ultimately ruled over the kingdom of Babylonia for 400 years, while the Babylonian state retained the written Akkadian language of the people. The Amorite rulers spoke a Northern Canaanite dialect (other Canaanite languages are Hebrew and Phoenician) while the state used the Sumerian language for religious worship. The Amorites started descending on Mesopotamia around 2400 BCE and were considered primitive people by the Sumerians and Akkadians, lacking civil behavior and social know-how with no knowledge of farming and consuming raw meat. Starting 1894 BCE under king Hammurabi during who's reign Babylon expanded to become a major capital city of the short-lived Babylonian empire which disintegrated right after his death. Interest in astronomy was methodical and rational. The heavens are seen from a physical perspective as opposed to the Egyptian view, which was only metaphysical. The movement of the sun, the moon and Planets and times of rising and setting was meticulously recorded. As early as 1921-1901 BCE that is about 4000 years ago, records of solar and lunar eclipses, comets and shooting stars were precisely recorded using water clocks and sundials. By 1260-1232 BCE, other instruments for transit determination were used in observatories made from the top of their temples, called Ziggurats (Ronan, 1982). These temples were built by the Sumerians, the Babylonians and Assyrians throughout the ages in Mesopotamia starting over 4000 years ago until the middle of the last millennium BCE.

During the beginning of the last millennium BCE, Babylonia was ruled by Akkadians and Assyrian neighbors. Following the Death of the Assyrian king Assurbanipal in 627 BCE, the Assyrian empire disintegrated in a civil war, and a short-lived Neo-Babylonian era rose in 626 BCE under Nabopolassar who's Chaldean tribe had migrated from the Levant to southeastern Babylonia at the turn of the millennia. In alliance with the ancient Iranian Medes people of Media (today's North-western Iran and south-east Turkey), who migrated while under an unstable period of a weakened Assyrian empire (1365-1020 BCE), the nomadic Scythians, another culture occupying today's southern Iran, and the nomadic Cimmerians from further north of the Caucasus, he went to occupy Nineveh in 612 BC, and restored the rule of Babylonia for the first time since the death of Hammurabi in mid-1800's BCE. This was another short-lived period of progress and economic prosperity in agricultural production, economic growth, architectural construction, science, astronomy, and arts. This era ended with the reign of Nabonidus in 539 BCE who saw the Neo Babylonian empire fall to the Achaemenid Persians who also grew strong under Cyrus II of Fars, known as Cyrus the Great. Starting this era until the deportation of the Babylonians in 63 BCE, records spanning five centuries of first visibility of the lunar crescent were retrieved. These records will be discussed in Section 10.1 from the perspective of the linear model of Chapter 9 and the work of modern investigators.

Babylon remained under the rule of the Achaemenid Persian Empire for two centuries, which grew fast to gulp four empires and became the largest empire the world has seen. Cyrus the Great and his successors did not impose the will of the state on other religions and did not interfere with other cultures. Cyrus was the only Messiah outside the house of Abraham. It was under the Persian King Darius, born in 550 BCE, and reined from 522 BCE to 486 BCE that the empire reached its peak, accounting for about 50 million people, nearly half the world population of the world at that time. His empire was extending south to the Sudan and Libya, to the Indus Valley in the Far East, Central Asia and to the North Caucasus, the Black Sea, and the Balkans. Named as King of Persia and Babylon and Pharaoh of Egypt, he ruled with sound economic principles. He built roads and infrastructures, introduced a unified monetary system and standard weights and measures. He made Aramaic the official language of the empire. His religion was Zoroastrianism, one of the oldest religions characterized by messianism, heaven and hell, creation, and end of existence. Darius had no interest in the Babylonian temples and priesthood. He was however interested in their science, their legislative code writing, and in their scientists who made his calendar more accurate (Olmstead, 1966). The rule of the Persians over Babylon ended with the conquest of Alexander the Great 332-323 BCE.

The final period of the Mesopotamian astronomy corresponds to the Chaldean dynasty. Chaldean priest-scientists and astronomers did not get interrupted in their work from the period of the rise of the Neo Babylonian era all the way to the Persian occupation. In such stability and equipped with the knowledge accumulated by ancient Sumerians and Babylonians, they observed at length the moon and other celestial bodies. The Babylonian astronomers' frustration with the lunar cycle being out-of-step with the seasons had finally ended when they figured that seven years of 13 lunar months plus 12 years of 12 lunar months would equal 19 solar years with a rounding error of only a few hours. This is the Metonic cycle, named after the Greek astronomer who "introduced" the concept derived already by Babylonian astronomers.

To reach an understanding of the motion of the moon, Babylonians astronomers followed a pure scientific methodology of data gathering and measurements. During each lunar month, they have kept meticulous records of six periods of time: (i) sunset to moonset whenever they sight the first crescent of the new month after sunset, (ii) moonrise to sunset when the almost full moon rises before sunset, (iii) sunset to moonrise during the evening twilight on the following day, (iv) moonset to sunrise on the last morning twilight that the almost full moon set before sunrise, (v) sunrise to moonset on the following morning, and (vi) moonrise to sunrise on the last morning before conjunction that the crescent moon was visible. These periods of time were measured, and whenever it was cloudy, calculations were made with fair accuracy (Stephenson, 1997). The Babylonians devised several units of time. The most important unit corresponds to 4 minutes of time. That is 15 units amount to one hour. This unit of time corresponds to the time it takes the celestial sphere to rotate 1 degree and was the source of the partition of the circle into 360 degrees.

The Babylonian Calendar is lunisolar. The new year starts at the Spring Equinox, and from the year 311 BCE, the year is comprised of 12 months, 29 or 30 days long, totaling about 354 days. Every two or three years, an intercalary month is added right after the sixth month or the 12 months to bring the calendar in step with the seasons. The months in the Babylonian Calendar are referred to as:

Nisannu, Ayaru, Simanu, Du'uzu, Abu, Ululu, Tashritu, Arahsamnu, Kislimu, Tebetu, Shabatu, and **Addaru.**

At the end of each month, either the 29th or 30th day, a watch is kept for the new crescent moon after sunset to determine if the new month has started. This is probably the reason why the day starts at sunset.

The Babylonians did not have enough time to formulate laws of motion of celestial bodies, even though they had the basic means to do it through their invention of Algebra. After Alexander the Great ended the Achaemenid Empire in 332 BCE, Babylon was under the rule of the Seleucid Empire from 312 BCE to 63 BCE which built a new capital, named Seleucia 40 miles north of Babylon. The ruler Seleucus I Nicator deported the Babylonians to the new city which was dominated by immigrants from Greece, sealing the end of Babylon. Babylon remained a depressed city until about 200 CE and was ruled by the Iranian Parthian and Sasanian. The astronomy work of the Chaldeans did not develop planetary motion laws but attempted to produce prediction based on prior observations. They tried multiple speeds for the heavenly bodies and gradually varying speeds to predict future motions and therefore produced genuine scientific work. Later they were erroneously blamed for introducing astrology and predicting fate and magic. Astrology prospered in the third century BCE and was developed by the Greek Egyptians.

It is widely acknowledged, that the Babylonian calendar formed the basis of the Hebrew Calendar and the first Greek and Roman Calendars (Ronan, 1982), in construct, form and other aspects. The work of the Chaldean astronomers was forgotten for centuries until about two thousand texts from cuneiform tablets were actively removed from Iraq during a period of a disintegrating rule of the Ottoman Empire between 1876 and 1882 CE. The lion share of the known tablets outside Iraq were brought to England and placed at the British Museum (Stephenson, 1997); they constitute 98 % of the trove. A very few Texts can also be found in the Staatliche Museum, Berlin and the Morgan Library collection in New Heaven. The few tablets left behind were disseminated together with the rich heritage held both in the historical sites and in the museums of Iraq in a domino effect of looting and destruction which followed the 2003 invasion of Iraq. The tablets revealed an incredible amount of work carried by Babylonian astronomers and testify to a civilization that propelled mankind out of darkness.

2.6.3 The Mayan calendars

The ancient Maya people are the inhabitants of Mesoamerica who developed a separate culture that dates back to 7000 BCE and built an advanced magnificent civilization raising to prominence between 2000 BCE and 250 CE Their heritage accounts for multiple large cities over an extent of thousands of kilometers, encompassing the geographic areas of what is now Honduras, Guatemala, El Salvador, Nicaragua, Costa Rica, going all the way to Central Mexico in the north.

Mesoamerica was the home of one of handful ancient civilizations where mankind independently developed knowledge in science, culture, medicine, agriculture in the world. The Maya, as the collective populations of Mesoamerica that interacted and exchanged goods and traditions over centuries, accounted for near 8 million people at the arrival of the Spanish. They are believed to be descendants of Mongolians and other Asian groups who migrated to the Americas via Beringia, a land bridge which connected the American continent to Asia about 13,500 years ago as illustrated in Figure 2.6-2. The group of migrants may be about 70 or so as revealed from genetic studies. The blood type of the Maya is 98 % of the type O, prevalent in all ancient populations of the Americas.

The Mayan nation had dozens of languages which derived from a proto-Mayan language, probably spoken around 5000 years ago. They developed their own written records using a complex script with no relationship to other languages. Christian Europeans who arrived first viewed these records as heretical[4]. It was unfortunate that Bishop Diego de Landa, who oversaw the conversion of the Mayans following the Spanish conquest of Yucatán, destroyed many texts in pyres. Few documents were recovered in their original languages, while others were transcribed into Spanish providing a record of an outstanding ancient culture and knowledge.

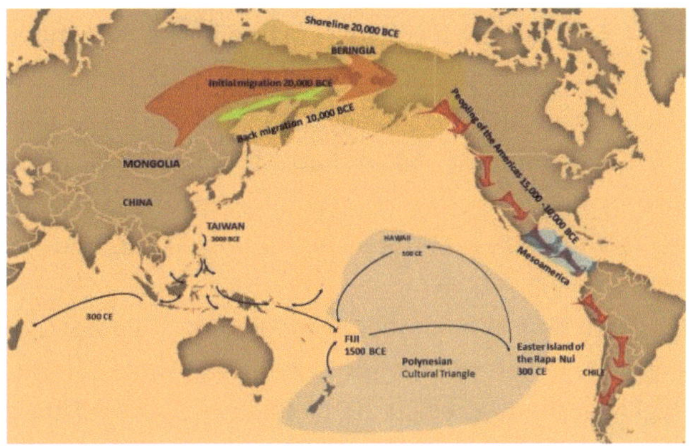

Figure 2.6-1: Early migration based on genetic ancestry mitochondrial DNA, linguistic, archaeological flow in and out of Beringia of the Native American founders and the Austronesian peopling of the Pacific.

American civilizations at the time of European encounter had achieved impressive accomplishments in astronomy and mathematics. The Mayan tracked time using a set of three separate calendars dwarfing all other ancient calendar developments. Bishop Diego de Landa did not know that the Julian calendar he used in 1549 was less accurate than what the Maya had developed (Duncan, 1999). They used positional indicators that permit an extended count of multiples without reliance on fractions and had already established the concept of zero. The Aztec city of Tenochtitlan, the ancient site of Mexico City was one of the largest in the world, with a population of 200,000. It is said that the Mayans may not be the original developers of the calendar system and that it came from another ancient South American

[4] Al-Biruni, mentions that when Kutaiba ben Muslim conquered Khawarism the second time after the inhabitants had rebelled, he "extinguished and ruined in every possible way all those who knew how to write and to read the Khwarizmi writing, who knew the history of the country, and who studied their sciences".

culture. Studies have shown that the calendar was in common use throughout Mesoamerica going back to at least the 5th century BCE, relating to calendars employed by earlier Mesoamerican civilizations (e.g., Zapotec and Olmec) and subsequent ones such as the Mixtec and the Aztec (Rice, 2007). The Yucatec Mayan believe that tradition and knowledge of the calendar were given to their ancestors by a higher being named ***Itzamna,*** identified with God of the sun ***Kinich Ahu***. It is interesting the same naming of "**Ahu**" was used by one of the most remote islanders of the world, the Polynesian Rapa Nui people (the closest coastal point is Chili, 3512 km distant), who also developed an independent civilization but had no relationship with the Mayans. The naming "**Ahu**" was used in reference to the platforms used for their famous figures carved from stones between 1200 and 1500 CE. Located at the southeastern point of Polynesia, which they settled in the first millennium CE and saw their population obliterated to about 100 people from a high of 15,000 inhabitants, had a different migratory path, origin, and chronology. Other references to similar names: ***heiau*** is a Hawaiian temple, for example, *heiau hō'ola* heiau to treat the sick, Ahu, altars in a heiau (Hawaiian temple), the name of old cities in China and Iran, and the name of Allah in Sufism. Despite these similarities in names, Mesoamerica was isolated, and its development was unique.

The Mayan people counted time in a number of moons. They did not use fractions but their observation for many years -uninterrupted for 522 years in some surviving records- allowed them to establish highly accurate predictions of the movement of the heavens. All Maya people of cities that erected monuments used similar calendars. At Palenque, the Maya used 81 moons = 2392 days giving an average lunar month of 29.53086 days, and at Copan and other cities using similar moon numbering the formula is 149 moons =4400 days. That is an average length of 29.53020, not bad for an ancient civilization! They were aware of the tropical year length and the 19-year cycle connecting the solar and moon cycles. At Copan, they measured the tropical year length using 19 years cycles, nearly equals to 235 moons and 149 moons is 4400 days giving an average of 365.2420 days in a year. Later at Palenque, a different formula gives an average of 365.2430 days, remarkably close to the Gregorian year of 365.2425 days. The Mayan people had accurate tables for predicting eclipses and information on planets. The Mayan solar calendar, named Haab' is composed of 18 months, twenty days long plus 5 "unlucky" days, totaling 365 days.

The Maya used a vigesimal system instead of the decimal system. The system may seem cumbersome at first, but if one has not been exposed to other systems, they will find it easy to use. For counting, we commonly use our ten fingers as a basis, while the Mayan used ten fingers and ten toes, and therefore had a basis of 20 for computation instead of 10. Place-position of numerals and use of the abstract concept of zero was also accomplished before it was discovered in the old world, and that is, by any stretch, a remarkable achievement. The achievements of the Maya in measuring time are remarkable, given the fact that they have arrived at the correct estimates the long cycles of the motions of the heavens. How such isolated culture could arrive at these results which were much superior to those of other connected Old Civilizations? The destruction of their heritage was certainly a tremendous loss. Their story is still unfolding with more archeological discoveries to be unearthed.

The day is referred to as the Kin. Twenty Kin make 1 Uinal and 18 Uinal make 1 Tun or 360 days. The Tun is a little less than a year and forms the unit of time. It is an arbitrary number not intended for a special meaning. The higher vigesimal unit is the Katun with 20 Tun make 1 Katun or 7,200 days, 20 Katun make 1 Baktun or 144,000 days. Twenty Baktun equals 8000 Tun and makes 1 Pictun, and so on -see Glyphs in Figure 2.6-2.

Figure 2.6-2 Glyphs for the six first orders of magnitude (from left to right: Pictun, Baktun, Katun, Tun, Uinal, and Kin) Glyphs (Courtesy of the Smithsonian National Museum of the American Indian)

Figure 2.6-3 Long day count (Courtesy of the Smithsonian National Museum of the American Indian)

Dates are expressed by an introductory glyph which identifies it as belonging to the Long Count system, followed by the full date. The Tzolkin is a date displayed as a number together with the Tzolkin day glyph, then the Haab date in the form of a number together with the Haab glyph, and ending with a glyph representing one of the nine deities of the Maya underworld given in the sequence

baktun.katun.tun.uinal.kin | Tzolkin | Haab | Lord of the Night.

The Maya kept a record of the lunar cycle in the above form. We will check this record to examine if they had the same issues with the start of the lunar month in Section 10.1. Typical glyph representing a Long day count is shown in Figure 2.6-3. Glance, such representation may seem complex. But they are simple once we get used to them.

2.7 Ensuing calendars

Except for the Mayan calendar which could have been destroyed together with most of the heritage of the Mayan, ancient calendars from the old world made their way out of their birthplace to be adopted by emerging new powers. Understandably, a newly emerging power would certainly need a good calendar to manage the day-to-day function of the state. Calendars evolved over the centuries and over large swaths of earth, first under the sponsorship of religion, and subsequently under the military or the office of weather prediction or independently in a designated office which oversees standards. The astronomical calendar was developed under a joined effort of the British and US naval powers in need for maritime navigation accuracy and other applications. Following the coverage of the Ancient original calendars in the previous section, we introduce next several ensuing calendars that developed later over the following centuries. Again, of interest are the historical perspective under which calendric development has taken place, and the motivation needed for adoption and implementation.

2.7.1 Ethiopian (East African) and Coptic calendars

Both Coptic and Ethiopian calendars are rooted in the Egyptian-Alexandrian calendar from which the Julian calendar was also derived. The Julian calendar and its revision, the Gregorian calendar, use 30 or 31 days in a month except for February which uses 28 or 29 days depending on whether it is a leap year. The Coptic and East African calendars use 12 thirty-day months with the thirteenth month of five or six days. The sixth day is added once every four years (365 days or 366 days in leap years) with no exception, like the Julian calendar and unlike the Gregorian calendar in which the years that are multiples of 100 are leap only when divisible by 400. The new year of the East African calendar corresponds to September 11 in the Gregorian calendar and on the 12th of September on leap years.

The East African calendar is quite important in the Orthodox religious communities in Ethiopia and Eritrea. It is used in determining celebrations of festivities, saints, certain prayers and readings of scriptures. The months of the Ethiopian calendar are:

Meskerem, Tikemet, Hidar, Tahsas, Tarr, Yekatit, Makwit, Miyaziah, Genbot, Sanne, Hamli, Nashi, and Quaggimi.

The Coptic calendar has the following names of the thirteen months:

Thuout, Paopi, Athor, Khoiak, Tobi, Meshir, Baramhat, Baramouda, Pashons, Paony, Epep, Mesori, and Nasie

There are four seasons in Ethiopia: Kiremt or Meher: occurs in June, July, and August characterized by heavy rainfall. Belg: September, October, and November: the spring season and time of harvest. Bega: December, January, and February constitute the dry season with frost in mornings, especially in January. And Tseday: March, April, and May are characterized by occasional showers. May is the hottest month in Ethiopia. Meskerem translates as September, but the name is drawn from the Geez word **MesQuer** which means a ship. The new year marks the celebration of the beginning of Spring and a celebration of the end of the great flood with Noah's ark coming to a rest on Mount Arafat which is the key location for Haj in Arabia -it is often said "the Hajj is Arafat" meaning that the scaling of the mount Arafat is the main ritual of the Hajj. We shall revisit this topic again in Section 10.5 when we discuss the origin of the names of the Arabic calendar.

Etymologically, the first month in the Arabic calendar AL-Muharram and the name of the first month of the East-African calendar name, Meskerem is very similar. Also, the twelfth month, Nashi is the same as that of the thirteenth month of the Coptic calendar Nasie, it is also the word used for the title of the head of the Jewish rabbinical Supreme Court, the Sanhedrin (in charge of the calendar), and also used for the intercalation process in the Arabic calendar which was abolished in the Muslim calendar as it will be discussed in the next few pages.

The Coptic years are counted starting the year 284 BCE with the first day is August 29 in the Julian calendar. This is the year the Roman Emperor Diocletian mass executed Egyptians Christians, prompting the reference to the era of martyrs, *Anno Martyrum.*

2.7.2 The Hebrew calendar

The beginning of the month **Rosh Chodesh** or **Rosh Hodesh** (note the similarity between the old Hebrew: ראש חודש and Arabic: **Roes Chard:** راس شهر means the Head of the Month. The first day of every month in the Hebrew calendar is marked by the appearance of the new crescent moon. Again, this is not the astronomical new moon, which is not visible to the naked eye. It is the day the new crescent is observed and is a holiday for the Jewish people. In their early history, the Jewish people had no Calendar to follow to observe religious rituals because the beginning of the lunar month could not be predicted by astronomers and direct sighting used to be the only way to initiate rituals. The **Sanhedrin (**Hebrew:,סַנְהֶדְרִין Arabic: شوريين , "Shourein" meaning council), the Rabbinical Supreme Court, is the keeper of the timing of month. The statement *"...This chodesh shall be to you the head of the month,"* and in another version: "This month shall be for you the beginning of months. It shall be the first month of the year for you". (Exodus 12:1–2). The wording "shall be to you," was interpreted as a divine order to the Prophet *Moses* and *Aaron*, and to the leaders of the Jewish people thereafter to sight –the original words meant actual seeing- the new crescent light and the initiation of the *chodesh*.

An account of the early process followed in the initiation of the month as led by the early rabbinic Jewish intuition[5] is summarized next. At the end of the waning moon, men would look for the first sight of a new moon following sunset, and anyone who claims to have witnessed the new moon would come to the Sanhedrin to state that he witnessed the new moon and gets interrogated on what they observe. The members of the Sanhedrin are trained in new crescent moon sighting. They have knowledge on the particularities of where in the sky (how high) the crescent moon might be visible, its direction of descent, its orientation with respect to the Horizon, its thickness, etc. Declaration of the month depends on the acceptance of the testimony of two male witnesses. The rabbis of the Sanhedrin would check the testimonies of all witnesses even if they were already satisfied with a pair of claims and invite the witnesses to have dinner. Upon satisfaction with the Testimony, the Sanhedrin announces the sanctification (*"Mekudash,"* Arabic word "Mokadass"). On the following night, the second night of the month, large fires on mountain ranges are started to communicate that Rosh Chodesh was the night before. The Sadducees, a sect of upper-class Jews who did not believe in rabbinic authority, or laws of the Sanhedrin or their interpretation of the Torah were blamed for lighting fires on the "wrong" days. That was the end of the usage of fire-on-mountaintop, and instead, messengers were dispatched which took longer; a delay which has *halachic* implications with regards to observance of the second day of holidays. In the case where no visibility was possible, and thirty days have elapsed since the last Rosh Chodesh, the following day is declared Rosh Chodesh. The Jewish month is either 29 days or 30 days. The practice was completely disbanded in 358 after Theodosius I dismantled the Sanhedrin, the head of which, Hillel II (also known as Hillel II, the Nasi) established the first Jewish calendar to avoid prosecution. An attempt to resurrect a clandestine Sanhedrin resulted in the execution of the last supreme Rabbi Gamaliel VI by Theodosius II in 425 CE.

The Jewish calendar is lunisolar; its months are lunar, but its years are solar. By this, we mean that the calendar's mean year length is close to the length of a solar year, so its years stay (approximately) in step with the seasons. A year begins on average around the time of the September (southward) equinox, and Passover, which occurs just after midyear, occurs shortly after the March (northward) equinox. The calendar's months, on the other hand, are synchronized with the phases of the moon. Each month begins close to an astronomical New moon and Full moon coincides approximately with the middle of the month. This is because the calendar's mean month length is very close to the mean length of what astronomers call a synodic lunation (a full cycle of lunar phases).

---

[5] http://www.chabad.org/library/article_cdo/aid/1209/jewish/The-Lunar-Files.htm

The present-day Jewish calendar has not fundamentally changed since 358 CE. As stated earlier, it is a rule-based, fixed arithmetic calendar, as opposed to its predecessor, which was an empirical one regulated by observation and prone to errors. Over long periods of time, the Hebrew calendar maintains accuracy due to the near-perfect rhythm of the lunar orbit but fails to predict the first day of the month like any other lunisolar or lunar calendar due to the non-local manifestation of the new crescent. Molad (plural Moladot) is both a Hebrew and Arabic word which means "birth" and refers to the time at which the New moon is born. Except for the month of Tishrei, moladot are announced in synagogues on the last Shabbat of the elapsing lunisolar month. The molad of the month of Tishrei determines the date of the New Year Day (Rosh Hashanah), which is subject to postponements of either one or two days depending on certain rules.

In modern time, the Molad is announced as a time given in hours, minutes, and 18ths of a minute starting from mean sunset. Because the Hebrew calendar day starts at sunset, attempts to local time zones and to convert to civil time has led to errors and confusion in announcements of Moladot. Errors can generate problems with the celebration of Rosh Hashanah. The months of the year are associated with themes and count either 29 or 30 days, according to the following rule:

1. **Nissan**—30 —Month of Redemption
2. **Iyar**—29 —Self-improvement
3. **Sivan**—30 —Related to the giving of the Torah
4. **Tamuz**—29 —Reflection on the Sin of the golden calf
5. **Menachem** Av—30 —Comfort and reflection on God's fulfillment of promise
6. **Elul**—29 —Repentance
7. **Tishrei**—30 —Strength
8. **Mar Cheshvan**—29 or 30 —Month of the great flood
9. **Kislev**—29 or 30 — Month of rest
10. **Teve**t—29 —God's grace
11. **Shevat**—30 —Month of the tree of life
12. **Adar**—29 —Month of good fortune

In leap years, which occur every two to three years, that is seven times in the 19-year lunar cycle, Adar is called **Adar I** and is followed by an intercalation month called **Veadar** or **Adar Beit** -30 days)

Kislev and Cheshvan can both be 29 days, in which case the year is said to be chaser), or 30 days with the year is referred to as malei, or if Cheshvan is 29 and Kislev is 30, the year is k'sidran). Therefore, there are three types of year: chaser, malei, or k'sidran. The rules of the calendar ensure that the first day of Rosh Hashanah will never take place on Sunday, Wednesday or Friday.

The names of the months are Babylonian and point to the origin of the calendar which the Jewish people started in 586 BCE. Nissan is the first month of the Calendar and corresponds to the beginning of spring around March 21 (spring equinox), at which time Jews celebrate Passover. However, the Jewish New year starts the first day of Tishrei (Autumnal equinox) around September 22 and not in Nissan. The dates vary from year to year depending on the lunation of the year.

The Jewish calendar went through multiple developments over the centuries and was not this sophisticated in its early history. Today, the rule-based calendar does not have the beginning of the month at the correct time and therefore was not acceptable to many faithful people, e.g. People of Yahweh. They believe that divine "appointed times," i.e., Holy Days as stated in Leviticus, Chapter 23 constitute of the Weekly Sabbath, and three days assigned in reference to the "Chodesh": 1st, 10th and 15th day of the Month (verses 24, 27 and 6, respectively.) They argue that keeping these appointed times in the correct days is essential and should not be substituted by man-made calendar dates. For the question of calculations: "Can't these dates be calculated and just put on our secular calendar?" Their response is "This has been tried over the centuries, but the moon cycle is complex for predicting a visible crescent moon. It will always be a 29 or 30-day month, but which one it is will occasionally only be confirmed by a sighting. "

The new crescent visibility law and the photometric calendar which are presented later in the book may be a substitute for direct sighting. Some conservative opinions still require direct visibility of the new month. The latter groups may find The Witness System a vehicle for the faithful to document the new month. The system responds to their need to visualize the new moon crescent each month. It has been argued that looking for the new moon with other people adds to the unity of the body of believers and introduces action in one's belief system. Some people go out as a family to a location with an unobstructed view of the horizon and blow the shofar when they see the new crescent and confirm calendar dates as the year progresses. The Witness system transcends religions and can bring people from all religions together to search for a common crescent, as the very same matter stands out in other religions and cultures. The single Witness System could break barriers and at least get people from the conservative sides of separate religions to talk and interact.

2.7.3 The Arabic calendar

Before the seventh century and prior to Islam, Arabia was a tribal society without a central power. Some prominent Meccan tribes formed the glue in the Arabian Peninsula by hosting economic fairs, entertainment such as organizing competitions in poetry and spiritualties by maintaining the Kaaba, a destination for pagan worshippers[6]. Arabia had multiple lunisolar calendars which were also adapted from the Babylonian calendar. The maintenance of the Meccan calendar, including the announcement of the intercalation month, was proudly provided by one of the leaders from the tribe *Banu Kinana* of Mecca (Ilyas M. , 1993). Leaders of the Meccan tribes needed a unified calendar to regulate the time of the fairs and commerce associated with the Hajj (i.e., pilgrimage) to the shrines of Mecca.

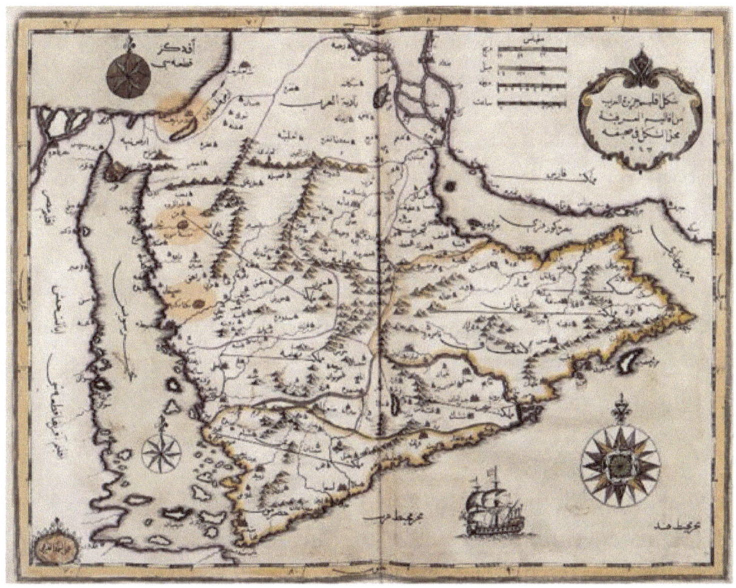

Figure 2.7-1: Drawing by Ahmed al-Kırımî of the Arabian Peninsula illustrated in *Displaying the World,* by the Ottoman Historian and geographer Katib Çelebi's 1609-1607. (Highlights are added here to show Mecca (most south), Medina (middle) and Jerusalem the most northern highlight). The top right scale is about 200 miles).

Arabian tribes attended successive major fairs which were scheduled to coincide with the season of pilgrimage in which safe travel is guaranteed through customs of forbidden aggression. They traded goods such as agricultural products, stock, birds, honey, olive oil, wine, perfume, spices, and clothing. There used to be a unique tradition of temporary adoption of newborn babies. Nursing mothers from the desert came to Mecca and offered to nurse and care for newborn babies back in their homes away from city-prone diseases. These babies, usually from wealthy families, grew healthier and articulate in the pure Arabic language of the desert. The Arabic language is one of the most advanced that mankind has

[6] According to Islam, the early monotheism teaching of Abraham degenerated to idolatry over the centuries. Arabs worshiped multiple stone and wooden gods which filled the premises of the Kaaba. There used to be nearly 400 statues including one of Jesus and one of Mary.

ever developed. Arabs meticulously recorded events, and other information they deemed important, such as lineages, using poetry. In addition to common poetry found in other languages, Arabs devised unique genres such as the *Fakhr* for stating pride, the *Medh* for elevating or promoting someone or a group of people and *Heja'a* for verbally attacking someone or a group of people. The finest poems in the *Ukaz* fair competition used to be hung on the walls of the Kaaba, indicating the importance of the spoken word. In fact, when Islam arrived, the proof of the prophet Mohammed had been the Qur'an itself, challenging anyone to duplicate its novel construct.

The Arabic calendar had sacred months, devised for pilgrims to travel unharmed. This spiritual designation, agreed upon by all tribes, formed the backbone for security over a limited period and allowed for economic commerce and exchange of goods in a desert with a meager food supply. Outside the sacred months, travel and caravan cargo required protection. Through negotiations, Meccan traders secured escorts for their merchandise by employing the aggressors. The system was respected in general. Tribes ceased hostility during the sacred months, with the understanding that everyone is bound to respond to violations decisively. The leaders in commerce from Mecca gained an honorable status that allowed them to enter in negotiations with the rulers of Rome, Persia and the Habesh of the Aksumite Empire of Eastern Africa to safeguard trade routes outside Arabia.

The Mecca calendar was not in use throughout Arabia. Other tribes, such as *Mudar* and *Rabiaa* had their own calendars, sometimes with different names of the months, and different sanctity status. Medina, the second prominent city of Arabia, to which the prophet Mohammed migrated in the year 622 CE, had a separate Arabic Calendar in use which was probably identical to that of the local Jewish tribes, at least when it comes to the start of the new month. It is not known if the Jews of Medina used the rabbinic system of fire-lighting atop the mountain ranges, discussed earlier, along with the mountain ranges highlighted in Figure 2.7-1, perhaps all the way to Jerusalem. The trade routes and other historical accounts clearly point to strong economic links between Medina and Jerusalem.

In the year 412 CE, fifty-four years after the release of the first Jewish calendar by Hillel II, a meeting of several leaders of Arab tribes led to the unification of the names of the months and their sequence. This meeting was needed to overcome problems with variation in the timing of the pilgrimage to the Kaaba, a practice that the Arabs were keen to do periodically since Abraham built the first house of worship with help from his sons. Before then, the names of the months and their arrangement varied among the Arab tribes. This landmark meeting of the Arab leaders which took place during the time of Kilab ibn Marrah, the fifth-grandfather of the Prophet Mohammed, resulted in the instauration of a unified Arabic lunisolar calendar that all tribes used for centuries before Islam and adopted later by Muslims after the abolition of the intercalation to become the Islamic calendar which is in use today. Four of the twelve months are sacred in which aggression, vengeance and any form of violence including tribal warring were forbidden. The same names were adopted in the Islamic Calendar, subject of Section 2.7.4, and their sanctity was further reinforced in Islam. The months in the Arabic calendar are named as follows:

1. **AL-Muharram**: (***Sacred***, means 'of Forbidden) which is the first month of the Islamic year, and the sacred month: so-called forbidden because the Arabs before Islam forbid warring during this month.
2. **Safar**: (origin word is sifr, i.e., zero, or nothing) this month is called Safar because Arabs left their homes without people to get provisions in expeditions named Safariyya.
3. **Rabīʿ al-Awwal**: (Rabīʿ I mean First Spring) **n**amed because of the coming of flowers and blossoms from the fall of dew and rain. Arabs refer to the fall season in the northern hemisphere as Spring. The Habasha in Africa also refers to this season spring.
4. **Rabīʿ ath-Thānī**: (Rabīʿ II means Second Spring) so named because it constituted the second phase of the spring, which is referred to as Autumn in the west.
5. **Jumādā al-Ūlā**: The first month of parched land, or for its freezing water
6. **Jumādā ath-Thāniya**: The second month of parched land and its freezing water.
7. **Rajab**: (**Sacred**, means of honor): The people are set to travel in security. Arabs stop fighting because of the sanctity of the month and honorable reasons.
8. **Shaʿbān**: Because this is a separation month between Rajab and Ramaḍān, also said the name means dispersion to seek water. In another opinion, Arabs separated for war after a pause in the month of Rajab.
9. **Ramaḍān**: So named because of the heat and intensity of the sun roasting stone.
10. **Shawwāl**: named for the time when camels become meager and dry from milk. It is also said for the oscillating heat.
11. **Dhū al-Qiʿda**: (***Sacred***, means 'of truce) Arabs refrain from warring during this month and stayed home.
12. **Dhū al-Ḥijja**: (***Sacred***, means 'of pilgrimage): Arabs before Islam visited the Kaaba in Mecca for pilgrimage "the Hajj" and performed pagan spiritualities. The yearly tradition was replaced by the reinstatement of the original worship of the God of Abraham and the elimination of pagan rituals. The Hajj in Islam takes place between 8-13 Dhū al-Ḥijja.

2.7.4 The Islamic calendar

No one cherished the moon as did the Muslims. They went as far as adopting the crescent as a symbol for their religion. The crescent moon is found in their flags, official buildings, houses of worship and common signs. In fact, the crescent is widely used as a symbol of their identity. Like it is the case in the Jewish faith, the sighting of the new moon is also a divine matter for Muslims. The following verses from the Qur'an are a statement for the Prophet Mohammed to follow the sighting of the new moon for time reckoning and for determining the time of pilgrimage, the practice of which is one of the pillars of Islam.

Concerning the New Moons
Say: They are but signs
To mark fixed periods of time
In (the affairs) of men
And the pilgrimage

(Qur'an: 2; 189)

Arabia had a few calendars in use by the main tribes. Two calendars were especially important, one in Mecca and the other in Medina. As the new religion of Islam took over Mecca, a calendar was needed, and the subject had to be addressed during the last year of the prophecy. The decision of whether Muslims should follow the Meccan or Medina calendar was probably a politically charged matter due to several reasons. The prophet had to choose one of the two calendars, but neither one appealed to the new religion. The Medina calendar had a strong connection to another religion as it was maintained by the Jewish people of the Medina. The people of Medina were wary that they may lose status and the prophet may even relocate now that Mecca is under Islam. The other side of the matter is that the Meccan calendar was associated with Arabian pagan practices which the victorious Muslims were working to end. Also, Muslims needed Mecca as the main site of worship and therefore keeping its time reckoning structure and geographic infrastructure was logistically important. It was the Meccan calendar that was retained at the end, and even this choice did no last too long. Management of the calendar was a specialty which was handled by members of an Arabian tribe which, like other key tasks of the time, such as the possession of keys and management of the Kaaba shrine, was in the hands of certain specialized people. The strategy of the prophet after his peaceful takeover was not to disrupt the lives of the Meccan population. He kept everything as is. Except for the stone and other idols counting to over 365 had to be destroyed, except the Jesus and Mary status which were buried somewhere near the Kaaba in respect to the people of the Book. The keys and management of the Kaaba were kept with the same family and the leadership of Mecca was kept unchanged. The calendar must respond to the new religion's celebrations and observances, however. Even though the Kalammas in charge of the calendar has adopted Islam, something had to be done to ensure the discipline of worship time and observance in the new religion. The topic of the calendar was addressed when the followers of the prophet asked questions about what to do with the timing of the hajj. As pointed out by Ilyas (Ilyas M. , 1993), during the Prophet Mohammed's last Pilgrimage to Mecca in the year CE 632, about ten years after his migration from Mecca to Medina, the topic received divine intervention through Qur'anic verses:

The number of months
In the sight of Allah
Is twelve (in a year)
So ordained by Him
The day He created
The heavens and the earth;
Of them four are sacred;
That is the straight usage

(Qur'an: 9; 36)

Verily the postponing
(of a prohibited month)
Is an addition to Unbelief:
The Unbelievers are led
To wrong thereby: for they make
It lawful one year,
And forbidden another year,
Of months forbidden by Allah
And make such forbidden ones lawful

(Qur'an: 9; 37)

The Prophet passed away three months later, and Muslims were left with two directives in this matter:

1. Number of months is twelve, four of which are forbidden (sacred)
2. Postponement of forbidden months is prohibited. The word "deferral" was used in another translation.

Following the above instructions, Muslims adopted a modified version of the Arabic calendar. The process of intercalation was dropped altogether without recorded protests, questions, or comments. The year of Hijra (migration) of the Prophet Mohammed from Mecca to Medina was chosen as the beginning of time reckoning. This had resulted in the introduction of the naming, *Anno Hijrae* (AH). The first use of the calendar is said to have taken place two and a half years into the succession of the second Caliph *(successor to the prophet) Omar ibn al-Khattab*, during the month of Rabīʿ al-Awwal of the year 17 AH. The Islamic calendar adopted the first day of the month AL-Muharram, of the year of Hijra as the first day of the new calendar, corresponding to on Friday, July 16th, 622 CE in the Julian calendar, i.e., Julian day 1948439. The start of the calendar corresponds to the time of Mohammed's flight from Mecca to Medina, reckoned as the first day of the first month of year 1 A.H. For religious purposes, the months are defined to start with the first observation of the crescent of the new moon.

Observances of the hajj, fasting of the month of Ramadan and other rituals started drifting through the seasons. The twelve months have either 29 or 30 days. The lunar year ends about 11 days earlier than the solar year. The days are organized in weeks, which are sets of 7 days, independently from the lunar day count in a lunar month, with al-'ahad (Sunday) as the first day, starting the previous day at sunset and ending at the subsequent sunset. The following days are named al-'ithnayn, ath-thoulatha', al-'arbi`a', al-khamis, al-jumou`a, and as-sabt. The names simply mean the first, second, etc.

Muhammad ibn Ahmad Al-Biruni, illustrated in Figure 2.7-2, is a medieval Muslim scholar who wrote extensively on multiple technical subjects including the calendars of ancient civilizations. He is considered one of the greatest pioneers of science. A translation of his technical writing on Arabian, Islamic, and other chronologies and traditions can be found in reference (Al-Biruni, 1879)[7]. His work is valuable not only because it came about three and a half centuries after the passing of the prophet Mohammed, but also because it addresses the pre-Islam Arabic calendar and the shift to the Islamic calendar. According to Al-Biruni, beside the importance of the Hijra in Islam, one of the main reasons the choice of the Hijra as origin was made is that it was witnessed by many and therefore there are no doubts about its correctness.

Figure 2.7-2 Muhammad ibn Aḥmad Al-Biruni (Abū Rayḥān) 973–1050 CE, born in Khawarism, Khorasan in what is today western Uzbekistan, Kazakhstan, and northern Turkmenistan, part of the greater Persia. He contributed to astronomy, mathematics, physics, medicine, Natural science, psychology, ethnography, anthropology, and geography. *Source:* Soviet-era stamp, public domain, Wikimedia Commons freely licensed media file repository.

The names of the months in the Arabic (and Islamic) calendar refer to the seasons and therefore indicate either a lunisolar origin or a solar one. However, such reference did not fit with the seasons, during the Islamic era and prior to Islam, with the months referring to cold, hot, and spring seasons wandering around the solar year. It makes complete sense that the Islamic calendar would have the months shift at a rate of about 11 days per year, but the Arabic calendar should not shift by more than 4 weeks throughout the years. For example, it can be shown that the 1st of the month of Dhū al-Ḥijja varied between Feb 06 and Mar 06 during the life of the prophet Mohammed, prior to the advent of the Islamic calendar. By construction, a lunisolar calendar is made to be in sync with the seasons, remaining in-step

[7] Edward Sachau in the preface of the translation said Al-Biruni betrays a strong aversion to the Arabs, the destroyers of Sasanian glory, and a marked predilection for all that is of Persian or Iranian nationality.

with a fairly accurate solar calendar and therefore there is no reason for major drifts whereby a month referring to the cold season, such as Jumādā al-Ūlā comes in the summer. The origin of the dilemma has remained a mystery which would have been forgotten if not to the emergence of the Islamic calendar which carries the same names, inheriting the same issue but overshadowed by its construct to float throughout the seasons.

According to Al-Biruni, there are several opinions on the etymology of the Arabic names for the months. The names were said to be given to certain tribes by their ancestor, and which have been referenced in Arabic poems. He also mentioned that pagan Arabs had originally let their months wonder around the year in a way like the Muslims have been doing, and therefore their pilgrimage shifted around the seasons. This had taken place until about 200 years before Hijra when they learned from the "Jews of their neighborhood" and applied intercalation, two centuries before the Hijra about the year 423 CE. This date is close to the time of the meeting of the Arab leaders of 412 CE, which led to the unification of the names of the months and their sequence. Could it be that the year 412 CE was the time in which Intercalation was initiated? Arabs did so to make the Hajj occur at an agreeable time of the year when their merchandise was ready for commerce, and the weather cooperates. Like the Jews, Arabs adopted "normal" and "leap" years. The leap years occur every two to three years, that is seven times in the 19-year lunar cycle. In a leap year, Arabs postponed the month of Al-Muharram, which is the first month of the year, to make room for an intercalation month which is not named, effectively postponing the first month of Al-Muharram. Postponement of Al-Muharram was abolished after the death of the prophet, thus giving rise to the Islamic pure lunar calendar as a substitute. The postponement was carried over generations by the Kalammas, a member from the tribe of Kinana who comes out during the fair and around the time of the Hajj and proclaims the following month to be named by the same name as the current month, in effect postponing a month, a process they also called Al-Nasi. According to Al-Biruni, the process of intercalation went beyond the month of Al-Muharram and has in fact cycled through the year as follows. The first intercalation happened in the month of Al-Muharram and therefore, Safar is called Al-Muharram, Rabīʿ I is called Safar, etc. shifting all months of the year. The subsequent intercalation happens in Safar, and therefore, Rabīʿ I is called Safar and so on until all months have changed. They continued until the intercalation happens again in Al-Muharram, at which time they repeat anew, moving the month in which intercalation occurs from one month to a subsequent one. Al-Biruni adds that the change in the position of the month in the season is recognizable from the Luna Mansions. This had given them a way to gauge when adjustments are needed. He added that Arabs counted the number of times the intercalation cycled through the year. They refer to dates by how many times the years have cycled around. However, it should be mentioned that no other reference from the era (following Islam or prior to Islam) corroborates this process. In addition, such a system of intercalation would need an excellent communication infrastructure throughout

Arabia and is, therefore, improbable. The other sources discussing the matter of intercalation refer to its occurrence only during the pagan hajj. We shall revisit this subject in detail in Section 10.5 at the end of the book.

As mentioned earlier, many of the original names of the months refer to the seasons, and therefore indicate the lunisolar (or solar) origin of the precursor Arabic calendar, as discussed in multiple references, e.g. (Duncan, 1999), (Ioh, 2014). We shall discuss in Sections 10.3-10.5 the topic of the emergence of the Islamic calendar, the origin of the names of the months and events associated with the transition from a lunisolar calendar to a pure calendar.

2.7.5 The Chinese calendar

It was only in 1960 that the West started understanding the contribution of Chinese heritage to science. Before then, doubts were often cast about what originated in China versus what was learned from the rest of the world. The lack of clarity was attributed to the language barrier and communication deficiency amongst scholars.

When it comes to calendars, the absence of immediate communication over a large swath of land together with stable administrative governance probably delayed the use of a standard calendar throughout China. We tend to erroneously think of China as a unified entity; it was instead ruled by separate powers for periods of time, with several political systems, multiple social structures, and a count of diverse dialects.

Chinese rulers throughout the different dynasties employed astronomers to keep a close watch on the sky as early as 800 BCE. Astronomical events were held as important omens. These astronomers kept an extensive record of solar and lunar eclipses, sunspots, novae, comets and other manifestation such as conjunctions. Under the order of one Emperor, wise Chinese thinkers started experimenting with a calendar as early as 2357 BCE, according to a legend. The calendar eventually became Metonic, adding seven months every 19 years (Duncan, 1999). In the west, a reference to the use of the Babylonian calendar by early Chinese dynasties is the default history for lack of other information. Excavation in Northern China unearthed many bones and other objects with inscriptions from the Shang dynasty which ruled over China around 1500-1050 BCE. There is also reference to the prior Xia dynasty between 2070 and 1600 BCE leaving written records, but no calendric documents or other science-related information can be traced back to the era. Texts from the Shang era dating between c. 1350 and 1050 BCE in the form of oracular texts inscribed by diviners on animal bones and other prepared media. Inscriptions are made using hot needles to generate cracks which are interpreted by the diviner; a practice widely used during that era. The same medium contains an old script with information on the divination. The bones which were buried to prevent desecration preserved valuable texts and records of solar and lunar eclipses in the form of a day number in a sexagenary cycle referred to as the Shang calendar.

As discussed earlier, uniform usage is needed for a calendric system to take hold. There are, however, several calendar variants in China. The Chinese calendars used in early history are lunisolar with many variants and evolving intercalations. The Chinese used numbers to identify the different months but used the zodiac to name the years in a cycle of 12. These are the Rat, Ox, Tiger, Hare, Dragon, Snake, Horse, Sheep, Monkey, Rooster, Dog and Pig.

Today, besides using the Gregorian official calendar, Chinese people still use a lunisolar calendar for social and traditional activities. The month starts at the astronomical new moon and does not relate to the phases of the moon.

Figure 2.7-3 Oracle bone from the Shang dynasty c.1500 -1050 B.C.E. Credit: The Chinese University of Hong Kong Library[8] -Creative Commons License terms.

2.7.6 The South-Asian calendars

South-Asia refers to the region of the sub-Himalayan territories of the Asian continent, including what is today Afghanistan, Bangladesh, Bhutan, India, Maldives, Mauritius, Nepal, Pakistan, Siri Lanka, and Tibet. South Asia is home to well over 20 percent of the world

[8] http://archives.lib.cuhk.edu.hk/repositories/5/archival_objects/209867

population and has seen some of the earliest calendric developments, given the confirmed advances in science, astronomy, and mathematics.

Figure 2.7-4 Map of South East Asia

Among others, early calendric information must have been lost or has yet to be unearthed. There has been a mention of a serious lack of proper dating and availability of literature together with a need for a massive undertaken to bring to light the history of some of the earliest vibrant civilizations. Uncertainty on the true nature of contacts with other neighboring civilizations and questions on the independence of the development of ideas have prevented western historians from drawing final conclusions on South Asian achievements.

The Soanians are the earliest settlers of the Soan Valley of Punjab in what is today Pakistan. Early hominids, including Homo erectus Soanians, used stone tools during the Lower Paleolithic, approximately 500,000 years ago. The Indus Valley region which covers most of Pakistan, Northern India and Afghanistan saw Homo sapien activities that culminated in multiple ancient cultures starting with the Mehrgarth settlement where agriculture and communities settled. About 6500 BCE, and perhaps earlier, some form of molar drilling seems to have been practiced marking the first practice of dentistry. Civilization in South Asia dates to that of Near East of China, to about 4000 years ago. Complex town planning, engineering, and trade date to 2300 -1750 B.C.E in Mohenjo Daro and Harappa. Where sophisticated and technologically advanced urban culture with paved streets and a developed economy even more advanced to that of Mesopotamia in certain ways was unearthed. This civilization left a written script language in the form of seals. Following this early prosperity, the Vedic Civilization emerged, mainly in Gandharan, Taksasila (in today Taxila) in Punjab (1500-500 BCE) and were driven by the invasions of the Indo-Aryan cultures from the North that ultimately overrun the entire subcontinent and put in place a precursor to the Caste system. This system partitioned people in four distinct groups, warriors, priests, merchants, and laborers. The heritage of the Vedic belief system was developed, written down, and transferred through generations and was ultimately used for the development of Science and Hinduism and science. Following the Vedic civilization, several empires ruled over the cradle of this ancient civilization, notably the Persian Achaemenid Empire about five centuries before common era, two centuries later it was part of Alexander the Great's empire, succeeded by the Maurya Empire until 185 BCE, and for a short period thereafter by the Indo-Greek Kingdom.

At about 1000 BCE, a unique approach was used by the Vedic civilization to reckon time. The position of the moon is tracked against the fixed stars, yielding a 27 or 28 –day month (an actual cycle is 27.32 days). The calendar adopted a 12-month, and a 360-day year requiring intercalation of some sort. They tried two approaches: adding days and adding one month at a regular interval, settling on the latter around 100 BCE. They also seem to have shifted from the 27-28 day month towards a 29-30 day month (Ronan, 1982).

In 1952, over thirty calendars may be accounted for in the different regions of South Asia. The two leading calendars which have seen many evolutions over the centuries: The Vikrama and the Shalivahana are lunar calendars. The former is adopted in Northern India and Nepal, while the latter is mainly used in the south. There is also a separate Tamil solar calendar and a well-developed lunisolar Hindu calendar.

There are two ways to recon the month, one counting from Full moon to Full moon and the other from New moon to New moon. The start of the month has also been a challenge for many Hindu priests in charge of tracking the month. Even a full moon can seem to persist over two and sometimes more days making it hard to determine the end and the beginning of the month. Certain versions of the calendar account for a lunar month of 29 to 32 days. There is also a cycle of sixty years in some of the Hindu Calendars.

Yoga is a name used in the West to refer to several ancient practices which originated in ancient India several centuries BCE and maybe even earlier. These practices rooted in Hinduism, Buddhism, and Jainism involve physical movements, meditations, and spiritual activities. Yoga arrived in western societies in the 20th century, gaining traction with individuals who struggle to cope with the daily stress of metropolitan life.

The moon cycle is important to most yoga teachings, and yoga practitioners are often attuned to the natural cycle of the moon. For example, Ashtanga, a form of Hatha Yoga, one of the two most known forms of yoga, discourages the practice of physical exercises on both new and full moon days, which are considered holidays. The force of Prana is considered the greatest acting upward to reach the head on a full moon. The fullness of the moon parallels the end of a breathing cycle and illustrates a feeling of maximum energy and emotion. The new moon is associated with the end of a breathing cycle in which the force of Apana is greatest, acting downwards giving a sense of groundedness and calmness.

Figure 2.7-5 Julie Peters is tuning to the energy of the new moon through a yoga pose called "Child's Pose" (self-reflection). Credit: with permission from Julie Peters[9]

The phases of the moon are tracked across the different Yoga schools. In general, the full moon holiday is associated with abundance and generosity, while the new moon is a time of quietness and meditation. The waxing moon is the time for a new beginning with strengthening forces and energy and are deemed good for starts of new activities. Exceptions to these rules and guidelines occur for w major holidays.

The common denominator in tracking the lunar moon in Yoga practices is the alignment of personal activities to coincide with the manifestations of the natural world. The moon is a major player in our world, and its phases matter a great deal in Yoga practice.

Finally, the question remains that while old yoga practitioners adopted the first crescent as the beginning of the month, today's practitioners follow the astronomical calendar, which states the conjunction as the start of the month. Obviously, the controversy is present for the yoga practice as well.

2.7.7 The Hellenic calendars

Early Ionian Science, starting the 7th century BCE is hailed by scholars as to the beginning of something new for its investigative nature which is separate this time from gods and

---

[9] http://jcpeters.ca

superstitions: "the Scientific Age" has begun (Ronan, 1982). Marshall Clagett, professor of the History of Science at the Institute for Advanced Studies at Princeton, President of the History of Science Society 1962-1964 and editor of Isis, the principal journal in the field stated that historians, in recognition of the Gulf that separates the approach of this natural philosophy emerging from Ionic states from that of the cosmology of the Egyptians and Babylonians, have called this phenomenon the "Great Miracle" (Clagett, 1976). Ronan explained that there is no special reason, whether racial or geographical, except there is new settler living in a new political environment of their own who started asking questions and providing different answers that are not bound by social or other constraints. They lived in a port hub where commerce thrived from the east and south-east, from the Fertile Crescent and from far away IRAN, India and even China. The Ionian lived in such thriving environment such as Miletus. It has been about a century since the Phoenicians were on the move. They settled in North Africa and started building Carthage, they have also resettled in Miletus at about the same time. It was where the Milesian school of philosophy and science was born led By Thales whose parents are of Phoenician origin, Anaximander and Anaximenes, which focused on the origin of physical processes instead of the supernatural intervention.

Thales (624-524 BCE) was statesman, a philosopher, a business tycoon, a mathematician, and astronomer. He was considered as one of the seven Wise Men of Greece. Thales was credited of all sorts of mathematical achievements including measuring the height of pyramids and measuring distances at sea. Ronan (Ronan, 1982) wrote that Thales's fame rests with his astronomy and that he is said to have predicted the total eclipse of the sun of 28 May, 525 BCE. He is supposed to have used the Saros cycle familiar to the Babylonians, which he learned during his trip to Egypt. "Most historians are satisfied that he could not have done so, at least not as the fifth-century Greek historian Herodotus described, for Herodotus states that Thales was able to tell within the year' when it happened."

During Classical Greece, in the fifth and fourth centuries BCE, much of present-day Greece had the same fate as that of the Neo Babylonian empire; they were under the influence of the Achaemenid Persians. The Greeks were scattered with alliances of city-states engaged in exhaustive fighting and did not have unifying political governance to align their civil affairs, let alone their calendars. The Greeks ended with many regional calendars. The Persians were not interested in interfering with the inner fabrics of the different people they occupied. Therefore, the lack of a unified calendar was due to the lack of agreement between the different city-states which were effectively under different political rules. The political will to enforce a calendar for the greater good across the city-states was absent. The idea that classical Greece was a unit is only in the minds of western historians who kept looking for an origin for the western civilization. As a matter of fact, it was erroneously and commonly reported that efforts to bring the lunar calendar in sync with the seasons culminated in the Metonic cycle, proposed by Meton in 425 BCE, and this cycle was made use of for several centuries thereafter, e.g. (Lindberg, 1992). It is well established that the 19-year cycle was identified earlier by the Babylonians (see Section 2.6.2).

Even when Alexander the Great of Macedonia (356–323 BC) set to fulfill his late father Phillip II's Dream of conquering all of Greece and then liberating it from Persian dominion, It can be argued that he was absorbed by the Military aspects of his military conquest and not managing the vast empire which extended from what is western Libya today to the edges of India. His life was dedicated to military action and not in building the infrastructure for a civil society that could make use of a good calendar. As a matter of fact, his empire was another short-lived one and died with him in 323 BCE. It was the end of what historians call the classic era (also called Hellenic era) and the beginning of the Hellenistic era.

Historians consider the Hellenistic era to be the period between the death of Alexander the Great in 323 BCE, when the partitioning of his empire by his military generals and the rise of the Roman Empire identified to be either marked the occupation of the Greek motherland following the Achaean war in 146 BCE or the Battle of Actium in 31 BCE which immediately led to the fall of Ptolemaic Egypt.

Ronan (Ronan, 1982) wrote: "The Greeks claimed to have received their Mathematics from Egypt by way of Thales; Herodotus, Aristotle and his pupil Eudemos, who wrote a history of Mathematics, all claim that Thales, after a trip to Egypt, brought this study to Greece". Achievements in Astronomy did not happen overnight. There is a need for lengthy observations and recording. Both happened to have occurred with the priests of Babylonia and Sumerians before them and the Egyptians. It took time and dedication; that work has led to the discovery of the lunar, lunisolar and solar calendars. The Greeks did not set up observatories to do astronomy for an extended time. They have certainly propelled humanity in many fields of science and their great achievements in geometry and mathematics are with no doubt impressive. There is however a missing link Greek historian choose not to be clear about. History did repeat itself with Western historians referred to Muslims as only translators when absorbing the Islamic achievements in Spain during the massive translation era.

2.7.8 The Persian calendar

Persians may have not invented the calendar, but their astronomers perfected several aspects of it, the most important one is accuracy. The use of advanced calendars goes back to at least the later Achaemenian period (650 to 330 BCE). Inscriptions show that early Persians used a 360-day Babylonian system of calendars. They are also behind the development of the original Civil Islamic calendar, at least its algorithmic and computational aspects.

The Persians have long used an accurate solar calendar with an origin at the vernal equinox (midnight between the two-consecutive solar noons between which the instant of the Northern spring equinox occurs). The first day is a day of celebration and is called Norouz (meaning new day). The calendar is observation-based and not rule-based with accurate but complex leap year system. The names, sequence, and lengths of the months are as follows:

Order	Days	Month name	Order	Days	Month name
1	31	Farvardin	7	30	Mehr
2	31	Ordibehesht	8	30	Aban
3	31	Khordad	9	30	Azar
4	31	Tir	10	30	Dey
5	31	Mordad	11	30	Bahman
6	31	Shahrivar	12	29/30	Esfand

The week starts on Saturday and ends on Friday. The days are Shanbeh, Yekshanbeh, Doshanbeh, Seshanbeh, Chaharshanbeh, Panjshanbeh and Jom'eh. Earlier versions had names for each day of the month. The length of the months varied each year, and a month could even count 32 days.

Like other calendars, the Persian calendar has evolved over the centuries, and culturally, they had a preference to measure time and keep track of the seasons by relying on the sun instead of the moon. With the arrival of Islam in the 7th century, the Iranian had to switch to the lunar Islamic calendar. And it was not until the reign of Sultan Jalaal ad-Din Malik Shah Seljuki in the 11th century that the latest form of the solar calendar was devised. It is often referred to as the Jalali calendar. However, the Islamic calendar kept being used until the beginning of the 20th century. During the Pahlavi era, in 1925, the Islamic lunar calendar was officially replaced by the Iranian solar calendar, reviving the ancient Persian names, which are still in use today.

2.7.9 The Zodiac calendar

The Babylonians came up with subdivision of the year into 12 equal parts and the corresponding zodiac. The early Babylonian zodiac entered the Greek culture together with some Egyptian influences to give rise to what is now known as the western zodiac which is traced to the 12th century of the translations from Ptolemy's work of the 2nd century BCE. The zodiac partitions the plane of the ecliptic into 12 equal 30 degrees sectors, as shown in Figure 2.7-6. These parts are given symbols and Latin names, as shown in Table 2.7-1. The partition starts at the point of Aries. Month after month, the sun appears to move from one sign to another over the course of a year, and so does the moon. In two to three years, two moons show up in the same sign. The full moon in that thirteenth moon is called Blue Moon in the Old Farmer's Almanac.

Table 2.7-1 The longitude divisions of the Zodiac name and symbols

Aries	Taurus	Gemini	Cancer	Leo	Virgo
♈	♉	♊	♋	♌	♍
Ram	Bull	Twins	Crab	Lion	Maiden
0°- 30°	30°- 60°	60°- 90°	90°- 120°	120°- 150°	150°- 180°

Libra	Scorpio	Sagittarius	Capricorn	Aquarius	Pisces
♎	♏	♐	♑	♒	♓
Scales	Scorpion	Archer	Mountain Sea-goat	Water-bearer	Fish
180°- 210°	210°- 240°	240°- 270°	270°- 300°	300°- 330°	330°- 360°

Figure 2.7-6 The belt of the Zodiac is the apparent path of the ecliptic across the celestial sphere (adapted from Leo Reynolds on Visual Hunt / Permission Creative Common)

2.8 Western calendars

It was on October 1, 1949, that the Gregorian calendar was set for use worldwide after Chairman Mao Zedong announced that China would adhere to its use. Today's Gregorian calendar is adopted by everyone for business, administrative work, and international transactions. All computers have their operating systems equipped with this calendar. In this section, we shall go through the progress of the calendar in western civilization all the way to the emergence of the Gregorian calendar.

2.8.1 The Roman republic calendar

The origin of the solar calendar goes back to the Roman republic calendar which in turn was based on primitive Roman and Greek calendar which dates to the 8th century BCE. The early

Roman calendar numbered the years *ab urbe condita*, that is, "from the founding of the city", that is the City of Rome in 753 BCE and clearly based on the naming of the months (September, October, November, and December), its original version had only ten months and by the first century BCE two months were added. The months were based on the Synodic month and alternated between 29 and 30 days to give the average of 29.5 days. The lunar year adds to 354 days, while the tropical year adds to 365.25 days. The difference between the two scales. The difference amounts to an out-of-step of a month every three years or so. It is then decided to add a thirteenth month every two years to synchronize with the seasons. The priests who oversaw the declaration of the announcement of the leap year abused the count to keep their allies in charge for a long time. This controversy led to a miscommunication, and people end up having a disagreement over the applicable year. This has led to the reform introduced by Julius Cesar in the year 46 BCE.

2.8.2 The Julian calendar

In the year 46 BCE, the emerging Roman Empire needed a good calendar to replace a dysfunctional Roman republic calendar which suffers from being out-of-step with the seasons among other problems. The other problem is the same as the one with any other lunar calendar problem: the Kalendae (Latin name for the first day of the month, the origin of the word calendar) and the difficulty of its determination.

After defeating the pharaoh's forces in 47 BCE and installing his sister-wife Cleopatra as a ruler over Egypt Alexandria, he adopted the Alexandrian Aristarchus' calendar of 239 BC, which consisted of a solar year of twelve months and of 365 days with an extra day every fourth year. Multiple references state that Julius Caesar consulted the Alexandrian astronomer Sosigenes. Little is known about Sosigines beside the reference from Gaius Plinius Secundus (CE 23 –CE 79), a Roman author who wrote "In addition to the three main schools, the Chaldean, the Egyptian, and the Greek, Julius Cesar added a fourth one with the assistance of the astronomer Sosigenes who brought the separate years back into conformity with the course of the sun". A good account of the relationship of Julius Caesar and Sosigenes who comes from the lineage of knowledge of Alexandria is addressed in (Preston, 2009). As far back as 238 BCE, a reform of the Egyptian Calendar was ordered by Ptolemy III Euergetes, the third king of the Ptolemaic dynasty in Egypt (reigned 246–222 BCE). Ptolemy III's stone contains decrees about priestly hieroglyph orders to use of a leap day which must be added to the calendar of 365 days and the associated changes in festivals.

In addition to dropping the lunar synodic month as a basic unit of the calendar, and adopting the tropical year of 365.25 days, Sosigines's calendar design has 365 days in a year, with a leap year of 366 days every year that is divisible by four. The additional day is added to the month of February. The average year would then be 365.25 days. January was assigned to be the first month of the year instead of March, rendering January 1, 45 BCE the first day of the Calendar.

The insertion of a leap year is equivalent to the adjustment made to the lunar calendar through intercalation of a thirteenth month every two or three years to bring the lunar calendar in par with the seasons. Because the vernal equinox had fallen out of its –then conventional date of March 25 in the Roman Republican calendar, Sosigines intercalated three extra months in the year 46 BCE, making it 445 days long. That year was called the "last year of year of confusion." The above constituted the Julian calendar in which the names of the months (for exmple September, November, and October) inferred an erroneous numbering. There were no subdivisions as weeks or parts of a month. After Julius death, the Roman Senate changed the fifth month, Quintullus (also known as Quintlis), the month of Julius birth, to July in his honor. The Roman Senate did the same again renaming the sixth month, Sixtilis, to August in honor of the Augustus Caesar, the successor of Julius Caesar.

As the Church inherited the Roman Empire's institutions, it had to address lunar dating in relation to the Julian calendar. The first milestone in the record was during the historical Council of Nicaea (Nice) in 325 CE. The dates of observance of Easter and other events were addressed. It was agreed that Easter falls on the first Sunday after the fourteenth day of the lunar month (near full moon) that occurs on or after March 21. In 325 CE, March 21 was the date of the Vernal Equinox. The choice of Sunday after the full moon was chosen to prevent possible coincidence with the Jewish Passover. Over time, the vernal equinox has actually gradually slipped from March 25 to March 21 because the Julian year of 365.25 days is 674 seconds longer than the tropical year of 365 days and 0.24219 days (5 hours 48 minutes and 46 seconds), and from 45 BCE to 325 CE this discrepancy accumulated to about three days.

2.8.3 Computus and Easter Tables

At the closure of the antique period with the fall of Rome in 455 CE, the middle ages settled in Europe for about five centuries. It was a barren period in the history of European civilization (Mason, 1979) time characterized by a lack of written records which lasted until the intellectual reawakening under Pope Sylvester II (999-1003). Two Roman statesman and authors from the sixth century who were instrumental in transmitting some elementary Greek knowledge to Western Europe were Boethius (ca. 480-524) and Cassiodorus (ca. 488-575). Neither one was a scientist, but they played an important role in the early history of science (Clagett, 1994). Boethius was born just four years after the last Roman Emperor was deposed. He was a Roman senator, consul, and a philosopher known for the writing of ***Consolation of Philosophy*** during his imprisonment. He had several translations and commentaries on Aristotelian logical writing and the mathematical Pythagorean topics. He also compressed the work of Nicomachus, ***Introduction to Arithmetic,*** to Latin as well as other translations from Greek treatises including the work of Ptolemy, and the theory of vibration in music. His work, especially in arithmetical concepts, is considered as the first in Latin early middle age. He produced two other translations which were lost: one in geometry and the other in astronomy. Cassiodorus was less intellectual than Boethius but had a role in the preservation of written manuscripts from monks and other sources. From his influential positions in the entourage of Ostrogoth kings, he encouraged learning and collected manuscripts, probably a first in the time.

Later encyclopedias and epitomes were assembled by others, including Isidore of Sevile, who took over the responsibilities of his brother the Bishop of Sevile for 35 years. Even though he did not have the mastery of the Greek language, Isidore produced two documents: O***n the Nature of Things***, a short work on several topics including elementary astronomy, cosmology, geography, and meteorology. He also wrote on liberal arts, medicine, law, chronology, and other topics before he died in CE 636. He wrote in a brief descriptive form, never based on observation and lacked understanding of the subjects. Nevertheless, several authors followed his footsteps. The famous English church historian Bede, (ca. 673-735) was one of them. He was an English monk at the monasteries of St. Peter and St. Paul in England. Credit for his education goes to the Greek monk Theodore of Tarsus and to Hadrian, former abbot of a monastery in Italy. He is known as "the father of the English History" for his work ***Ecclesiastical History of the English People,*** but more relevant to the topic of the moon, he was able to make remarkable contributions to ***Computus:*** a collection of materials used in the development of ecclesiastical calendar, that is, for the preparation and understanding of Easter tables. These tables address the 19-year lunar cycle, the day of the lunar month, what sort of day and moon is the Pascal solemnity, how Easter is to be computed, and the ascension and descending of the sun. Before discussing the history of the ecclesiastical calendar, we continue the history of the Middle Ages and stop by another milestone in the history of knowledge in western Europe which set the stage for the most effective proliferation of science and emancipation of mankind.

2.8.4 The French Roman Emperor's calendar

The French King Charlemagne (742-814 CE) never learned writing or reading at a young age. He struggled to acquire such knowledge at a mature ruler. Charlemagne became King of the Franks in 771 CE, a Germanic tribe that settled over a large swath of present-day Western Europe encompassing The Netherlands, West Germany, Luxembourg, and France. He waged a savage war on the Saxon, another Germanic tribe. He intended to unify the Germanic people and force their conversion to Christianity. His wisdom and perhaps also his literacy handicap made him appreciates knowledge. He had two tutors to help him overcome his lack of reading and writing but could not progress much, even though his entourage was not literate. His military success received immediate attention from the Pope in Rome who saw in him the potential to restore the obliterated Roman Empire and therefore offered him the title of the Holy Roman Emperor which he accepted in the year 800 CE.

Knowing Charlemagne's fascination with Hourglasses, (there was a disputed report that he had a very large one that can last a twelve-hour period), the caliph Haroun al-Rashid in Bagdad send him presents including the earliest geared clock as he saw in him a key person to convince to adopt Islam, but the Pope in Rome won the competition.

Charlemagne provided generous stipends to scholars who came from all of Europe in completion with Haroun Al-Rashid's Beit El Hekma. Even scholars from Muslim Spain joined his court. He also ordered that all children receive educations in Monasteries, In 789 CE, he ordered: 'let those who can, teach.' 'Let the ministers of God's altar... collect and associate with themselves children ...that there may be schools for reading-boys. Let them learn psalms, notes, chants, the computes, and grammar in every monastery and bishop's house' (Duncan, 1999). Charlemagne, who knows of the ways of the Moslem's education in teaching the memorization of the Qur'an and other forms of knowledge, wanted to duplicate what they had in the Christian backward and illiterate world.

Being a decisive Barbarian warrior, he acted as if he was faced with a war to propel his owns to catch up. His death and subsequent defragmentation of his empire did not see a continuation of what he started. But it ignited the first light in a world where life expectancy was 35 years, and half the children died before the age of five. A plague hits every quarter of a century.

Because emperors come with a calendar, Charlemagne needed one. He named the months of January through December: Wintarmanoth, Hornung, Lentzinmanoth, Ostarmanoth, Winnemanth, Brachmanoth, Heuvimanoth, Aranmanoth, Witumanoth, Windumanth, Herbistmanoth, and Heilagmanoth.

Emperor Charlemagne died in 814 CE after a reign of 47 years. His affection toward science died with him. The learning centers that he started were closed, the scholars in his court were dismissed, the political order he managed to impose through his decisiveness vanished and his empire was fragmented among his fighting heirs.

2.8.5 Ecclesiastical lunar calendar

The Ecclesiastical calendar (also referred to as the Church calendar) is based on the lunisolar cycle and is used by Christian churches in determining the dates and periods of fasts and festivals. Reference to the ecclesiastical lunar calendar in the western world was as early as the middle of the 13th century, during the fragmentation of the Islamic empire took place. However, its existence must have been in practice much earlier. The Ecclesiastical lunisolar calendar of the medieval time begins with the first day the crescent is visible. Because the crescent new moon could not be seen for certain, errors are also common. Roger Bacon (1214-1292 CE) addressed the issue of the announced dates of the lunar month being off by three to four days.

Figure 2.8-1: Statue of Roger Bacon (1214-1292) earliest European thinker and philosopher, Oxford Museum of Natural History. Source: Wikipedia, Photography by Michael Reeve, Creative Commons Attribution-ShareAlike 3.0 unported license.

In his treatise, Opus Majus, Bacon also proposed a calendric reform similar to the later system introduced in the Gregorian calendar. He referred to the Julian calendar as "intolerable, horrible and laughable." The length of the solar year of 365.25 days has led to gross error over the course of centuries which has meant the date of Easter slipped by nine days since the First Council of Nicaea in 325 CE. Bacon proposed to drop one day every 125 years and not to rely on fixed equinoxes and solstices. His complaints were addressed three centuries later by Pope Gregory XIII in 1574 CE. It must be noted that Roger Bacon was one of the earliest Europeans who understood the importance of experimental science, philosophy, and arts. He was impressed by the translations of the work of Abu Ali Hasan Ibn al-Haitham (Alhazen) in optics and astronomy, from which practical results were emerging at the time, such as the manufacturing of eyeglasses in Italy and telescopes that peer in the heavens. He was influenced by Yakub Al-Kindi[10] in philosophy, ethics, and theology. Roger Bacon was the first to write the formula for gunpowder in the western world[11]. Bacon referred to Islamic thinkers as "both philosophers and sacred writers" and promoted the integration of Islamic philosophy into Christian learning.

The Ecclesiastical calendar is based on movable and immovable feasts. Easter is the principal movable feast to which most other movable feasts refer to, while the principal immovable feast is Christmas and is fixed to December 25. As mentioned earlier in Computus, Latin name given to the calculation of the day of Easter which was fixed to fall on the vernal equinox, fixed to 21st of March, and the full moon is fixed at 14 days after the beginning of the ecclesiastical

---

[10] Scholar from the House of Wisdom of Kufa

[11] , information probably obtained from Arabic translations as well in his treatise Opus Majus of 1267. This 9th century Chinese invention made it to the Islamic world with the Mongolian invasions and led to multiple experimentations such as those documented in the middle of the 13th century by the Arab chemist and engineer Hasan Al-Rammah

new moon. The vernal equinox does not necessarily fall on March 21, and the full moon is not necessarily 14 days from the ecclesiastical new moon. The new moon in the Easter date Tables, which are derived based on the Metonic cycle, is considered day 1. The new moon, as intended in medieval time was the traditional visible first light. When Roger Bacon specifically mentioned the discrepancy[12] involving the ecclesiastical new moon, he was certainly referring to the observed new moon and was not referring to the onset of conjunction. Some churches announce the lunar date daily.

2.8.6 The Gregorian calendar

In the year 1570, the Italian astronomer Aloysius Lilius (Luigi Lilio) made a proposal to adjust the outdated Julian calendar and correct for the accumulation of the yearly seasonal slip of 674 seconds. A papal committee took over his ideas, but it took several years for the ecclesiastical administration before something concrete came out. In the meantime, Lilius had died and the German Jesuit Christopher Clavius had taken over his work.

In 1582, the yearly seasonal slip of 674 seconds had accumulated nearly 13 days in the Julian calendar and if no action is taken, the first day of spring will come in the winter season. One option was to gradually straighten out the differences by carrying out a small correction over several years, but Pope Gregory XIII (born Ugo Buoncompagni) opted for a drastic measure. The result was that the day following Thursday, October 4, 1582, was re-dated as Friday, October 15, 1582.

To provide a lasting correction to the slip in the seasons, and manage the difference between the tropical year length of 365.24250000 and the actual value of 365.242199 which accumulates to a full day every 128 year, while maintain simplicity, it was decided that three leap years in four centuries which are divisible by 4 and not by 400 revert back to common years. This makes the year 1700, 1800 and 1900, which are divisible by 4 become common and renders the average Gregorian year 365.2425 days (correct to 1 day every 3300 years). Today, the Gregorian calendar has been further tuned to make it accurate to 1 day every 20,000 years by making the year multiple of 4000 a common one.

The day of Easter which was defined to be the Sunday following the full moon succeeding the Northern Spring Equinox (referred to as Paschal moon), which was fixed to fall on the 21st of March, and the full moon is fixed at 14 days after the beginning of the ecclesiastical new moon. With the corrections introduced in the Gregorian calendar, the 21st of March is now April 3rd in the Gregorian calendar. Therefore the date of Easter which varied from March 22 to April

[12] Christianity was not immune to the new moon controversy either

25 inclusive in the Julian calendar, now varies between April 4 and May 8 in the Gregorian calendar. Eastern Christianity still use the Julian calendar today

Figure 2.8-2: Pope Gregory XIII and the Decree of the Gregorian calendar of 1582. Source: public domain, Wikimedia Commons freely licensed media file repository

Pope Gregory XIII studied law in Bolognia, taught law and was appointed as a judge. After his appointment as Cardinal-Priest, his legate to Philip II of Spain produced a strong connection with the Spanish monarchy. In addition to his legal expertise, he was also a highly productive manager who could get things done. He was decisive an independent. The influence of the cardinals waned during his tenure, and as a result, he brought sound reforms, one of which was the triumph of the Gregorian calendar. He produced an advanced edition of the Corpus juris canonici, compilations of law in the Roman Catholic Church, the main source of ecclesiastical legislation which was in effect until 1917 when it was replaced by the Codex Juris Canonici (Code of Canon Law). He also replaced the Cardinals Consistories, with many new colleges of the Jesuits throughout Europe, The Roman College of the Jesuits became the most important center of learning in Europe, known then as the University of the Nations. It is now named the Pontifical Gregorian University. Pope Gregory XIII founded numerous seminaries for training priests, beginning with the German College at Rome, and put them in the charge of the Jesuits. He commissioned artists, to decorate the Vatican, including the commission of the 120 m Gallery of Maps.

2.9 Eastern Orthodox calendar

The Eastern Orthodox Liturgical Calendar organizes all religious observances of the Eastern Orthodox Church. Detailed and complex commemorations rules involving fasting/feasting, prayers, and ceremonies are scheduled depending on the day and time of the year.

The Eastern Christian Orthodox Church uses the old Julian calendar in the determination of these celebrations. Feasts in the Orthodox Church calendar can be classified as fixed and movable. Fixed feasts occur on the same calendar day every year, while movable feasts vary from year to year. The moveable feasts relate to Pascha referred to as Easter in western churches. Pascha is central to the Orthodox Church and as is for Western churches commemorate the event of the resurrection of Jesus. The word *Pascha* (Greek: Πάσχα), is a transliteration of the Greek word, which is itself a transliteration of the Hebrew Pesach, which means Passover. The Jewish people celebrate Passover, the day of their liberation from slavery in Egypt. Some Orthodox Christians discourage the use of the word Easter, as the English name is related to the pagan spring festival of fertility. The origin of the term Easter comes from the Germanic name for the month in which the Christian celebration usually fell and is called Eostur-monath. Eostre is the pagan Anglo-Saxon goddess of spring and monath is the old English name for the month, which has the connotation of the moon.

Pasha is often celebrated at a different date than Easter in the West. Even though the rule for the determination of the date is the same in the Eastern and Western Church, the use of a different calendar is a source of controversy.

The Eastern Church did not adopt the Gregorian calendar. It took almost four centuries to have a proposal to synchronize the Eastern Orthodox Church calendar with the seasons. During the 19th century, the entire Eastern Church still used the Julian calendar (due to Sosigenes of Alexandria). The German astronomer Johann Heinrich von Mädler (1794-1874) made accurate determinations of the length of the tropical year and proposed to Tsar of Russia, and the Orthodox Church to drop 12 days in the year 1900. He suggested a 128-year rule for an additional common year. And if the years which are multiples of 3200 are not leap, the duration of the mean year will then be 365.2421875 days very close to the real duration of the mean tropical year which lasts 365.24219 days. So that a very little error of only 0.6-hour error in 10,000 years will then be made. This would have been a great improvement on the Sosigenes tropical year estimate of 365.25 days. The Tsar and the Orthodox clergy end up not accepting the recommendation from Germany. Following the Bolshevik Revolution, Lenin of Russia adopted the western Gregorian calendar, with its conventional tropic year of 365.2425 days (i.e., 365d. 5h 49min. 12s) due to Christopher Clavius (1537-1612).

In 1923, Serbian astronomer Milutin Milanković, who is well known for the Milanković cycle, proposed a variant to the von Mädler leap rule. He suggested applying the 128-year adjustment at the end of the century. Every year which divides exactly by four would be a leap one, except years which divided exactly by 100 are common, unless when divided by 900 they gave 200 or 600 as remainder. The leap rule of Milanković puts the mean year within two seconds of the then-current length of the mean tropical year.

In May 1923, what was referred to as Pan-Orthodox Congress of Constantinople was only attended by the followers of the Patriarchs of Constantinople, and Serbia. Represented were also Croats and Slovene eastern churches and absent were the Patriarchates of Jerusalem, and Alexandria and the Russian Orthodox Church. In this meeting, the Milanković proposal was adopted. It synchronizes a new calendar with the Gregorian calendar by specifying that October 1st of the Julian calendar would be October 14 in the new calendar, thus dropping thirteen days. Easter was proposed to be Sunday after the first day (midnight-to-midnight) of the full moon following the occurrence of the vernal equinox at the meridian of the Church of Holy Sepulcher in Jerusalem. If the full moon occurs on a Sunday, Easter will be delayed one week to the following Sunday. All Eastern Orthodox churches rejected this rule and continue to use the Julian calendar to determine the day of Easter. The Finnish and Orthodox Churches use the Gregorian Easter. Today, those Orthodox Churches that advocate for a revised calendar are called sometime the new Calendarists, e.g., Orthodox churches of Constantinople, Alexandria, Czech and Slovak Orthodox Church, and most of the Orthodox Church of America. The old-Calendarists are Orthodox Christians who rejected updates to the Julian liturgical calendar to keep it on par with the seasons. These proposed updates, in addition to their similarities to the Gregorian calendar, have Orthodox rules for the celebration of the Pasha (Greek for Easter). In the Orthodox Church, old Calendarists went as far as breaking ties with the New Calendarists over the matter. Anathemas against the Gregorian calendar issued by Patriarchal and Pan-Orthodox Synods in Constantinople were issued following the proclamation of the Calendar revision, as well as various condemnations by multiple local synods. And since, another major controversy between Eastern and Western Churches has dominated Christianity. Emissaries of Pope Gregory to convince the Orthodox to accept the New Calendar did not succeed. The rejection is because the date of Easter would change in violation of the injunctions of canon 7 of the Holy Apostles, which states that

> "If any Bishop, or Presbyter, or Deacon celebrates the holy day of Easter before the vernal equinox with the Jews, let him be deposed."

According to the ruling of the First Ecumenical Synod in 325, Easter Sunday should fall on Sunday which follows the first full moon following the vernal equinox. If the full moon happens to fall on a Sunday, Easter is observed the following Sunday. The day taken to be the is March 2, the date of the Vernal equinox of the time.

According to scholars, insistence on the vernal equinox is so that the celebration of the event does not take place twice in one year. And the reference to the Sunday after the Passover of the Jews, on which date the slaughter of the lamb symbolizing death precedes the resurrection afterward that follows. Note that the Jews celebrate Passover on any day that happens to be the fourteenth of the Eclectics moon, The Apostles also say whenever it so happens that the legal Passover falls on a Sunday, Easter is not celebrated on that day, but on Sunday next thereafter, to avoid celebrating along with the Jews.

The basis for the refusal to implement the necessary correction to the old calendar is that the canons were ratified by an Ecumenical Synod that knew only the Julian calendar and only an Ecumenical Synod could institute such reform. As a result, except for the Church of Finland, all Orthodox Churches observe the event of Resurrection on the same day, Easter is therefore celebrated as much as five weeks later in the Orthodox church. Today, when it comes to Christmas, the Orthodox Churches which have adopted reforms to the Julian calendar observe Christmas on December 25, while the Orthodox Churches which have not adopted reforms celebrate Christmas 13 days later, on January 7. Easter in all Orthodox Churches is however determined using the old calendar. Finally, because of some irregularities of the sun's course, the vernal equinox does not occur always on the same day. During the time of the Holy Apostles, it occurred on the 22nd day of the month of Drystrus, that is March for the Romans, and it occurred on the 21st day of March in the Julian calendar at the time of the First Ecumenical Council.

2.10 Calendric unification efforts

As discussed in Section 2.8.5, the date of Easter was set in CE 325 at the council of Nicaea. The council, convened by the Roman emperor Constantine, declared that Easter should come on the first Sunday following the 14th day of the Paschal – or ecclesiastical – full moon. Consequently, the day of Easter falls somewhere between March 22 and April 25 in the Gregorian calendar. The use of different calendars, such as the Julian calendar leads to a different celebration date. Early in 2016, the Archbishop of Canterbury, Justin Welby announced that he was engaged in discussions with other dominations to unify the date of Easter within 5 to 10 years and solve a 1600-year-old disagreement between Eastern and Western Christianity. The drive to unify the observation date of Easter goes back to the 10th century, and this renewed attempt will probably have a major chance for success. Archbishop Welby entered in discussions with the Coptic Pope Tawadros II in Cairo, Pope Francis -who in 2015 signaled openness to change the date of Easter in the west- and the Ecumenical Patriarch of the Orthodox Church and signaled that they were heading in the right direction. Pope Tawadros has mentioned the suggestion that such a celebration could be either the 2nd or 3rd Sunday of the month of April. The legal foundation already existed for such idea in England where the Easter Act 1928 has already proposed setting the date on the Sunday following the second Saturday in April. However, the act has never been implemented, and successive governments have left it to churches to agree on any such change.

Muslim scholars have been trying to address a lunar calendric unification for several decades. Except for a handful of non-technical meetings, Muslim governments have unfortunately ignored the topic altogether. The Kuwait Conference of 1973 recommended that difference of matali' (i.e., initiation of the lunar month) must be disregarded between locations sharing part of the night. Five years later, in the well-referenced Istanbul Conference of 1978, the principle of unifying Muslim occasions was finally adopted. But, despite the success of the meeting, no binding technical criteria could emerge. And even a new effort materialized by the Mecca and Cairo conferences of 1986 did not yield tangible advances to the topic. In the US, the non-governmental IIIT Lunar Calendar Conference was a venue for astronomers and thinkers to discuss technical aspects of the topic of moon sighting and the development of a lunar calendar. After this meeting, a conference on the topic was organized in Malaysia, one in Tunisia and another in Morocco. In 1998, the issued guidelines of the Islamic Society of North America and other Muslim Groups were adopted, and the majority of Muslims in North America celebrated the end of fasting 'Eid al-Fitr' on the same day. This Unification was ,not only short-lived but also restricted to North America and some pockets in Asia. Not much improvement can be reported thirty years later. Muslims are still in disagreement over the dates of Ramaḍān, Eid al Fitr and the timing for the Hajj. Reports of new moon sighting that are scientifically impossible kept coming from certain Muslim countries. Ahmad (Ahmad, 1998) and others place the blame on certain Muslim countries, as many Muslims in North America and around the world tend to follow these two countries. In addition to inconsistencies in the criteria used to determine the start and end of observances, many documented erroneous and astronomically impossible sightings continue to be announced.

2.11 Secular calendric proposals

2.11.1 The French Republican calendar

The leaders of the French revolution knew very well that they had accomplished a major historical event framing their achievement as a product of the era of enlightenment set forth by several great thinkers and philosophers. They also knew that every power in the history of humanity required a calendar of its own, a testimony of their accomplishment, civilization, and identity. Nothing was wrong with the Gregorian calendar from a technical standpoint. However, the association of the names of the historical Roman emperors with the names of the months did not fit their newly found principles. Instead, they represented symbols of despotism and clergy and had to be removed. The month names of Juillet (July) and Août (August), after the names of the Roman emperors Julius and Augustine in the Gregorian calendar, had no place disgracing the French Calendar (Garraty & Gay, 1972). The Republicans in power at the time had no doubt about their self-importance and their rise to take the moral higher ground proclaiming September 22, 1792, the first day of the first month of the year I of a new French Calendar representing the new era of freedom, liberty, and fraternity between fellow man. The first day was chosen for commemorating the beginning of the "Republican

Era," the day the French First Republic was proclaimed following the convention which abolished the monarchy the day before.

The Republican calendar was short-lived; it was adopted by the Republican government for about 12 years, between 1793 and 1805, and was abolished by Napoléon on January 1, 1806. It was conceived amid a drive to perform sweeping social and administrative changes that replaced the old institutions. Some of these changes ultimately became global such as the world's metric system of units. A commission which included famous politicians, scientists, mathematicians and astronomers, such as Joseph-Louis Lagrange, a geographer, a poet, and prominent artists.

The year starts the day of the autumnal vernal equinox starting midnight and comprises 12 months, each of which has three-ten-day weeks referred to as decades. The day of rest is the tenth day of the week instead of the Sunday of the seven-day week. There are five or six extra days added at the end of the year to bring the calendar in step with the solar tropical year. A Leap day is added once every four years. The day also follows a decimal partition with a ten-hour in a day, each hour is partitioned into 100 minutes and each of which has 100 seconds. Clock builders started making decimal clocks but were not successful. Ultimately the decimal time was abolished in 1795 with some sporadic usage in certain French communities the following few years.

The names of the months are made-up and derive from either French, Greek, or Latin origin and relate to the local seasonal weather. The autumn season months are: Vendémiaire, Brumaire, and Frimaire; the winter months are: Nivôse, Pluviôse, Ventôse; the Spring months are: Germinal, Floréal and Prairial and the Summer season months are: Messidor, Thermidor, and Fructidor. The months start somewhere between the nineteenth and the twenty-fourth of the Gregorian months. The ten days of the week are primidi, duodi, tridi, quartidi, quintidi, sextidi, septidi, octidi, nonidi and décadi.

It is debatable why the Republican calendar did not catch and failed before having a chance to migrate out of France while the same time the S.I. standard became global today.

2.11.2 The International Fixed calendar

Industrial heavyweights from the new world who dominated world markets and amassed exceptional new wealth and powers during the 20th century also had to have a new calendar. George Eastman of the mighty Eastman Kodak Company that filled shelves around the world with photography films during most of the 20th century has embarked on a completely unexpected mission, the promotion of the International Fixed Calendar (IFC) which he has enforced in his offices and has been in effect from 1928-1989. The IFC is a solar calendar of thirteen months each consisting of 28 days was developed by Moses Cotsworth (1859-1943) to provide a fix to the lack of uniformity of the Gregorian calendar in the number of days and weeks in a month. From an accounting perspective, working for the North-Eastern Railway Company, Cotsworth used his analytical skills to develop a calendar that responds to business accounting needs. The IFC has been known as the Equal Month Calendar, the Eastman Calendar, the Cotsworth calendar or simply the 13-Month Calendar and was first proposed in 1902.

The fixed calendar consists of 13 months, each consists of four seven-day weeks. The Seven-day week starts on a Sunday and ends on a Saturday. The 13 months with 28 days total of 364 days in a solar year. The same Gregorian names for the months are kept with the addition of month 13, named ***Sol*** in honor of the sun, occurring between June and July. To keep in step with the season, the day following the month of December, called ***"year-day"*** is added as a Holiday and would not count as a weekday. In addition, the concept of a leap year of the Gregorian calendar is adopted so that every year which is divisible by 4 unless it is divisible by 100, but not by 400, has an additional day, is once every four years another day is added. This leap year day is introduced at the end of the Month of June and does not count as a weekday either.

Figure 2.11-1 Shown left to right Secretary of State Col. Henry Stimson receives the report of the Committee on Calendar Simplification which he will forward to the League of, Dr. G.K. Burgess, Director of the Bureau of Standards; George Eastman of Kodak Co. and Chairman of the Committee; Dr. Charles F. Marvin, Chief of the US Weather Bureau; Rep. Stephen Porter, Chairman of the House Committee on Foreign Affairs. Credit (AP) 1929.

The 13-month calendar did not get traction despite the push from heavyweight names in North America. George Eastman believed the calendar properly maintained records and was especially passionate about it. He printed many calendars in all shapes and forms and had them posted everywhere in the company and started the International Fixed Calendar League headquarters in Rochester's Kodak offices in 1929 after it was selected the best out of 130 calendars proposed to the League of Nations, a precursor to today's United Nations. He convinced businesses to switch to the new calendar as he did and chaired the National Committee on Calendar Simplification. His efforts came to an end when the League of Nations failed to approve the proposal in 1937.

There is no clear conventional half or quarter year due to the prime number 13. In addition, the day of worship for several religions did not fit with the week starting a Sunday, and a Sabbath would be erroneously observed. For others, the number 13 was simply a bad luck number.

Should such calendar received traction, it would have been a historical triumph of the world of business and industry and more importantly to the American spirit of innovation and ingenuity following the massive number of innovations in consumer products and technologies that flooded mankind from the new world in practically all sectors changing the world once and for all.

2.11.3 Hanke-Henry Permanent calendar

If it was just up to the United States, the Gregorian calendar would have been replaced by a new calendar by now. However, being unafraid of innovation is not the main requirement in this case, and it is mainly the consensus that is needed for a world calendar to take hold. In a world dominated by commerce and trade, one would expect a new calendar to be proposed for business reasons. The failure to drive the 13-month calendar did not prevent American innovators from proposing another alternative to the Gregorian calendar. After all, this is a nation that plans and listens carefully to its innovators. Visitors are usually amazed by the level of planning that takes place in the "new" world. In a non-related topic, When the United States joined other nations and finally adopted soccer as a mainstream sport, the first matter US officials addressed was to suggest modifications to the rules of the game, even though these officials knew that soccer rules have not changed since its early inception in England. In contrast, American football rules change on a yearly basis. The US is a country of invention and change. For example, in the field of education, courses are planned over a year ahead of time. Students know when and where classes are offered when exams are held, and even graduation day is scheduled. In today's world of globalization, the shift in the dates to do business from year to year requires constant revisions and rescheduling of activities. And that is not good for planning. For example, a class for a course must be redesigned on a yearly basis because the term starts on different days of the week. Planners do not like changes and want to work with uniformity, and several of them went to work to propose a solution. It is in the culture of the new world, and it had to happen. Calendar reformists are driven by bringing about better scheduling and planning for mankind. They understand the high cost involved with the initial rewriting of software and the updating involved in current products, terminologies education, informing the public, etc. but they argue that the price tag for changes will be amortized over a long era and therefore the final cost will be negligible in the long run. We will have a calendar once and for all free of ambiguities so that the prediction of work dates and holidays is straightforward.

In the year 2004, Richard Henry of Johns Hopkins University proposed that the world switches to a calendar known as Common-Civil-Calendar-and-Time, an adaptation of a proposal by another calendar reformist by the name Robert McClenon's. Henry's calendar would be identical every year by simple tweaking the computation of leap years of McClenon's calendar. He introduced a rule that parallels the Gregorian leap year rule. He proposed the addition of a leap week named "Newton" between June and July making the difference between the start of the Gregorian year nearly unchanged. First, he proposed the implementation of the calendar on the first of January of 2006. Henry's colleague, the economist Steve Hanke, proposed the addition of a leap week at the end of the year instead, calling it "Extra." Years that are divisible by 5 are leap years except those divisible by 40 but not divisible by 400. Consequently, the beginning of a new year can vary by up to 17 days relative to the Gregorian year. Under the Hanke-Henry calendar proposal, the weekdays are the same as in the Gregorian calendar, but months will have different lengths. January, February, April, May, July, August, October, and November have thirty days, while March, June, September, and December will have thirty-one days. This makes each quarter have two 30-day months and one 31-day month.

The proposed calendar is well thought through and is far superior to previously suggested reformed calendars. In fact, it does not break the cycle of the days of the week, that is except for the addition of the extra week every 5 or 6 years, the calendar is fixed and wouldn't change. It also has the same number of days in a quarter, rendering financial reporting consistent and eliminates troubles with earning reporting such as the Q4 reporting of Apple in 2012. Unlike the case of reforms that contain one or more "blank days," the proposed calendar would not alter religious celebrations because the cycle of 7 days between weeks is always maintained (i.e., from a Friday-to-Friday, Saturday-to-Saturday or Sunday-to-Sunday we always have 7 days). However, the calendar would have variation with the seasons; the equinoxes would shift more than in the case of the Gregorian calendar. In the same proposal, it is also recommended the adoption of the universal time for the entire world, making a date and a time identical everywhere on earth. The calendar is referred to as the Hanke-Henry Permanent Calendar; it was unsuccessfully pushed for universal adoption on 1 January 2017.

2.12 The Astronomical calendar

The moon, black sheep of celestial bodies, has finally revealed its secrets
Anonymous

The source of the Astronomical Lunar Calendar is the *Astronomical Almanac*, considered to be the world resource for standard astronomical data. The Astronomical Almanac is rooted in the British *Nautical Almanac and Astronomical Ephemeris*, a product of the Nautical Almanac Office (HMNAO) which was first published in 1766, and *The American Ephemeris and Nautical Almanac*, first published in 1852. The American Ephemeris is a product of the U.S. Nautical Almanac Office (USNO) which was established in 1849 by Congress to prepare and publish an official national almanac to avoid dependence of American scientists and seaman on foreign almanacs. The Nautical Almanac Office started in Cambridge Massachusetts, under the direction of Lieutenant Charles Henry Davis, and was later moved to the U.S. Naval Observatory, in Washington DC in 1893.

In 1852, the Nautical Almanac Office published the first American Ephemeris and Nautical Almanac for the year 1855, providing much-needed data for astronomers and surveyors and at the same time a display of national pride. It was also a display of the emerging scientific prowess in the US by new Harvard graduates from Harvard who acquired knowledge in mathematical astronomy. These graduates made the new office a center for research as they tackled the problems of those days. The main problem of the time was a thorny misfit of the two-century-old Newtonian mechanics to observations of lunar data. The orbit of the moon happens to be one of the most complex problems handled by astronomers all the way until these days and age due to the need for corrections representing multiple influences.

Simon Newcomb was one of the researchers who studied at Harvard. He was the Office's Superintendent in 1877 and had already derived new models for the orbits of the moon and other celestial bodies based on sound theoretical considerations which were verified through observational data. Newcomb became one of the most famous scientists in America known for bringing fame and high regards to the Nautical Almanac Office. Aided by the talented mathematician George Hill and other assistants called "computers," Newcomb worked on the moon and the planets Mercury, Venus, Earth, Mars, Uranus and Neptune. Hill, tackled the extremely difficult theories of moon's motion and within a couple of decades, and before the start of the 20th century, they developed sound theories for all planets used until at least 1960. With developments for Mercury, Venus, and Mars used all the way until the '80s. Born in 1866 in the UK and naturalized American Citizen, the mathematician-astronomer Ernest William Brown was introduced to the work of George Hill in 1888 while a postgraduate student at Cambridge, England. Brown had George Darwin, son of Charles Darwin for an advisor and an applied mathematician, with whom he kept contact and stayed with during his visits back from the US during his career in Celestial mechanics. It was Darwin who guided Brown to continue Hill's work, which Hill gracefully approved. Brown moved to the US, and within two years, he held a professorship position in Haverford College, Pennsylvania from 1891 to 1895. Brown connected immediately with Hill and Newcomb in Washington DC. Soon after this visit, Hill referred to Brown's early work on Lunar theory and recommended that the office of Almanac hires computers to work on Brown's theory. Later, Brown moved to Yale University, where he was very well funded and continued his work on his Lunar theory (Study of the Lunar motion). He finished his theory within one year and developed the well-known and precise *Motion of the moon* Tables in 1919, replacing the Hansen Tables which were in use since 1857. Brown's work was a massive undertaken focused on developing accurate ephemeris of the moon using three-body (earth-moon-sun) mechanics and accounting for perturbations from Jupiter and Venus and imperfections of earth and moon shapes. He had over 1 million in today dollars in support to do the job. Historical and technical accounts of his work can be found in (Wilson, 2010).

Figure 2.12-1 From left to right: Simon Newcomb, Superintendent of Nautical Almanac Office at the US Naval Observatory, courtesy of Library of Congress, George William Hill, American astronomer and mathematician, Naval Oceanography portal and Ernest William Brown, US naturalized Mathematician and astronomer of English origin, portrait by Maull & Fox, ca.1898 © The Royal Society.

Official collaboration between the U.S. and the U.K. on the almanacs started in 1912, following a U.S. congressional authorization of data exchange, and has continued until today. Several standards and advancements have resulted from this collaboration. In 1925, Civil time was introduced with day change begins at midnight instead of noon. In 1941, Almanacs adopted *FK3,* the International Astronomical Union (IAU) catalog of stars as the fundamental reference system. By 1955, it was agreed that that the 1960's publications of the American and British Almanacs will be identical. The unified ephemeris was also adopted in Germany, replacing the Berliner Jahrbuch, while key components were adopted in the Japanese Ephemeris.

The 1960 publication of the Ephemeris came with the release of the First Edition of the reference *Explanatory Supplement to the Astronomical Almanac* which explains the theory, algorithms and computational details behind the annual Almanac publication. The reference was jointly produced by HMNAO and USNO. The second edition of the book appeared in 1992 and contained contributions from The Jet Propulsion Laboratory (JPL), Bureau des Longitudes in addition to HMNAO and USNO. One of those contributions was the 1972 additional precision for tabulated longitude of the sun, half-day values of the horizontal parallax of the moon, and the moon's distance in units of earth's equatorial radius. Also included were the International Astronomical Union 1976 constants and FK5 (J2000) system which were adopted in 1984. The edition also had the Ephemeris Time (ET) replaced with the Barycentric Dynamic Time (TDB) and Terrestrial Dynamical Time (TDT). The basis for the ephemerides was changed to Jet Propulsion Laboratory's DE200/LE200. Also, the Supplement to the Astronomical Almanac 1984 was published; it detailed the introduction of the improved IAU System of Astronomical Constants, Time Scales, and Reference Frame into The Astronomical Almanac. Brown's tables were replaced with completely new computational methods with data from new observation and lunar laser ranging.

Figure 2.12-2 Le Roy E. Doggett Chief of the Nautical Almanac Office at the US Naval Observatory from 1990 to 1996 (after whom the Doggett Prize for Historical Astronomy of the Astronomical Society is named) 1941-1996. Credit: Library of the US Naval Observatory.

The Third Edition of the reference *Explanatory Supplement to the Astronomical Almanac* (Seidelmann, 2012), was made available starting mid-2012 is a complete revision of the 1992 reference; it contains the major advancements in positional astronomy including the ICRS replacing the FK5 system (a descendant of the FK3 system), the new precession and nutation theories and a new positional paradigm that is no longer tied to the ecliptic and equinox. Of relevance to our subject matter is the address of the visibility of the Crescent moon in the concise survey by Lee Doggett in Chapter 12 of the Edition (Seidelmann, 2012). It is reported that the New moon was sighted as early as 15.4 hours following conjunction (Schaefer, 1988). Schafer, Seidelmann, and Doggett, shown in Figure 2.12-2 and Figure 2.12-3 contributed extensively to the topic of new moonlight visibility during the 1980s.

Figure 2.12-3 Kenneth Seidelmann (left) and Bradley Schaefer (right) are contemporary leading scholars in the subject of new moon visibility.

2.13 Basis of Lunar Calendars: First visibility

There is a consensus that first visibility of the new moon is the basis of all traditional lunar calendars, see e.g. (Fotheringham, 1910), (Hasanzadeh, 2012) and (Gautschy, 2014) and its prediction has been proven to be a non-trivial task (Fatoohi, 1999). As a result, many empirical models have been proposed throughout history. Even today, new models are still being developed, and the problem has yet to be solved conclusively as pointed in (Gautschy, 2014).

Prediction of first and last crescent moon visibility goes back to the time of the Sumerians and the Babylonians who are credited to have first devised criteria to predict new moon visibility, see e.g. (Sultan, 2007). Many clay Tablets illustrate their usage of multiple methods for prediction together with continuous observations for extended periods of time. The Babylonian criterion that the new lunar crescent can be first observed when the separation between the moon and the sun is greater or equals 12° (Bruin, 1977) is baseless as discussed in reference (Fatoohi, 1999) and confirmed later in (Gautschy, 2014). Based on the study of Babylonian lunar ephemerides, Neugebauer suggested that the Babylonians may have used a criterion of the form: elongation + lag time between sunset and moonset (in degrees) greater than a constant (Neugebauer O. , 1955). Other criteria were used by the Babylonians either concurrently or at later times. There is a consensus however that the Babylonians used rules for prediction of first visibility with more complex criteria emerging during the Seleucid Era.

First records on calculation of first visibility tables are attributed to the 9th century Persian Scholar of Baghdad Muhammad Al-Khawarizmi (King, 1987). Muslim astronomers proposed several criteria for first-crescent visibility between 8-10th century, which revolved around relative sun moon altitude, moon lag, relative azimuth, and crescent width. These criteria are derived based on local observation and could not be valid everywhere. It was known from old texts that a new moon visibility varied from one geographic location to another.

Renewed interest in predicting crescent moon first visibility came from Europe at the end of the 19th century and continued into the twentieth century. After a non-successful attempt to compute the new and old moon by K .C. Bruhns around the year 1880 (Mommsen, 1883), it was Fotheringham who formulated an empirical expression for first visibility (Fotheringham, 1910). Based on measured data from Mommsen, Fotheringham suggested first visibility can be given by an empirical relation between the minimum altitude of the moon and the relative moon-sun azimuthal distance at sunset. Improvements on Fotheringham's work came from Maunder (Maunder, 1911). Twenty five years later, Danjon (Danjon, 1936) argued that the arc length of the crescent moon decreases with the elongation of the moon and using observational data, he concluded that at 7 ° elongation, the length of arc is nil. This value is widely known as the Danjon limit and is the limit at which the moon crescent is formed. Even though this limit is not strictly accurate, it happens to be close to the results obtained using other methods of analysis (e.g. Sections 5.3-5.4).

In modern time, computers and instruments added new information which led to sophisticated models to predict new crescent visibility based on celestial mechanics, astronomy, atmospheric optics, etc. Bruin derived a sighting criterion based on relative sun-moon altitude and crescent width (Bruin, 1977). From Malaysia, Ilyas proposed multiple criteria for new moon sighting (Ilyas M., 1988). He introduced a lunar dateline for first visibility around the world (Ilyas, 1984). Schaefer and Doggett proposed a criterion which accounts for light scatter and other atmospheric considerations (Doggett L.E. and Shaefer, 1994) and (Schaefer, 1996) drawing from observations by Doggett, Seidelmann and Schaefer (Doggett, 1988) and (Schaefer, 1988). With the arrival of modern astronomy to personal computers and the dissemination of computerized algorithms from astronomy centers in the US and England, criteria for first visibility were implemented using computer graphics. Notable is the early work of the Ahmed, who provided a much-needed computational tool (Ahmed, 2001) for predicting the beginning of the month of Ramadan around the world. His software *Moon Calculator* which he made available free of charge performs scans of the globe to locate first visibility using most criteria for new moon visibility.

Despite such advancements, decision-makers could not be convinced that the new moon is a global event which occurs at a single instant in time, irrespective of the position of the observer and ignored the call for synchronization of festivities. They often summon the population to look for the crescent over two consecutive evenings, calling one of them "the night of doubt." Some regrettably ignored science with a pretext of reliance on religious texts and consistently initiated and ended observance at the wrong time. Non-funded works of Muslim thinkers and astronomers often received no consideration from those who decide on the dates of celebration, triggering emotions to run high at times with some of the scholars characterizing the situation as chaotic (Afzal, 1988), embarrassing (Khan, 1988) and even finding the umma (i.e., Muslim world) has settled at its lowest point (Al-`Alwani, 1988).

2.14 Conclusions

As mentioned earlier, the history of the calendar was not initially planned to be part of this work. Also, the technical particularities of the calendar were considered outside the scope of this effort. Our chief goal is limited to when and where the first moon is detected so that festivals and observances are coordinated around the world. Midway in the documentation of our results, it was realized that there is a need to gather, organize, and connect historical developments of lunar-based time measurements. It is worth noting that the effort of gathering historical information on the subject has been rewarding as there have been lessons learned. Even though some of these lessons may seem to lack adequate backing and amount to a mere perception, they remain important and are therefore shared here.

Time-keeping has long been a need for humans. As we started digging for evidence on how we came to rise over the rest of the species, in addition of finding remains of our ancestors and their cousins, we stumbled on records left behind in a multitude of forms. So far, the earliest calendar was of African origin in the form of markings inscribed on a 44000-year-old baboon bone. Such markings probably constitute a tool for planning activities over several days; a fundamental feature that humans leverage in their interaction with the environment and achieve dominion over the rest of the world. As planet earth stepped out of the ice age, and the last cousin disappeared in the European continent, humans discovered agriculture and domesticated livestock in Egypt, Mesopotamia, Indus Valley, and China with minimal competition except between themselves. In this window of mild weather, which will probably not last for long, earth saw the unique ascent of man on earth. Scientists are warning that humans may be missing a unique chance for future survival beyond earth. Competition for resources, greed, egos, racism, nationalism, and warring are hampering the much-needed technological advancements to reach this goal.

It was not until the rise of Charlemagne that Europe turned the page on the middle ages and looked towards education and knowledge in churches. Africa has been under constant conflicts and even European occupation. Stability, education, and strong states are prerequisites for emancipation. It is ironic that the state of war and instability in the Middle East is currently delaying the advancement of the people of the region who started civilization in the first place. At the writing of these words, ignorance has risen again, and the world seems to run on bullying and conflicts.

It was not until the rise of the strong states of Egypt and Babylon that scientific and computation-based calendric developments took place, shifting time-keeping responsibility from the female gender to full-time state-supported priest-astronomers. The immediate outcome of this shift is the birth of the technical calendar derived from astronomical observations. The technical calendar was the science of the day; a strongly guarded secret which priests and religious leaders oversaw because of the need to define festivities and observances. Subsequently, and over several centuries, organized religion set foot in communities outside the major power center, leading the calendar and modern science, in general, to develop everywhere in the cradle of religion. Ultimately, science developed at a fast pace through innovation and the religious principles remained stationary due to their construct and had a tough time reinventing itself to keep in-par with parallel social developments triggered by new products and ways of life. The divorce was painful and marred with capital punishments in Europe, destruction of other nation's heritage in Mesoamerica and even recently infringement on the scientific body through the introduction of creation as a science. The Muslim world is currently in a state of confusion on how to define the start of the lunar month due to a conflict between scientific evidence on when the new moon can be seen and an artificial "requirement" for local visibility claimed by some interpretations. This is not a conflict between science and religion. It is rather a conflict created by men and entities not wanting to give up control over the people they have exploited for a long time. In fact, the triumph of the Gregorian calendar would not have happened without the push from people like Bacon and the wisdom of Pope Gregory XIII. Infringement on science came from the political establishment and not from religion; it just happened that politics, much like science, was inside the religious establishment. In the west, the most recent example of infringement on science is in the matter of environment and global warming. Deniers want to convince the world that "there are alternative facts" to scientific measurements for narrow-minded economic and selfish reasons. Even current attempt to teach "creation" as a form of science takes root in politics and has nothing to do with religion. The same is also prevalent in the Muslim world when it comes to the topic of the moon and the start of religious observances. The states, operating in the background behind religious figures, would not spend the necessary funds to lift the discrepancy and act responsibly. As a result, observances end up driven by pure ignorance and egos.

Throughout history, emerging civilizations had a need for a functional calendar. The Egyptians, the Mesopotamians, the Mayan, Persians, East Africans, Indians, Chinese, Romans and recently Americans in the new era of industrialization and global influence, have had a need for time measurement to run their civilizations. Initially, there were not enough good calendars to go around. For example, the Romans imported a calendar from Egypt, while the Persians adopted the Babylonian calendar, and all have introduced refinements and corrections. However, not all calendars get to adopted, even if they are technically superior. The main ingredient for successful implementation is either consensus or forced usage. The triumph of the Gregorian calendar was due to a combination of early religious enforcement and subsequent economic needs which came with the advent of international commerce and the digital revolution. At the turn of the past century, the Americans took the baton from the British and solved the thorny moon problem for maritime navigation purposes and certainly did not stop there. The proposal for a new accounting-friendly International Fixed calendar came rushing from Eastman Kodak following a global dominance in the camera film industry. More recently, the business-friendly Hanke-Henry calendar surfaced, and finally, the ultimate development of the atomic clock has disconnected mankind from the need to rely on the diurnal cycle for time measurement. Earth rotation around its axis has many issues. For those who are familiar with instrumentation, measurements need to be carried independently of what is being measured, which is not the case here. Mankind worked hard to fit the clock to earth's rotation, which is not perfect to start with. Small errors accumulate over time resulting in gross inaccuracies.

Finally, even though people from all nations and all faiths are social by nature; they want to celebrate together and at the same time. But because of their instinct and drive for competition, they have a tough time to get along. And because an agreement is a prerequisite for a calendar to take hold, civilizations could not help it but live with controversies as demonstrated in many cases over the ages. Efforts for the unification of the Day of Easter are blocked by some even though it is only a celebration and not a call for aligning opinions. Also, the start and end of fasting of the month of Ramaḍān could not be unified, even though there is only one moon shining a new crescent at a specific time following conjunction. Muslims are in need of an adequate lunar calendar, yet they remain in discord until these days and age. The date for Rosh Hashanah, the Jewish New year and anniversary of the creation of Adam and Eve could be celebrated as an instant of time around the world irrespective of geographic locations. The internet and social media are full of questions such as *"... two friends of mine and me celebrated Passover on the wrong day. We celebrated the Passover on Monday night, and Tuesday the 1st day of Unleavened Bread. If I remember well, on the week of the sacrifice, Jesus celebrated the Passover on the evening of Tuesday ...beginning on what we now call Wednesday evening*." And "...when is the Eid, I heard that Eid is Monday in Saudi Arabia...?

Could a calendar based on lunar photometry demonstrates that the cycle of the moon is unique and does get affected by states, religious and political establishments and overcome discord?

Chapter 3
New moon verification system

The computation according to the appearance of new-moon is a subject the exposition of which would be both of great length and difficulty and would require difficult calculations and numerous tables.

Al-Biruni

Having realized that there was no solution in the literature to the question of when and where the first light from the new moon reaches the earth surface and considering the social anxiety to have the question addressed, the concept of a hands-on system to tackle the issue was born. The solution consists of a system designed for the documentation and verification of new moon sightings. It is expected to eliminate confusion caused by erroneous claims of sightings around the world for those who need to confirm new moon visibility. Upon deployment, the system is expected to synchronize naked-eye and instrument-aided sightings with the photometric calendar. As a result, it will unify the dateline for major religious and social events and support the task of assessing and declaring official sightings. The system is expected to provide leadership in the effort to adopt a unifying approach to this age-old problem.

The latest communication, optical, sensor, and security technologies are planned to be used in a scalable architecture so that the verification system can be deployed over several sites and later expanded throughout the world. Observation stations capture the young crescent and transmit images to a control center for authentication. The control center analyzes the image and associated data, and upon authentication, an official new moon sighting is declared. Once in place, the system can be expanded to support other academic, government, and commercial uses. For example, the control center can include an observatory for astronomical studies, receivers for secure, satellite-based communication and fixed ground terminals for broadcast and media systems and high-speed network communications. Deployment can occur in two phases. The first phase is scheduled for 12 months and demonstrates a prototype observation station and a prototype control center to be housed in an existing facility. The second phase should also last approximately 12 months, deliver twenty production observation stations, and demonstrate the fully functioning of the system.

3.1 Description of the need

With billions of people worldwide dependent upon new moon sighting to establish the timeline of lunar months and festivals, there has long remained controversy regarding the adoption of a unified photometric lunar calendar. The difference in opinion on how to determine the beginning of a month (i.e., by naked-eye, instrument-aided sighting or sun-moon conjunction) has hampered the arrival at a universal solution. Occurrences of erroneous sightings around the world have only further added to the discrepancies between communities, regions, and countries.

In the Muslim world, the controversies around when to start and stop fasting in the month of Ramaḍān led to a meeting attended by scholars and decision-makers who produced guidelines on new moon sighting, known as the "Istanbul recommendations" which yielded some success. However, the controversy surrounding the declaration of the new month remains, even with the new guidelines. The new crescent occurs at a unique time but is independent of the location of the observer, varying from month to month. therefore, a global solution that transcends geographic boundaries is urgently required. Science and technology can be used to separate fact from fiction and bring clarity to the actual meaning of arguments presented in support of the different sighting opinions. And from a practical perspective, there is a need for a framework that brings clarity and convergence of new crescent moon sighting through adequate documentation of claims, standardization of reporting claim procedures, verification of sighting claims and transparency in reporting and verification processes. By resolving the issues surrounding the verification of the new crescent sightings, it will become possible to maintain a standard universal photometric calendar.

At no other time in history, we have technology, expertise, and desire so readily aligned to solve this problem. The system described in this chapter provides the framework to handle the multiple aspects of decision making and can be guided by a body of scholars. It is a verification platform designed to eliminate all confusion associated with erroneous claims of new moon sighting. The system operates under the direction of a Competent Authority, a body composed of technical experts and scholars which is expected to be the single reliable source that people everywhere will look to for the declaration of a new photometric lunar month. Once deployed, the control center can be expanded to include an observatory for astronomical studies, receivers for secure satellite-based communication and fixed ground terminals for broadcasting, media systems, and high-speed network communication. Observation stations can include sensors for identification and monitoring of air pollution and tracking of natural hazards. Adoption and implementation of the system should provide the framework for accuracy, understanding, and exchange of information. Once implemented, the system coined "*The Witness*" by our group of enthusiasts in the summer of 2008, should inevitably lead to a unified photometric lunar calendar.

3.2 System baseline

The witness system is composed of a set of distributed observation stations and a control center. Observation stations provide a high degree of certainty that the new moon will be observed and verified. Each station is operated by a trained technical staff member, while the control center is run by the Competent Authority, the body with final authority on sighting authenticity. The responsibilities of the control center include receiving requests for documentation of sightings from observation stations, processing documentation of sightings, and establishing sighting authenticity. The control center issues the official and final declaration of the Hilal sighting following an evaluation process of all sighting claims.

A typical observation station is equipped with several integrated capabilities including 1) an imaging system tuned specifically for the young crescent, 2) a global positioning system that determines the location of the observation unit anywhere in the world and the Universal Time of observation, and 3) a communication system with a secure link to the control center to request and acquire instant permissions for new moon documentation in the form of a coded digital stamp. The observation station has the processing capability to merge the digital stamp, time, and global position of the observing station with a digital image of the new moon. In addition, built-in scanning and tracking capabilities are used to direct the observer to the correct location in the horizon sky.

The control center has the capacity to visualize and evaluate the conditions of each observation station, including exact astronomical predictions of crescent position, crescent age, visibility and local atmospheric conditions as seen from the observation station. The control center maintains direct contact with several observation stations at a time as sunset moves westward from one longitude to another on the surface of the earth.

Figure 3.2-1 Proposed Observatory and Control Center

Authentication of new moon sighting occurs in three steps. First, an observation station operator sends a stream of data that includes GPS location, Universal Time, atmospheric condition, and visibility to the control center and requests a permit to document a potential sighting. Second, the control center acknowledges receipt of the request and assigns a unique electronic stamp to the observation station, following verification of its current observation elements.

The control center then awaits further notification from the observation station and monitors the progress of its data elements. The control center handles multiple stations at a time and continuously assesses the potential for positive sightings in all observation stations. In the third and last step, when a claim for positive sighting arrives at the control center, a notification of acknowledgment of claim reception is sent back to the observation with either an end of transmission signal or a request for further documentation.

During a sighting attempt, the observation station undergoes automatic calibration and tracks the predicted new moon path in the sky. A sighting aid guides the observer to the location in the sky where the new crescent will appear. When a decision is made by the observers that a positive sighting has been made, an electronic image is acquired by the operator. The electronic image is automatically integrated with an electronic stamp and forwarded to the control center. Further assessment and checks of the image are made at the control center, and an authentication level is assigned to the sighting. Following the reception of several sightings, the Competent Authority convenes and issues the authentic and final declaration of new moon sighting.

Following an open competition for a pilot control center and a lunar observatory organized by *The Witness* Organization, the winning architecture provides a very low footprint so that it integrates with the historical surrounding (see Figure 3.2-2 Frontview of a pilot control Center). The theme of the building, as explained by the architectural team, is of moving shadows expressing the progress of time and light. The building has a visitor center to provide information about ongoing activities and past work, an auditorium and a museum. The external walls are coated with traditional indigenous small bricks.

Figure 3.2-2 Front view of a pilot control Center(left) and a pilot Observatory (right)

The pilot observatory, shown in Figure 3.2-2 and Figure 3.2-3 can be remotely controlled from the control center. Its proposed location is in a dry mountain range at an altitude of about 1000 meters above sea level. This location has a sweeping view to the west over an angle of more than 180 degrees. There are however agricultural fields down in the surrounding plains west of the mountain range that may contribute to air turbidity. The substrate of the observatory site is made of solid rock and should be optically stable. The observatory is going to evolve as the needs of the technical staff grow in time. Historically, major astronomical observatories around the world saw their equipment evolve with the maturity level of the users and with the development of better technologies. It is usually recommended that users start with lower grade equipment and work their way to more and more sophisticated instrumentation as the needs and experience evolve. The requirement to track and observe the new moon will be met by installing a set of instruments in the observatory where continuous observations for an extended period will be carried. There will ultimately be unique optical capabilities specifically designed for observing the new crescent. The benefits of the lunar observatory include:

- Systematic development of independent data that accurately describes the position of the moon in the sky, accounting for temperature, pressure air turbidity. Such data is expected to become a world reference.
- Early and reliable new crescent detection. Its unique optical capabilities will allow the new crescent to be seen earlier than ever before and provide unequivocal documentation of sightings,
- Enabling students and astronomers to conduct research, either on-site or remotely using the Internet, in planetary science through ordered time booking schedules.

The vision is to first equip the observatory for moon observation, and later introduce a second architecture for solar observation. The observatory is designed with the intention to operate during daytime & twilight conditions and therefore differ from common nighttime observatories.

The observatory's major components are shown in Figure 3.2-4. The observatory structure includes a building, a dome, a mounting pier (to which the telescope is independently secured), an observation deck, control & communication room, and a dormitory. The support systems include those typically found in buildings but also includes special lighting and communication equipment. Special support equipment, such as hoists to move heavy equipment, is also included. The observing system includes the telescope, a tracking mount, a control system and a collection of instruments used for viewing.

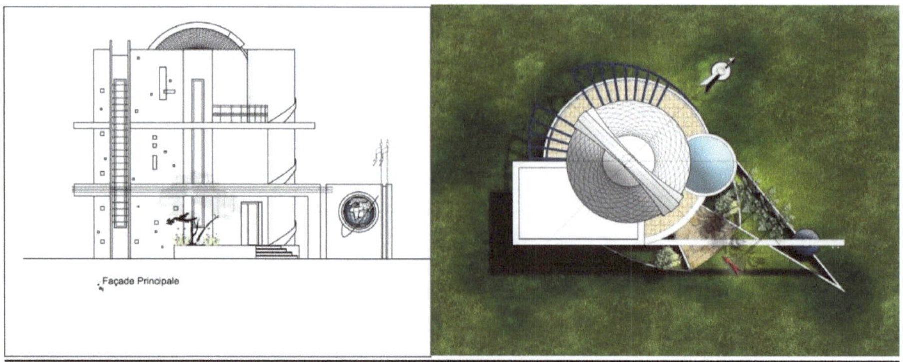

Figure 3.2-3 front and top views of the pilot observatory structure

The building is the main structure that encloses the equipment and supports the dome. The building can be constructed of normal materials, including concrete and steel. The walls are insulated on the inside to prevent them from causing convection currents in the observatory when they get hot. The resulting turbulence will blur the image, especially during twilight viewing. The dome sits atop the building on bearings that allow it to rotate. The dome has a door that opens to allow a viewing. The door is controlled so that it is always in-line with the telescope. The inside of the dome is also insulated to minimize convection currents.

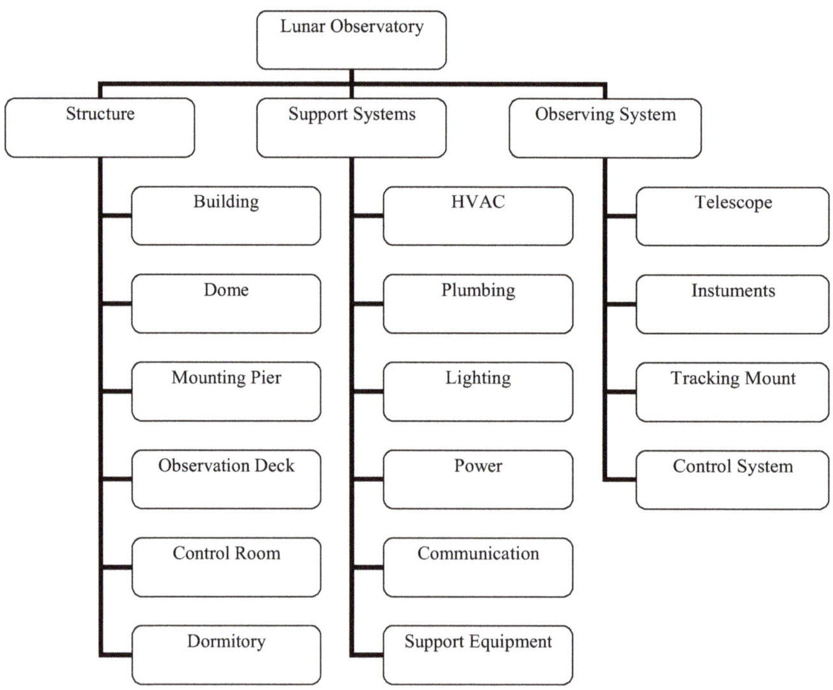

Figure 3.2-4 Major Components of the Observatory

Table 3.2-1 Recommended components for the Phase I Basic Lunar Observatory

COMPONENT	DESCRIPTION / RECOMMENDATION	REPRESENTATION
Location	The observatory is in a remote dry mountain range. Its position is minimally affected by light and pollution, traffic vibration, and city dust.	
Observatory Building & Dome	Standalone 5-meter diameter steel dome and building. A standard design is modified for near horizon viewing using up to 1-meter telescope. The observation deck accommodates up to 8 people.	
Telescope and mount	The telescope is 0.5 meter in diameter. It is either a standard Cassegrain reflector or a Ritchey-Chretien type depending on cost. The telescope mount and focus mechanisms are computer-controlled.	
Instruments	Several eyepieces and a filter wheel provide good viewing. A 35mm camera and an IR camera will be used initially to capture imagery of the moon, planet, and stars.	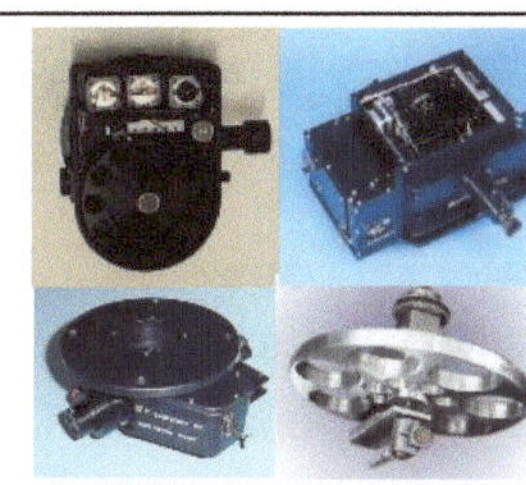
Control Room	Housed in a separate building, the control room has all observatory's electrical equipment.	
Communication	High-speed intranet and internet access, large data storage capabilities	
Visitor Center	The control building has a common area to receive visitors and serve users.	

The mounting pier is one of the most critical components for achieving good viewing. The pier must be isolated from the building to prevent vibrations from reaching the telescope. The observation deck is an elevated floor that brings the observer to the proper height on the telescope. The proposed observatory has the unique requirement of having to be able to view lower sky elevations.

In large observatories, the control room is typically located in the same building next to the observing room. However, with smaller observatories, the control room is in an adjacent building. The control room also contains as much of the electronics as possible to keep the amount of heat generated inside the dome to a minimum. The control room typically has one or more control computers with large flat panel displays and several workstations for data reduction and analysis.

Observatories which are in remote regions have dormitories to house users who are making extended observations. The dormitories typically include several bedrooms, a kitchen and a shared area with furniture and a television. The support systems are critical to the proper operation of the observatory. The heating, ventilation, and air conditioning system plays a key role in achieving good viewing. The observatory must be always environmentally controlled. Ideally, the conditions within the observatory would be maintained to match the outdoor conditions when viewing is likely to occur. This minimizes convection currents within the observatory when the dome is opened to the environment.

Two types of lighting are usually required inside the observatory, red and white. White light is used to work around the telescope and install instruments when no observations are being made. Red light is used during observations to illuminate the stairs and other areas.

Data storage and communication infrastructure are extremely important in today's observatories. Often, data is collected and quickly disseminated to several locations for reduction and analysis. High bandwidth communication is also required if the observatory is linked to an optional planetarium or internet for real-time viewing.

The specialized support equipment is needed to repair or maintain the observatory. The telescope's optics will eventually need to be removed for re-polishing, and special equipment will be needed to safely handle them.

The observing system is the heart of the observatory. It generates the images used to study the moon, planets, and beyond. A telescope is used to collect light and focus it to a small spot to increase the brightness of the image so it can be seen by the naked eye or a sensor. The amount of light the telescope can collect depends mostly on its size; bigger telescopes collect more light than smaller ones. Another important aspect is the ability of the telescope to focus light to a small spot without distorting it.

The motorized telescope mount must meet stringent performance requirements to achieve good viewing. It must point very accurately and move very smoothly. It must be very stiff so that the telescope does not sag as the mount rotates and users don't move the telescope when viewing through the eyepiece. The mount must also provide easy access to the eyepiece. The control system can control all aspects of operating the observatory. This includes opening the roof, setting the telescope position and rotating speed, adjusting the telescope's focus mechanism, operating the instruments and monitoring and recording data. All control is made possible remotely from the Control Center.

Table 3.2-1 shows the minimum system components necessary to meet the objectives of the observatory. There are numerous combinations of sizes and capabilities for the structure, support systems, and observing system that meets the objectives. The Basic system has a 0.5-meter diameter telescope housed in a 5-meter diameter dome. The telescope is mounted to a computer-controlled tracking mount. The telescope has eyepieces for visual observation and cameras for recording digital observations. The control room is in a separate facility and has substantial data storage capability. The control room has high-speed internet access and a separate computer to host the observatory's website.

3.3 Instrumentation

There are major differences between classical optical telescopes conceived for use in astronomy to resolve faint celestial bodies against a dark sky and a special-purpose lunar observation system dedicated to new moonlight extraction during twilight conditions.

There is a pressing need for new lunar observation tools to be designed to aid in solving the problem with new moon assertions and resolve the silver thin crescent from scattered sunlight. The lunar observation system must be designed to filter out scattered sunlight in the layers of the atmosphere by taking advantage of the differences between its spectral properties and those of moonlight. Most telescopes are designed for viewing very dim objects in the night sky, such as the planets in our solar system and distant galaxies. Two performance aspects are important for these telescopes to provide good viewing: First is its ability to collect light and focus it to a small spot thereby increasing the brightness of the image so it can be seen by the naked eye. The amount of light the telescope can collect depends mostly on its size; the bigger the telescope, the more light it can collect. The second aspect is the ability of the telescope to focus the light to a small spot without distorting it. A distorted image appears blurry to the viewer. Telescope optics must be manufactured to an extremely precise shape using quality materials to avoid distorting the image. It is this requirement that is responsible for most of the price differences between low and high-end telescopes. The light-collecting surface in a telescope is either a lens or a mirror. Thus, there are two classes of telescopes: refractive and reflective. Refractive telescopes use a lens to focus the light, and reflective telescopes use mirrors to focus light. All commercial cameras and most low-cost commercial telescopes are of the refractive-type, and all large observatories have reflective telescopes.

The western account of telescopes states that a spectacle (frame) maker by the name John Lippershey presented a petition to the States-General of Holland in 1608, applying for a patent for seeing at a distance. When the story came public others claimed the discovery as well, but no litigation or hard proof of invention was presented. The basic idea was to place a combination of lenses in the direction of an object, and the image becomes nearer. Within a year "Dutch trunks" –as they were called- were sold everywhere in Europe. All famous scientists of those days got interested and started experimenting with lenses. Galileo Galilei has built his own set of telescopes. His first produced a magnification of 3 diameters and encouraged by his success, he built two more and reached a magnification of 32 diameters. With his third telescope, modern instrument-based astronomy has begun. Looking at the moon, he realized that it was not a flawless sphere and later discovered the four brightest moons of Jupiter. Kepler, the author of the laws of planetary motion, improved on Galileo's instrument and produced a correction to by introducing a concave lens for the eye-piece. Later in 1637, to "theoretically" eliminate aberrations (circles and colored fringes forming around star images) which place a severe restriction on the size of the optics, Descartes proposed the use lenses with either elliptical or hyperbolic surfaces and suggested to lengthen the focal length, which is the distance from the surface of a lens or mirror to its focal point. And by 1650, telescopes around 2m in length were produced and polarized the interest of most scientists of the time.

Properties of convex glass were known way before the 1600s. One would expect that simple arrangements of lenses should have been established prior to 1608 CE. For example, there is an account that Thomas Digges in England produced an optical system having a convex lens and a mirror around 1570 CE, but was never claimed as an invention. It is well established that lenses and magnifying glasses were brought to Europe just before the start of the fourteenth century. Venetian glassmakers started making convex glass disks that could be worn in a frame and Marco Polo, while traveling in China in 1270 CE, observed the elderly using spectacles. The Chinese themselves claim that spectacles originated in Arabia in the 11th century. It is well established that corrective lenses were produced earlier by Al-Abbas Qasim Ibn Firnas in the 9th century; he polished round-shaped clear glass for viewing and were referred to as reading stones. The earliest evidence of a magnifying device, a convex lens that forms a magnified image," also dates to the Book of Optics by the father of experimental physics Abu Ali Hasan Ibn al-Haitham in his book of Optics of 1021 CE. Its Latin translation in the 12th century was instrumental to the manufacturing of eyeglasses in the 13th century in Italy. Refracting telescopes are known to be the earliest designs. They seem to have shown up in the Netherlands around 1608. Gallelio constructed a telescope of his own around the same time sometime in 1609 The principle of a refractor is simple. The light gathered by an objective lens larger than the human eye is focused at a point so that it gets magnified by an eyepiece giving the viewer a brighter and clearer virtual image.

Because of their interest in astronomy, the Huygens brothers improved the lens grinding technology to produce longer focus and reduce aberration and in 1655 made glass with about 4 m focus which led to the discovery of the Rings of Saturn and its largest moon Titan. They ultimately produced a near 8 m long telescope with an objective lens of about 6 cm in diameter. A race has started and until 1722 CE lenses with longer and longer focal lengths were produced in Europe producing telescopes between 50 m and 65 m long. They were hard to use, and all they contributed to was plotting positions of planets and stars. During this same period, some have started early experiments on collecting light using reflectors (mirror) instead of lenses was underway. James Gregory drew plans for a setup, termed reflector in which a reflective light-gathering mirror with a hole in the center and a concave mirror outside the focus reflected light back through the hole back to the eye-piece. Opticians of the day were unable to produce such a mirror. In 1672 CE, Isaac Newton presented a telescope made from a spherical reflective mirror and a small mirror within the focus of the large mirror but reflected light to the side of the telescope. This design is named a Newtonian reflector and is still in use today. In 1723 CE, John Hadley presented a slightly modified Newtonian telescope with about 15 cm reflective paraboloid mirror and a focal length of about 1.6 m. His telescope performed better than the 40-meter refractor telescope owned by the Royal Astronomical Society in England. Hadley was generous with his technology and spread his knowledge on how to make paraboloid mirrors, ending the age of awkwardly long telescopes. On the refractor side of design, a 7.5 cm achromatic lens was finally produced in 1733 CE by Chester Hall and later improved and mass-produced in 1755 by John Dolland and his son using good flint glass and better design. In 1772 CE William Herschel found a way to grind large mirrors. Large mirrors suffered from the bending of speculum metal under its own weight, creating distortion, and the mirrors had to be reconfigured every several months. This gave a chance to the Swiss Guinand and the German Franhaufer to produce fine 30 cm optics for refractors in the early 1800s. But then in 1850 Ce, Liebig and Steinheil in Germany and Faucoult in France produced silvered mirrors of Glass, and in the past 150 years, we had good reflectors and good refractors. In the pursuit of compactness and telephoto effects, the Cassegrain reflector was proposed by Laurent Cassegrain in 1672. The telescope combines a concave primary mirror and a convex secondary mirror. The telescope delivers a long focal length and convenience of locating the focal length behind the primary mirror, eliminating the need for a ladder system to perform observations, like in the case of a large Newtonian system. The shape of the primary mirror of a classical Cassegrain is parabolic while the secondary mirror has a hyperbolic surface. In the early 20th century, George Willis Ritchey and Henri Chrétien combined a hyperbolic primary mirror and a hyperbolic secondary mirror to eliminate optical coma and spherical aberration. Such design proved to be attractive to professional astronomers and serious amateurs. In fact, the Hubble Space Telescope and, many ground-based telescopes (e.g., Keck, VLT) adopted this design. There are several other variants of the Cassegrain telescope which are either easier to manufacture, cheaper or both. One design is the Dall-Kirkham telescope which was built by Horace Dall in 1928 and uses an elliptical primary mirror and a convex spherical secondary. This design is easier to manufacture but introduces off-axis coma which, depending on the application, may be acceptable at longer focal lengths. Other Cassegrain variants use off-axis and tilted mirrors to avoid shadows from the mounting of the secondary mirror and dealing with having to drill a hole at the center of the mirror at the expense of introducing other aberrations. Other designs variations of the Cassegrain combine reflectors, usually in the form of spherical mirrors to reduce cost, and refractors to correct for aberrations. Such designs are referred to as Catadioptric Cassegrains. The Schmidt-Cassegrain, designed from the original wide field the Schmidt camera, invented by Bernhard Schmidt in 1930 and The Maksutov-Cassegrain, named after the astronomer Dmitri Maksutov are two popular Catadadioptric Cassegrain designs. Both start the optical path by introducing a correcting optical element. The Schmidt uses a corrector plate to eliminate the spherical aberration of the primary mirror. The first patent for such telescope design appeared in 1946. The well-known Kepler observatory which identified a number of exoplanets uses a Schmidt camera design. The Maksutov design uses a corrector lens and uses a spherical secondary built into the correcting lens.

There are no front runners when it comes to telescopes. In fact, many designs can provide plenty of power for the new moon task. A good telescope designed to track and collect the first light that reflects off the moon has to be very sensitive. This requires the telescope to be large enough, yet not too large to have the image subjected to optical distortions introduced by daytime/twilight hot air turbidity. It must also be portable and rugged so it can withstand the harsh environments of wind, heat, cold, humidity, and rain that are found in the open. Under normal conditions where overcast may be a problem, it can be shown that portability needs to be a key requirement for observation stations. Such requirement would, in turn, place some limitation on weight and power.

The use of the two-element lens system called achromat makes refractors aberration-free images. They are also quite rugged and suitable for the new moon task. The telescope size should be between 80 and 12 cm in diameter with a 15 cm diameter (6 inches) is the largest practical telescope that can be manageable. The relatively new use of apochromatic lens elements delivers chromatic-free free images at a higher cost. The telescopes are well appreciated by the high-end and semi-professional amateur community.

Based on the above information, there can be several options that one may choose from. For immediate results, an observation system with extended capabilities to collect infrared and near-infrared can be used. Radiation is collected in a similar way that visible light is collected. The collected radiation is a combination of radiation from the moon, from the background and from the telescope itself. To minimize the radiation from the telescope itself, the detector may be actively cooled, and the telescope itself may be cooled. The background radiation is eliminated by continuously measuring the radiation of the surrounding and subtracting it from the collected image. Such operation is made several hundred times per second with a rugged fast steering mirror.

During new moon observing time, overcast skies due to clouds or sandstorms may be a problem that needs to be dealt with. Their portability was found to be a key requirement for the observation stations so that observers have the flexibility to move from one point to another on short notice. Based on weather predictions obtained several hours before sighting from satellites and land-based radar, observers need to be ready to displace by land or even fly to remote locations and catch open skies. Limited experiments in the United States have shown that portability is essential to ensure successful verification and positive sightings.

The requirement of portability places limitations on the weight and size of the telescope, its mount, power supply, and other accessories. Having a lighter system means that one must limit the size and stiffness of certain components. The latter is essential for good tracking and imaging. Lack of stiffness makes the instruments shake and deliver weak performance. Hence a conflict arises, and optimization is required to find the optimal solution with the best combination of minimal weight, maximum stiffness, and proper overall size.

The system must also be exceptionally rugged, so it can withstand transportation and the harsh environments of wind, heat, cold, humidity, and rain that are found in open environments. Experience has shown that the necessary transportation containers and other gear need to be part of the equipment

Shutter speed and exposure time need to be adjustable to combat the first two types of seeing. Generally, *Type I* seeing is caused by heat from the observer's body, air currents in or above the tube or poor telescope design. Tubes with low heat conductivity and heat capacitance are better at eliminating thermal gradients inside the telescope, especially during daytime and twilight observation. Separating the observer from the telescope and automating the mount helps eliminate some of *Type I* seeing. Moving the observation system to a different location can help with *Type II* seeing. Air movement is usually caused by thermal gradients in the area, which results from changes in topography or soil composition. Consequently, some locations present better air stability than others, so it is important that the lunar observation system be portable. There is practically nothing that one can do to eliminate seeing of *Type III*. However, adequate photography is still possible with *Type III* seeing conditions.

Most telescopes have magnifying eyepieces for viewing. This is acceptable for viewing dim objects at night in the visible wavelengths. The new observation system will view the moon at visible and infrared wavelengths. Infrared light is not visible to the human eye, so an LCD screen is required to display the infrared image of the moon. There is also a safety concern with an eyepiece. The system will operate in bright conditions and will view the moon when it is close to the sun. An observer viewing through an eyepiece would be instantly blinded if the sun accidentally came into view.

Telescopes do not have an integral means to store images or record video. The new system will have built-in capability to do so, as well as transmitting images to a predetermined location via a communication link.

Multiple images can be processed in a way that imitates the way humans process images observed with the unaided eye. This method will help the observer focus on the moonlight and discriminate against the unwanted background.

A suite of special instruments can be designed to use with the new lunar telescope to help observe the new moonlight during daytime and twilight. We now know that the visible universe is only a tiny part of reality. Each band of the electromagnetic spectrum reveals physical manifestations that are not seen in the visible light - much like the X-ray band reveals the structure of the denser bone material covered by softer tissue. By extending human vision to the other regions of the light spectrum, a new lunar observation system could observe the moonlight at new frequencies that reveal the moonlight against the background of scattered sunlight and allows the new crescent to be seen with certainty. We also know that the twilight sky is variably polarized and the polarization of scattered sunlight by molecules and aerosols in the atmosphere differs from that of moonlight. This leads to another discriminator that can be used in the design of the new lunar observation system.

The motion of the moon in the sky is unique and differs from other celestial objects. The moon moves quickly across the sky, covering a distance equal to its diameter in about three minutes. The special lunar observation system needs a special mount with a fine tracking device to accurately follow the moon path and properly and image its collected thin crescent light.

New instrumentation for daytime coronography and occultation can be adapted to observing the new crescent. Coronography is capable of attenuating the light of a central star by several orders of magnitude, thus revealing a dim light of an off-axis object like a planet. While coronograhs have been used to study high contrast light sources in some disciplines in astronomy, artificial occultation may be also considered to observe the new moonlight at small angular separation from the sun, even during daylight.

Unlike other observation systems, because information resulting from new moon observation needs to be immediately documented and communicated so that announcements can be made to the public, special data and voice links may need to be added to the design and made integral to the observation system.

Special solar cornographs for daytime astronomy can be designed for the sole purpose of extracting the new light of the moon. A coronograph is an instrument able of attenuating light from a central star (several orders of magnitude) without modifying the light from an off-axis object like a planet. From a historical perspective, early coronographs were produced to observe the solar corona in the thirties and were later adapted to stellar studies.

The difficulty in imaging the new moon lies in the important contrast between sunlight and new moonlight —more than 1 billion from visible to near-Infrared spectrums and about 1 million in mid-infrared- and the small angular separation from the sun and the moon. Special phase-difference is used in coronagraphs to further improve performance. By dividing the collected light into small domains, and using phase difference, to eliminate the background through destructive interference, images are shown to be much less sensitive to atmospheric disturbances.

Occultation is the passage of a celestial object across a plane between an observer and another celestial object, like in the case of a solar eclipse when the moon passes between the earth and the sun. The progressive blocking of light from a planet or a star has been very important to astronomers for a long time for important measurement could be made during those instances. Occultation can be ground or space-based. A long tubular structure can be designed to aid in daytime IR observation, and an artificial steerable light-weight large satellite (50-100 m) can be used in an elongated elliptical orbit that periodically coincides with the new lunar month. Such a system will give an unprecedented capability for superior contrast imaging of the new crescent and help to detect new planets around faraway stars.

3.4 Scientific impact and expansion capabilities

The Witness System will have an impact on scientific advancement in critical areas such as environmental monitoring. The system will jump-start a new scientific culture in optical engineering, atmospheric science, atmospheric physics, computer engineering and telecommunication and last, but not least, in systems engineering.

The inherent use of experimental physics and hands-on characteristics of the system is well-aligned with the interest of young and future generations in astronomy, space, and telecommunication. Should many mobile stations be deployed and made accessible to students, the system will inevitably keep young people engaged with applied science and constructive discoveries.

It is a fact that engineers and high-tech business owners in technologically advanced parts of the world grew up involved with science, instrument usage, and technologies. They are part of a trans-generation technical culture whose origin is usually traced to certain needs that manifested at some point in time. *The Witness* is designed to turn the old need for a uniform calendar into hands-on and practical scientific programs capable of impacting a substantial number of engineers, and scientists across multiple disciplines. The modularity of the system is such that an underlying teaching vehicle is provided to users with endless resources for discovery.

Once the observation units are mass-produced, *The Witness* will provide a good platform for the exchange of information and dissemination of know-how in multiple fields of science and engineering including astronomy, weather science, and optical engineering to name a few. The exchange of information and experience among users will become a standard practice due to the sheer number of topics and details. It is expected that there will be on-line science discussion groups where users share their latest findings and improvements to the hardware. This interaction will culminate in regular international meetings that will focus on subsystems.

The Witness system is an instrument-based standard designed for verifying new moon sightings using science and technology. It is conceived to witness and substantiate claims for traditional naked-eye sightings and instrument-aided sightings and is therefore in full conformity with all voiced opinions on the matter of new moon sighting. *The system* is structured as a baseline system with multiple conceptual options. Choices on what options to retain can be selected either from the start of the execution of the project or after completing the baseline system.

The Witness is conceived to initially work in the visible light spectrum. However, advanced technologies are available to expand the network's operation into the near-infrared and long-wave infrared regimes. Verifications will then be carried closer and closer to conjunction.

The Witness can serve as the backbone for further technological expansion. The control center can be expanded to include other critical capabilities including government and administration services, broadcast and media services, telecommunication system networks, and enterprise support systems. The control center can also be the center for astronomical studies and exploration.

The control center can provide the government and public organizations reliable and encrypted communication for connecting to remote sites. It can serve as the primary ground station for the acquisition of Satellite Remote Sensing (SRS) data for earth resource surveying, mapping, environmental monitoring studies, and for meteorological studies.

The control center can also provide communication from an e-school to a large "class" of students. The control center's forward link can provide sufficient capacity to distribute pictures, speech, and text to a large population. Interactivity is achieved because each user terminal can communicate back to the control center via satellite. A small group of teachers can, therefore, provide education services simultaneously to hundreds of students spread over a large region.

The control center can be expanded to enable astronomical studies and advance space science. Its communication capabilities will enable it to acquire and analyze shared scientific data from orbiting observatories. Depending on its location, the control center can be equipped with an observatory and multiple sensors to explore the sky over a wide range of the electromagnetic spectrum.

Additional expansion may include inter-government data exchange. This exchange may allow for interconnectivity between countries.

3.5 Lunar laser synthesis and ranging option

This option consists of building an observatory capable of detecting laser-pumped light from the surface of the moon. The observatory is optimized for near-horizon observation during twilight and typical line-of-sight to locations on the moon from which moonlight emanates in the early hours following conjunction.

The solar-powered lasers emit pulses from crescent moon radiation which would otherwise be bound for earth. Once implemented, this system would effectively bring the signature of a crescent new moonlight right from its origin to the observatory, bypassing traditional limitations on crescent visibility. New month confirmation is achieved even in poor climatic conditions, and even if the crescent is expected to be first-seen in remote locations on earth. The observatory operates either as a lunar laser ranging facility, traditionally known to bounce photons on special retroreflectors positioned on the moon surface or as an inverse laser terminal.

The idea is to drop and orient solar-powered lasers with retroreflecting capabilities in selected locations on the moon surface. In addition, the observatory can also bounce its own laser on the existing Apollo and Lunokhod reflectors with superior efficiency to what is available today, bringing world scientists to join and study current and future pressing topics in multiple disciplines. The observatory would operate on a 3.6 m meter telescope, supported by detectors, timing electronics, gravimeters, metrological arrays, etc. to provide local data to allow for full calibration of measurements.

Lunar Laser Ranging has produced remarkable discoveries in recent years including some of the best tests of General Relativity and a new understanding of the physics of the moon's crust and deep interior. This has been accomplished by repeated measurements and analyses of the distance to the moon with incomparable accuracy. These applications often lead to major discoveries, including the most accurate lunar orbit. The proposed new lunar laser system is expected to become the world premier observatory providing data for science topics such as the analysis of general relativity, the rotation of the earth, lunar physics and, of course, the timing for the accurate lunar calendar. The facility would provide dedicated multiple observation times per day to study topics, such as:

• General Relativity: Fundamentally, General Relativity (GR) and Quantum Mechanics (QM) cannot both be correct. Because QM has been tested to greater accuracy than GR, the current community belief is that at some level GR must break down and lead to a reformulation. Thus, measurements of great accuracy are very important to understanding our world.

• Lunar Physics & Astronomy: Multiple topics can be addressed through new and continuous precise measurements. These include as a minimum:

1. Verification of the earth-moon drift (when formed, the moon was much closer to earth, and since has been drifting farther away).
2. A variety of scientific activities such as the study of earth's gravity and plate tectonics, earth's axis tilt, lunar and solar system dynamics, mars future missions, providing high precision orbits, mapping missions, relativity theory, and astronomy.
3. Investigation of moon formation theories such as the moon was formed from two halves that fused at some point in time.
4. The observatory will be able to determine the earliest time of the new moonlight and observe the moon at the exact start of the new lunar cycle.

3.6 Time-keeping & broadcasting option

Time-keeping and broadcasting of time tables and schedules such as daily prayer time is a unique option of the system with no comparable system capabilities. Such an option would include, for example:

1. Universal Prayer Time (UPT): a novel protocol that allows anyone to receive broadcast of:
   a. Accurate Local Atomic Time: The local time, accurate to within a second every few million years.
   b. Local daily time tables such as prayer time: An atomic watch that receives its time from a broadcast station and equipped with a GPS receiver will deliver standard prayer times.
2. Beginning and end of religious events such as announcements signaling the start of the month of Ramaḍān, Moladot, etc.

The atomic clock sends information by broadcasting a high power and low-frequency radio wave to receivers for synchronization. The broadcast information is binary-coded using a number of bits and several separators at the rate of one bit per second. It would take, for example, sixty seconds to send a complete code which includes the year, date, and time. The atomic clock is synchronized with the UTC and can be housed in a special Time Keeping Laboratory.

Radio transmitters would broadcast the time signal for all to use. Anyone within about 4000 to 5000 km from the antenna's location can pick up and interpret the time information. The combination of atomic time and GPS is used to establish local time schedules such as standard local prayer time. This transmitted signal can be broadcast at 60 kHz and carries information in the form year|day|hour|minute|second, followed by indication on daylight saving time and leap information.

Atomic clock-time and GPS are combined to generate time schedules such as prayer time. This is achieved with the use of special hand watches, clocks that integrate miniature GPS, radio receivers, and cell phones. Thanks to this combination, a smartwatch or cell phone displays, for example, prayer-time within the nearest second, taking into consideration the receiver's geographic location.

The system may be used for emergency announcements to instruct the masses or emergency personnel to proceed through predefined public and/or private action codes in areas with frequent earthquake, tornados, and other emergencies and hazards.

3.7 Sighting from high altitude

There have been opinions voiced about new moon sighting that can be only made from ground level and only right after sunset; some have even coined ground sighting after sunset as "legal sighting" as opposed to other sightings. Without getting into a discussion about sighting validity and from a purely technical perspective, there are several advantages and needs to rise above the atmosphere. First, there is no difference between sighting from ground and sighting from space. As earth rotates from point B to point A, between noontime and sunset, the actual movement of the moon is not appreciable, but for an observer on earth, his/her position moves from A to B as shown in Figure 3.7-1 and therefore the distance to the moon increases by about the radius of the earth which is 6350 Km. By analogy, the distance between an arbitrary point of observation on the ground and the moon varies only 350 Km over a period of 15 minutes. Therefore, *sighting from a near-earth satellite from a typical orbit of 350 Km altitude is equivalent to performing two sightings from the ground at a mere 15-minute interval.*

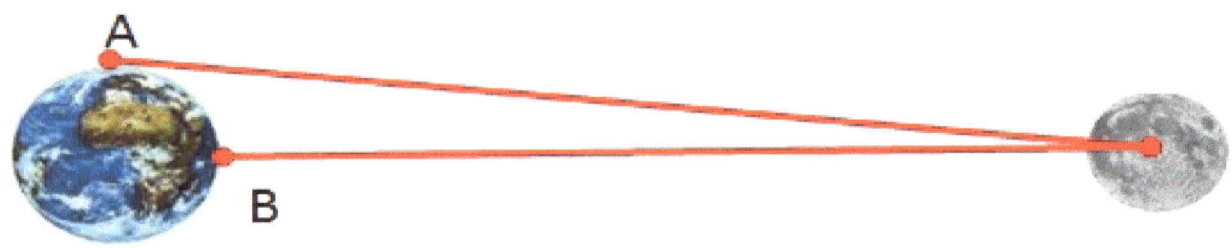

Figure 3.7-1 Change in distance between an observer and the moon as the earth rotates

Second, looking at the dim moonlight is no longer a process similar to the pre-industrial era. If it happens that you live in or outside a large city, you are probably submerged in a thick haze of man-made aerosols and dust that greatly affects the dim moonlight visibility. From satellites imagery, we also know that volcanoes have dimming effects lasting a few years after an eruption. Particulates emitted from volcanoes literally circle the globe affecting visibility everywhere.

Atmospheric dust near populated areas makes the visibility of lower altitude objects such as the crescent moon (less than 10-13 deg) quite difficult. Pollution from plants, vehicles, airplanes, heating furnaces produce particles called aerosols which get to be suspended in the air. These particles scatter, reflect, and absorb sunlight as well as moonlight. Imaging the new moon outside Santa Barbara on May 21, 2012, California was hampered by city pollution particulates and other aerosols as illustrated in the above photograph. Later, the detailed technical assessment revealed the tiny silver moon buried in the fog.

Recent studies on global warming, carried by NASA, have shown that atmospheric optical thickness (reduction of light reaching earth surface) can increase several folds following volcanic eruptions. Satellite observations of troposphere aerosol optical thickness show that volcanic eruptions such as the one of El Chichón in 1982 and the one of Mount Pinatubo in 1991 produced a large rise in aerosol concentrations, with impacts lasting 3 years or more as Figure 3.7-3. It is less known that the massive volcanic eruption of Samalas, Indonesia in the year 1257 CE, resulted in the world climate change of 1258 CE, as it darkened the sky around the globe causing crop failure, and resulting in the dissemination of the population of London, and probably bringing down the flourishing city of Bagdad together with the Islamic Empire.

Third, dust storms are common occurrences in and around desert regions with effects lasting days at a time and extending thousands of kilometers around the globe. Thanks to satellite photography, today we have a good understanding of the extent of the effect of such storms and their regional effect reaching thousands of kilometers. Satellite images of the storm of early June 2012 are shown in Figure 3.7-4 providing a clear view of the extent of such events taking place over a large swath in the Middle East and parts of Asia.

Figure 3.7-2 Pollution and other man-made particulates scatter light and hamper visibility from the ground. The crescent buried in the lower left of the right image taken outside Santa Barbara, California on May 21, 2012

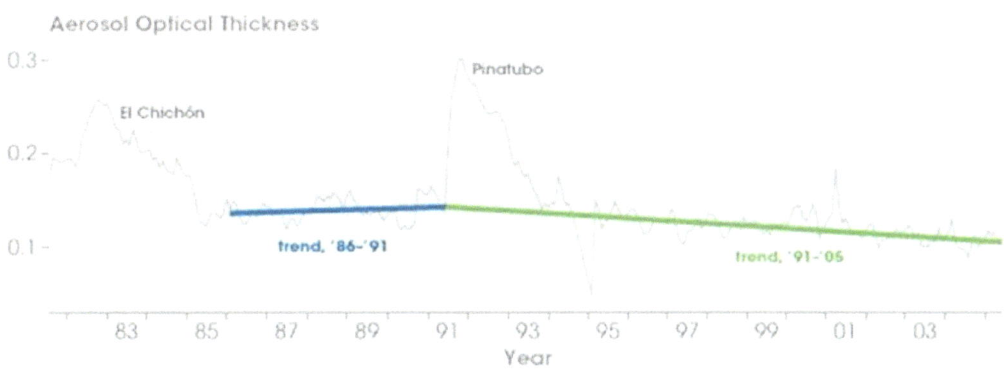

Figure 3.7-3 Increase of aerosol optical thickness following volcanic eruptions). - Courtesy NASA Earth Observatory

Natural cloud cover is the worst problem that one encounters when attempting to locate a new crescent, especially at lower moon angles following sunset. Even if the sky is clear at the zenith, clouds stack up when one looks at the horizon. Everyone who sought a new Hilal would tell you how disappointing to face such adversity. Figure 3.7-5 shows the superposition of eight days of sky cloud fraction around July 19, 2011, used to assess site selection for the July 19, 2012 observation carried by our Team.

Figure 3.7-4 Dust Storm over Iran, Pakistan and Afghanistan June 04, 2012 (left); Arabian Gulf Jun 02, 2012 (right) - Courtesy NASA Earth Observatory

Figure 3.7-5 Cloud fraction stacked over a period of 8 days around July 19, 2011
-- Courtesy NASA MODIS

Cloud seeding and weather modification have emerged for economic and military purposes. There is a major topic that needs to be addressed separately, as what is being done in the skies, without public permission, amounts to hazardous and dangerous environmental hacking. Contrails from aircrafts such as those shown in Figure 3.7-6 produce artificial clouds that obscure the horizon and prevent sighting of the new moon near cities and along major airline routes. Chemicals break light into oily brown color when scattered around these artificial clouds. Other chemicals in the form of fluffy particulates similar to what is produced by cottonwood trees and filaments fill what used to be dark blue clear skies Colorado's Rocky Mountains. These particulates can be seen floating when observing near the sun direction while its glare is blocked by a building.

Figure 3.7-6 Contrails blocking moonset visibility

The option of sighting from controlled balloons involves the build of retrievable observatories to fly on controlled balloon probes to fly at 20-30 Km altitude, pointing to the horizon during twilight. Most of the moisture responsible for scattering sunlight and providing an obstacle to extracting the dim light of the moon occurs below 10-12 Km above ground. Major pertinent aerosols of concern are in the 2-4 km altitude range. Therefore, clearing these altitudes provides a tremendous advantage in document Hilal sightings. We have preliminary designs that can be exploited for such purpose. We have already been involved in building an experimental observatory to minimize risks. The cost of such systems should be very affordable, and the time required to build them is very short.

The disadvantage is that new moons usually present opportunities for sightings in different regions of the earth and negotiations must precede the launch of such observatories to comply with the rules and regulations of local authorities.

The International Space Station (ISS) is a major collaboration between 16 countries to maintain a long-term human presence in outer space. *The Witness* proposed to coordinate efforts to use ISS as a platform to document Hilal sightings from space. The ISS could provide an important link with ground stations to perform new tasks never imagined before. The moon of Shaʿbān was photographed from the ISS and is shown below.

Figure 3.7-7 Waning moon of Shaʿbān 2012 from the Space Station (crescent moon can be seen in the blue haze of the atmosphere in the middle of the photograph - Courtesy NASA.

For example, on July 19, 2012, starting 14:28 GMT, we had seven opportunities to sight the crescent of Ramaḍan from the ISS in a span of nine hours. Figure 3.7-8 illustrates those possibilities. While earlier sightings were possible, the first time the ISS crosses a geographic location, where a separation between sunset and moonset is about six degrees occurred over the southeast of the African continent. The second opportunity occurred about one hour and a half later, at 14:28. The third and fourth opportunities occurred at 16:01 and 17:33 respectively over the South Atlantic Ocean. The next two opportunities were over the eastern and western coasts of Chile at 20:38 and 22:11 GMT. These two opportunities were important

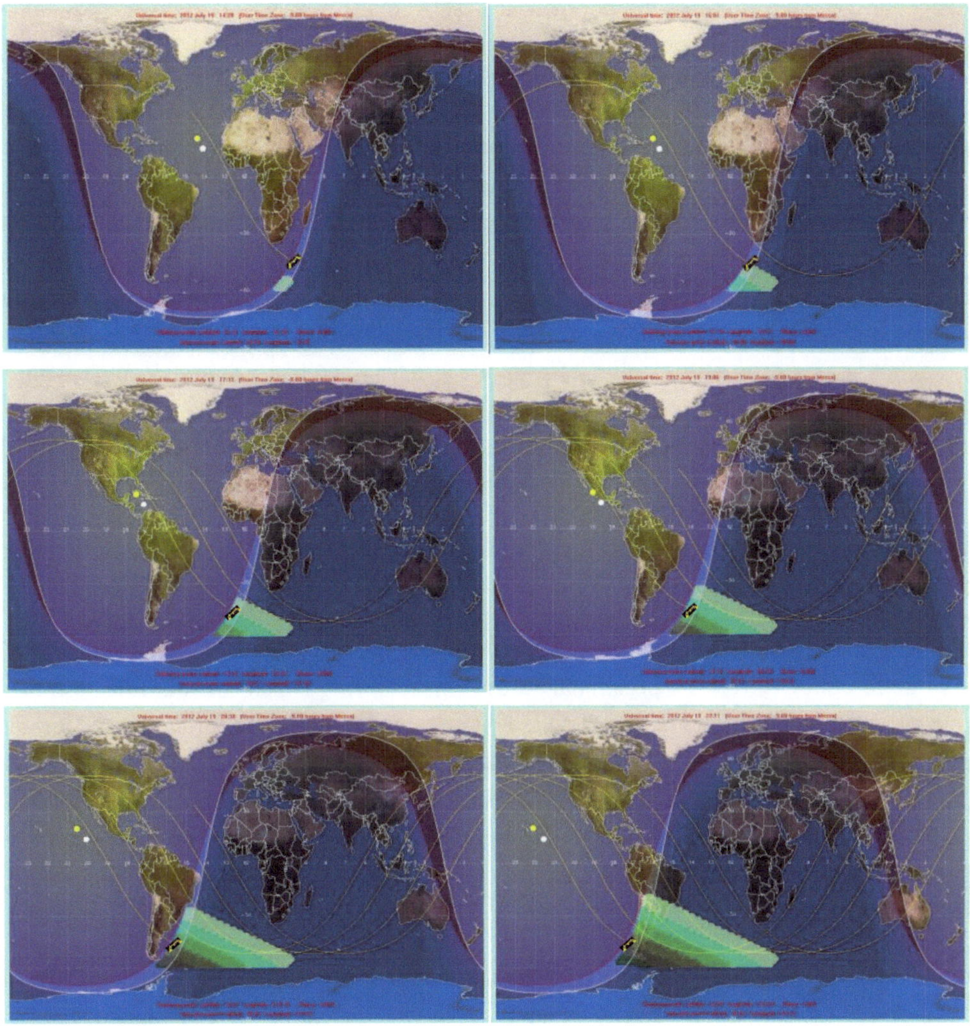

Figure 3.7-8 ISS ground tracks and 1st through sixth sighting opportunities during the month of Ramaḍan 2012

to *The Witness* team because we had an observation station in the Atacama Desert that we could use for comparison to evaluate the effects of earth shadow and air quality. Finally, the last opportunity occurred over the South Pacific Ocean at 23:42 GMT.

As we have illustrated in the case of sighting from ISS, a near-earth satellite would give a discrete number of opportunities to document the new crescent. One can think of about 7 to 10 opportunities depending on the month and the season if we choose a mission design with a period of 90 to 110 minutes. Such orbits are economical and are expected to provide unprecedented opportunities to confirm the Hilal over multiple regions of the globe. Depending on the mission's objectives, such satellite would be in the 150-300 Kg class with a life span of > 5 years.

Low altitude retrograde satellites with orbit inclinations of 96-100 deg and Molinia satellites with a semi-major axis orbit around 25,000 km and an inclination of 60-65 deg are more complex missions but can provide a superior coverage time. In addition, such orbits require advanced attitude control and maneuvering. The other advantage such orbits present is in their capacity to clear the field of view from the atmosphere at smaller angles of separation from the sun. Because there is no precedence to such missions, it is recommended to approach the development as part of a space program with a coherent administrative, educational and industrial framework in which the first mission would be exploratory in nature. For example, the first mission would carry multiple experiments to assess the efficiencies of several types of telescopes and detectors.

In addition to carrying sightings in the visible regime, an IR telescope would provide a major advantage in early moon detection. Such capability would make the mission more complex, but the return on investment may prove to be quite important in terms of reliability.

The satellite option requires a preliminary study phase during which the identification of the top-level mission requirement is carried. Following such phase, a preliminary design is proposed to demonstrate the capabilities of the satellite at the analytical level with reference to heritage in key hardware selection. Recommendations would result from such preliminary design review, and the detail design would then proceed. Once the design phase is complete, a second review takes place, and upon agreement, manufacturing and procurement of components begin.

Depending on the retained satellite system and depending on whether procurement relies on off-the-shelve components, execution may take up to 30-45 months. Starting small will not only reduce the cost, but it will also shorten the schedule.

3.8 Implementation plan

The Witness is intended to be developed in three phases. The baseline project is set to be fully operational in 24 months. The first phase lasts 12 months and demonstrates a fully functional observation station and a control center. This includes a custom-designed optomechanical system with high resolving power and embedded tracking capability. Workstations and other communications hardware can be integrated into an existing facility to create the control center. The communication architecture and information technology structure is made to support a large number of observation stations. On-site training on operating the observation stations and control center has been developed. The construction of the control center and observatory are carried in parallel during Phase I and Phase II and should last 18 months Phase 2 complete the development and deployment of *The Witness* system in approximately 12 months. Such schedule assumes a total of two dozen of deliverable observation stations and all the hardware needed for the observatory.

From a project management and system engineering perspectives, coordination between the organizations involved will be essential to the successful deployment of *The Witness*. Accordingly, the plan must include teleconference meetings shortly after the start of the project to identify the roles and responsibilities of the organizations and to establish a shared vision for the system. An outcome of such meetings is a set of baseline requirements for the system that will include those needed for basic system operation and those specifically identified to meet the needs of the party in charge of operations. Complete set of requirements should be drafted and approved within one month of the starting date.

Coordination between the organizations must occur frequently throughout the demonstration phase. Primary contacts for each organization must be established. These contacts will direct information flow within their respective organizations. Email can be the primary method of communication, but teleconferences must be used when needed.

The design of the observation station is relatively mature, but it must be customized to meet all of the final requirements. The design effort should focus on modifications to three functional areas: software design, mechanical and electrical design, and optical design. The final design is then presented approximately four months after beginning the design effort. The final review must be scheduled to present the overall architecture of the system, design details, and operating procedures.

Fabricating the observation station will take seven months. This "prototype unit" should closely resemble the final form and should be fully capable of demonstrating system performance. Assembly, integration, and testing of the observation station and the control center computer servers should take an additional four months.

In the demonstration phase, the control center must be simulated by computer servers. An internet connection is used to establish communication between the observation station and the computer. The control center computer servers contain *The Witness* software necessary to verify a sighting. This includes two-way communication, image download capability, data analysis, and authentication functions.

After the communications system design is finalized, the necessary equipment is procured and installed. This includes system servers, computer workstations, and communication equipment.

The information infrastructure should support the expansion of *The Witness* sighting verification software to allow the control center to process and validate data from multiple observation systems. An information technology (IT) architecture is implemented to enable high-speed data processing and long-term data storage along with broadcasting capabilities through a website.

The demonstration phase ends with the delivery of an observation terminal and a demonstration of *The Witness* system. Systems engineering provides on-site training to host personnel in the operation of the system. Testing is used to identify desired improvements to be implemented into the design at the beginning of the implementation phase.

In this second phase, the design of the observation stations is upgraded to incorporate the necessary changes identified in the demonstration phase. The modification should be completed within two months, and production must commence immediately thereafter. The two dozen units would be delivered to the system operator approximately eight months after the start of the Implementation Phase.

Most of the effort in the Implementation Phase is focused on completing the control center and the Observatory. The baseline plan is to use INMARSAT as the communications provider for the remotely located observation terminals.

The Implementation phase ends with *The Witness* system deployment and on-site training. The observation stations would be set up at various agreed-upon locations to demonstrate the effectiveness of the system. Ideally, system training should occur at the same time as a Hilal sighting so that a complete system demonstration can be conducted under real-world conditions. Other auxiliary options, including the Time Keeping System, must be implemented over a separate track for two reasons: First, the technology involves a particular electromagnetic spectrum band which requires different types of engineering and hardware vendors that may distract from the core *The Witness* system execution. Second, the system does not require much adjustment to incorporate its capabilities within *The Witness* System, once the decision is made to carry on with execution.

3.9 Concluding thoughts

The vision of The Witness is to start the system with a few observation units, and later mass produce the units to make them available to anyone. Interest in acquiring these units will come from academics, scientists, amateurs and others. As with any other widely used product, the introduction of *The Witness* should bring a positive economic impact in multiple sectors such as job creation, manufacturing, trade, and infrastructure development.

Impact on the economy should reach remote regions as testing measurements, and experimental work need to take place in remote geographic areas far away from cities and developed regions. This requirement is driven by the need for clean atmospheric conditions.

The control center can function as an attraction point for year-around local and international tourism. During a visit, young students may have hands-on classes to build, for example, their own observatories and earth-moon experiments. Display of ancient astronomical tools and ancient discoveries may be placed in a small museum situated within the control center facilities. Displays of *The Witness* system workings and efforts that led to its deployment may be included in the museum. There are other ideas on how to generate return using hands-on activities for children and adults.

The Witness system provides a unique catalyst for the reinforcement of several key technology sectors. It will contribute to the growth of the culture of knowledge in multiple disciplines such as astronomy, weather science, environmental science, optics, controls, and telecommunication. This culture of knowledge is the fabric around which sustained economic growth takes place. Important advancements in the history of mankind are the result of some needs that were either forced upon us or developed because of our innate drive for discovery. The need is very clear in the case of new moon sighting documentation, and The *Witness* can turn this need into economic growth.

In addition to Hilal sighting verification and documentation, *The Witness* incorporates state-of-the-art technologies that can handle expanded capabilities to support academic, government and commercial applications. Due to its modular nature, there is no limit to what such a system can do. As an example, instantaneous vital physical parameters such as temperature, pressure, humidity, and aerosols may be gathered for desertification evaluation and analysis.

There are several options that can augment the capabilities of *The Witness*. The proposed time-keeping and broadcasting system provide a unique achievement with no existing comparable capabilities offered elsewhere. This system will broadcast an official prayer time utilizing an atomic clock and provide the ability to ensure that prayer times occur precisely everywhere in a standard way.

The inherent use of experimental physics and hands-on characteristics of the system align well with the interest of young and future generations in astronomy, space, and telecommunication. Should many mobile stations be deployed and made accessible to students, the system will inevitably keep young people engaged in applied science and constructive discoveries.

Once the observation units are deployed, *The Witness* will provide a good platform for the exchange of information and dissemination of know-how in multiple fields of science and engineering including astronomy, atmospheric science, and optical engineering to name a few. The exchange of information and experience among users will become a standard practice due to the sheer number of topics and details. It is expected that there will be on-line science discussion groups where users share their latest findings and improvements to the hardware. This interaction will culminate in regular international meetings that will focus on disciplines.

It is a fact that engineers and high-tech business owners in technologically advanced parts of the world grew up involved with science, instrument usage, and technologies. They are part of a trans-generation technical culture driven by societal needs that manifested at some point in time. *The Witness* is designed to turn the old need for a uniform lunar calendar into hands-on and practical scientific programs capable of impacting many engineers, and scientists across multiple disciplines. The modularity of the system is such that an underlying teaching vehicle is provided to users with resources for discovery.

Chapter 4 Calendric moonlight

The world is a book and those who do not travel read only a page.

Saint Augustine

In this chapter, we describe the shifting light from the moon during a lunar cycle. Of interest to our topic is the manifestation of moonlight from both temporal and geographical perspectives as perceived from the earth surface.

4.1 Moon Phases

The new moon is born when its position in the ecliptic plane is between earth and the sun and is therefore said to be in conjunction with the sun, as seen from earth. At the time of conjunction, there is no sunlight reflected off the moon that reaches earth. That is, the side of the moon facing earth is dark, and the opposite side which faces the sun is illuminated. As time passes, the moon moves away from the direction of the sun, and when a certain separation angle is reached, a thin crescent can be seen for the first-time during twilight. Observation of the new crescent is then possible because, at twilight, the sun drops sufficiently low below the horizon causing the earth shadow to cast over a good portion of the lower layers of the atmosphere in the line of sight, where sunlight scatter is dominant.

In the first few hours following conjunction, when the angular distance between the moon and the sun is less than 10 degrees, the sun and the moon rise together and set together. During the daytime, even though traveling through a short cross-section of the atmosphere, the new crescent moonlight is lost in the glare of the sun, overcome by the scattered sunlight due to its proximity to the sun. In the following days, the separation angle grows larger and larger, and the moon lags the sun as they travel in the daytime sky. The silver moon crescent becomes visible before sunset. As the moon phase increases, the moonlight can be seen during daylight. At about 45-degree separation, half the moon appears illuminated by sunlight, and we have the first quarter moon. Around the middle of the lunar cycle, when the separation angle between the moon and the sun is about 180 degrees, the full moon rises from the East around the time of sunset, crossing the night sky. As time elapses, the separation angle decreases until it reaches 45 degrees, at which time we reach the last quarter of the cycle, which is characterized by a half-moon. During the daytime, the moon seems to be ahead of the sun in the sky and sets earlier than the sun. As the moon is separated by a few degrees from the sun, the crescent moon of the ending month could only be seen before sunrise. Otherwise, it is again engulfed in the haze of the scattered sunlight. The moon continues catching up with the sun in the sky until conjunction, at which time we have another new moon.

The phases of the moon designate the proportion of the moon's illumination. There are eight commonly adopted phases of the moon, which are shown in Figure 4.1-1. The moon, like any other spherical celestial object, has half of its outside surface illuminated by the sun, while the other half is in darkness. The eight inner half disks in the Figure illustrate the position of the moon at the distinct phases of the month, while the corresponding eight outer moons show the degree of illumination as seen from earth.

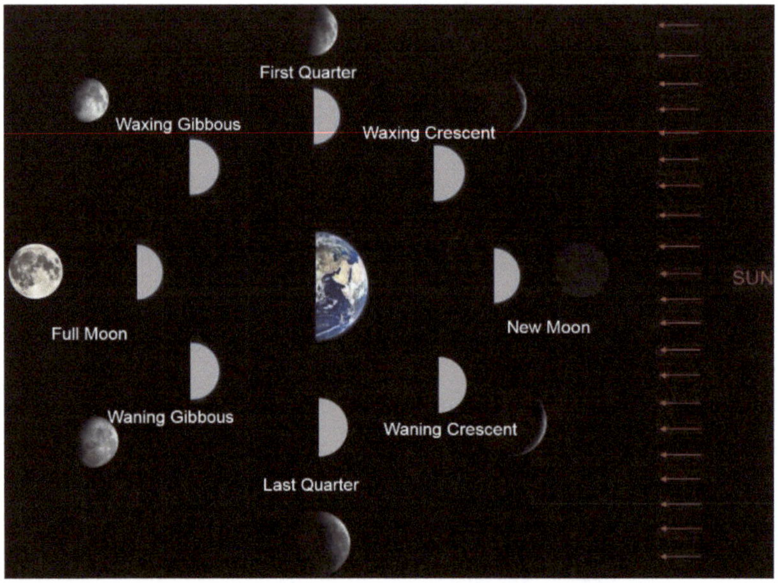

Figure 4.1-1 Phases of the moon

The New moon, First Quarter, Full moon and Last Quarter are called the primary phases of the moon, and they correspond to when the apparent longitudes of the moon and the sun in the ecliptic plane are separated by 0, 90, 180 and 270 degrees respectively. Such a definition is technical and does not relate to how actual the visibility of the moonlight is. The phases correspond to accurate Universal Times which are independent of the location of the observer and are published in common almanacs and calendars. The astronomical lunar calendar, an example of which from the US Naval Observatory is given in Table 11.1-1 for the year 2010, shows the Universal time and date of each primary phase of each lunar month of a solar year.

It remains that all social and religious events relate to moonlight visibility and not to the astronomical definition of the phases.

Early astronomers have precisely established the lunar cycle but could not come up with the correct prediction of when the first light from the moon is visible. As we will see next, modern astronomers have put together the right framework for predicting first visibility but failed to provide a solid mathematical rationale for when a new moon could be seen for the first time.

4.2 Alternative lunar month

In the absence of clouds, sand storms and aerosols, we may routinely document the first radiation emanating from the surface of the moon during daytime using electronic detectors –especially those geared to sense a good portion of the deep red and near-infrared. We take this opportunity to mention that we cannot emphasize enough how extremely dangerous to look in the direction of the sun for fear of immediate blindness. Humans and animals, being creatures of the earth, have an innate reflex that prevents them from focusing on the sun to avoid blindness. Staring in the direction of the sun is much more dangerous if someone looks through an optical device which collects light from an aperture larger than that of the eye. Optical devices which are geared for consumers carry large warning signs not to point them in the direction of the sun. Back to our topic, and a few days past conjunction, when the angular separation between the sun and the moon is large enough, one can see the moon partially illuminated in broad daylight. But then the new moon is no longer "new," and the window for detection of a new crescent has already elapsed. In the pursuit of establishing an origin for the first light from the new moon, one seeks detection as early as possible right after conjunction. Because the moon is near the sun in the sky, obviously there is no new moon crescent observation during night time, because both the sun and the moon have already set in the early hours following conjunction, and as a consequence, daytime detection would seem to be an obvious place to consider. There are several advantages to tracking the new moonlight during the daytime, some of which are:

- The light from the moon has a shorter path through the atmosphere in opposition to looking near the horizon after sunset, and therefore, extinction is minimized.
- The window of opportunity for tracking the moon is longer than the 45-60 during twilight. If the sky is cloudy at a certain time, a window of opportunity could open, and a chance to perform observations is possible.
- The time since conjunction would be the shortest possible and therefore responds to our need for first light.

There are, however, several fundamental disadvantages to tracking the moon during daytime which eliminate such choice altogether:

- The new moon detection cannot be accomplished using a naked eye. There is a requirement to use special hardware for detecting the new noon near the sun, and therefore, it may be stated that the process is hardware-dependent.
- Even if we standardize the detection hardware, historically, the month for many is the visible new crescent as seen by the naked eye which does not respond to intended social, religious, and other observance requirements.
- Electronic sensors, such as those of the CCD and CMOS types used in detection hardware, do not duplicate the human eye; they collect radiation and convert it to electric signals, integrating multiple wavelengths outside those of the human eye capacity. For example, detected deep red and near-infrared signals represent the heat absorbed by the lunar body and re-emitted in the form of thermal glow and do not constitute a reflected visible light radiation.
- Daytime observation of the new moon requires expensive hardware, which is outside the reach of the masses and requires trained personnel due to the risk of severe injury and technical challenges.

Even if we decide that the advantages of daytime detection counterweigh the disadvantages for detecting the first light from the moon, the effort will not lift the controversy. Some would still argue that their texts, scripture, or tradition pertains to the first observation of the new moon. And looking in the direction of the sun does not respond to their needs or is not in-line with tradition. The only agreeable, natural and logical origin for the lunar month is, therefore, the time of first visible light from the moon as seen using either the naked eye or aided by instruments during twilight following conjunction.

We learned from the previous survey of calendars that calendric adoption comes with an agreement. In fact, Section 2.2 refers to the Calendar as an agreeable system for organizing and measuring extended elapsed time. If we were to unify celebrations and observances, we need to use something simple that everyone accepts and can relate to. The lack of success of the Secular Calendars, e.g. The French Republican and The International Fixed Calendars of Section 2.11 is due to a lack of consensus surrounding their adoption, irrespective of their accuracy and efficiencies. Therefore, in the current effort, we seek a development that everyone accepts because it is useful, simple to adopt and intuitive. The proposal of a photometric new moon which we are about to define has such characteristics.

4.3 The photometric lunar calendar

The photometric lunar calendar keeps the lunar phases as traditionally defined and adds the new moon as a fifth phase. It should be noted that even current Lunisolar and pure lunar calendars could make use of such additional information, irrespective of intercalation. Each month in the photometric calendar starts with the first sight of the new crescent following astronomical conjunction and ends at the subsequent sighting of the new moon. Sighting is from anywhere on the earth surface. Computation is tabulated in the Coordinated Universal Time (UTC) and therefore, the length of a month is determined by the simple difference in UTC and will vary from month to month and from one year to another. As an option, the first light from a new moon is seen somewhere on the surface of the earth at a specific geographic location.

Photometry is the branch of optical science that quantifies the perceived light brightness by the human eye. This field is distinct from *Radiometry*, another branch of science that studies radiation in general and visible light in particular. Radiometry deals with energy power and relates to the photometric measures through *luminosity functions* which describe the relative human eye's sensitivity to brightness over the visible spectrum. Because the human eye operates differently during the day from the night time, there are two functions, one for day time adaptation, called *Photopic* and one for nighttime darkness adaptation, called *Scotopic*. These functions have been standardized by the *Commission Internationale de l'eclairage*, (CIE). Photometric quantities are weighed to the sensitivity of the human eye while the radiometric counterpart is absolute power. The photometric calendar applies the standard human eye model to the new crescent light detection to establish the onset of new crescent visibility. Sighting is in accordance with standard human visual acuity and assumes a clear atmosphere free from aerosols. The NASA standard atmosphere is the world standard atmosphere used today.

What kind of astronomy pertains to the twilight observation of the new moon? Classical astronomers carry observations at night, and when bad viewing occurs, they can come back and do their work the next evening; hence they have nightly chances to do their work. In opposition, those who track the new moon have only about one hour per month to verify procedures and test new methods of observation; this is if it happens that the above conditions for visibility (or equivalent ones) are met in their geographic locations. If it happens to be overcast, they will have to wait until the next month. Add to this the fact that new moon position is close to the sun, and the thin silver crescent is amid sunlight scatter. Basically, the new moon observer is given the wrong tool for the job: a telescope designed for night time and is pressed to report on a positive sighting if the computation predicts a chance for visibility. And even if certain new moon observers decide to develop tools to do the job properly, they have only a brief period to do their experiments and tests.

Differences between classical astronomical telescopes and instruments that can be conceived for twilight observation of the moon are many. From the basic purpose to operation and function, one immediately recognizes the need to rethink the way we want to observe and verify the light of the new moon. In the next chapter, we address the fundamental characteristics of classical telescopes and give a vision for an observation system that responds to the needs of new moon observation. The topic of new moon verification should open a new area of astronomy that could prove to be significant. It is noted that many recent technologies have just been developed for exploitation in the daytime and during twilight observation. Once we identify the need to at least understand what can be done, many solutions will further emerge. These solutions will compete, and some of them will prove to be more efficient than others.

4.4 Cylindrical projections

To respond to the question of the relative temporal/geographic setting of the sun and the moon, a normal cylindrical projection map of the earth is used for analysis. The terminology of "normal cylindrical projection" refers to projections in which the earth's meridians are mapped to equally spaced vertical lines. Similarly, the lines of latitude are mapped to horizontal lines. The mapping is produced by wrapping a co-axial cylinder around the earth, projecting the earth surface onto the cylinder, and subsequently unfolding it.

The intention here is not to provide accurate timing of the onset of sunset and moonset. Instead, we intend to focus on the geographical and temporal relative time of sunset and moonset and its evolution in the 24hr that follows conjunction.

As time goes by, the moon moves westwards in the sky and earth rotates. Using some reliable ephemeris, we obtain the exact position of the moon in the sky and its cylindrical projection. Assuming that (a) the moon is fully lit up and (b) in the absence of sunlight, the moon would light up the projection of the earth surface as shown by the lighter blue region in Figure 5.2-1. The far side of the earth that does not see the moonlight is illustrated by the darker blue shade. The moon is drawn as a light gray circle on the cylindrical projection of the earth surface and is located at the center of the lighter blue region as illustrated in the figure.

The lunar profile shown varies from day to day during a sidereal lunar month and from season to season. The trace of its projection also varies in longitude and in latitude as well to produce a cyclic trace on the earth surface. The bright side moves continuously from east to west. For illustration, we do not have to move the earth surface; all we have to do is adjust for its motion relative to that of the moon and take snapshot images.

Figure 5.2-1 also shows two terminators the western terminator is when the onset of moon rises, while the most eastern one is the onset of the moonset. The profile of the sunlight is also computed to reflect the projection of daylight and night time. Figure 5.1-2 illustrates the situation at some time in a typical February day. One may notice the short day and long night at any latitude in the northern hemisphere. The situation is reversed for latitudes in the southern hemisphere where summer is in effect. The situation changes in spring/summertime and the day becomes longer in the northern hemisphere

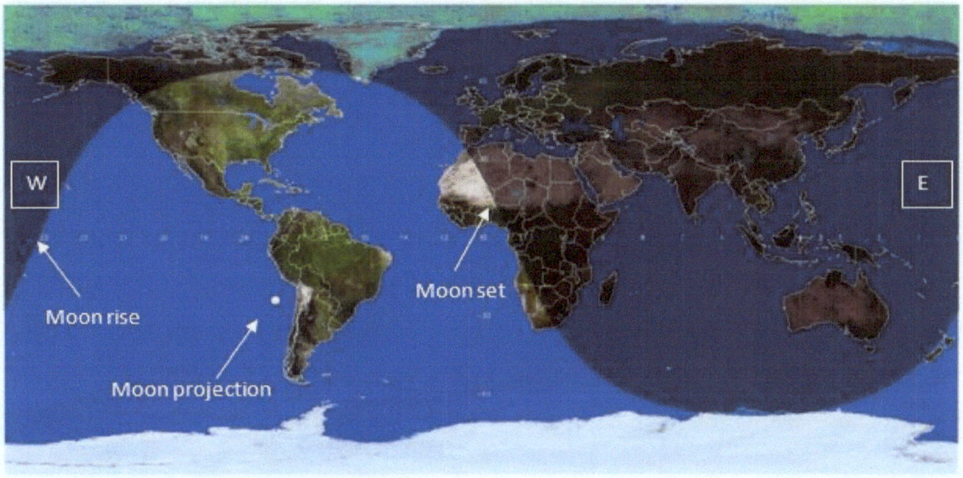

Figure 4.4-1 Cylindrical projection of moonlight at an instant in time

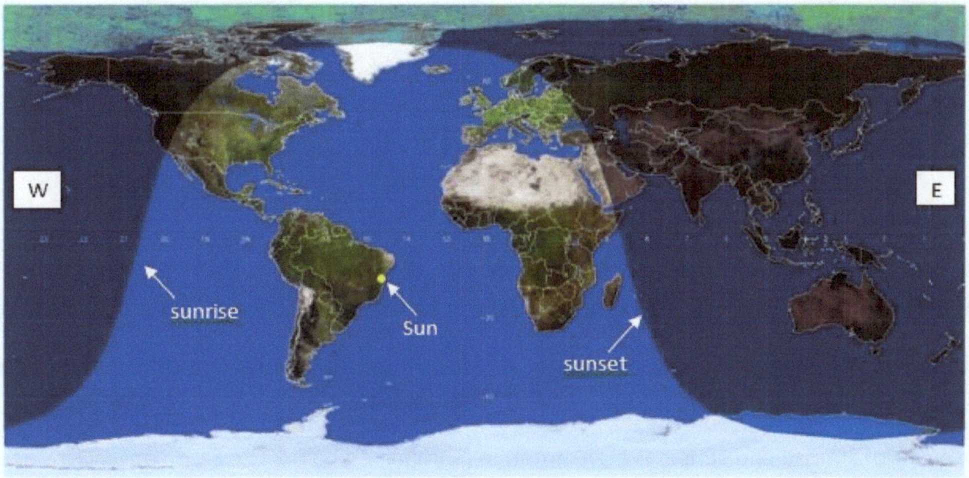

Figure 4.4-2 Cylindrical projection of sunlight at one instant in time in the month of February

The trace of the sun is displayed by a yellow dot and over a twenty-four hour period. The bright shape of the daytime moves also from East to West. We also have here two terminators: one for sunrise and one for sunset. The former is the most western one while the latter is the most eastern.

The introduction of the twilight phases to this study is critical to the understanding of the dynamics of moon sighting possibilities, as naked-eye new Hilal sightings occur during twilight. In a twenty-four hour period, the sky illumination undergoes drastic changes defined by astronomers as five separate phases (even though the boundary between some of the phases are not well defined and depend on several factors) highlighted by the contour lines from hot to cold shades in Figure 4.4-3:

- Daytime (yellow-orange-red): Sun is above Horizon,
- Civil Twilight (pink): Sun is 0 to 6 degrees below the horizon,
- Nautical Twilight (purple): Sun is 6 to 12 degrees below the horizon,
- Astronomical Twilight (Blue): Sun is 12 to 18 degrees below the horizon,
- Nighttime (Dark Blue): Sun is less than 18 degrees below the horizon.

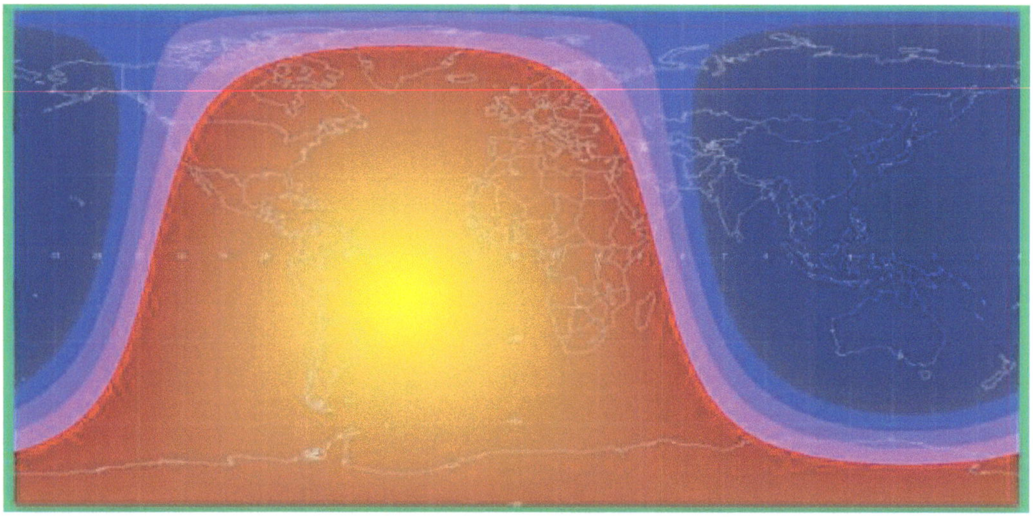

Figure 4.4-3 The five phases of a 24-hour period: daytime, civil, nautical, astronomical twilight, and night-time.

The onset of sunrise and sunset is highlighted in a bright red color. The same procedure used for the moon and for the sun is repeated for twilight periods. We have at any time, depending on the local position of the sun, an indication of whether the time is daytime, civil twilight, nautical twilight, astronomical twilight, or nighttime. These profiles are also dynamic, that is they change over a twenty-four-hour period and from season to season.

Twilight is an optical phenomenon caused by the refraction and scatters of sunlight by the different layers of the atmosphere. As direct sunlight gets weaker and weaker, the illuminance at the earth's surface drops by a factor of about a billion, allowing moon observation to occur. The moon does not have such a phenomenon because it has no atmosphere. Instead, at any time it is either night or day on the surface of the moon, and a sharp terminator is found on its surface. Back to earth, the duration of twilight is dictated by the speed of the sun dipping below the horizon, which is a function of the time of year and latitude. The duration of twilight is related to the width of the band shown in Figure 4.4-3. The wider the twilight band is, the longer the duration and the better the chances are for sighting.

Because we are looking at the entire globe at the same instant in time, astronomers suggested the use of a uniform clock to compute the time of events. The Universal Time which is related to the motion of the sun as observed on the Greenwich meridian gives enough accuracy for our particular topic, even though the earth is not a perfect timekeeper due to small effects that influence its cycle in its journey around the sun. However, these days a more accurate time measure is adopted thanks to the use of atomic clocks and the adoption of the Coordinated Universal Time. When looking at events in the distant past, another time measure which accounts for historical changes in the rate of rotation of the Earth is used. We shall discuus the different time measures and their usage in Section 8.4.

4.5 Sun-moon relative setting

The simultaneous manifestation of the moon and the sun against the twilight background is displayed in Figure 4.5-1. Where the onset of moonrise and moonset is highlighted by a light gray color and compared to the red colored onset of sunrise and sunset at a particular time giving an instantaneous global view of the relative moon/sun situation everywhere around the earth. The example of February 10, at 14:52 UT shown in Figure 4.5-1 illustrates all the elements of Figures Figure 4.4-1, Figure 4.4-2 and Figure 4.4-3 simultaneously.

The reader should first refer to the display of the projections of the sun and the moon to get proper orientation on where daytime is over the planet and where the moon is above the horizon. Then looking east, the reader needs to locate the sunset and moonset terminators (note that the terminology for moonset terminator is not strictly correct for we are not on the moon surface). Looking west, the reader should locate the terminators for sunrise and moonrise.

From the composite Figure 4.5-1, we consider the relative position of the red sunset and moonset curves displayed in red and light gray, respectively. On February 10, 14:52 UT (17:52 Mecca civil time), at the onset of sunset, the moon has already set. The situation will continue as the two terminators move west. The geographical area between the two moonset and sunset terminators gets a somber tone when moonset precedes sunset. When the opposite takes place, the area between the terminators gets a brighter shade as it is shown in

Figure 4.5-2 -. An estimate of the duration between sunset and moonset can be immediately evaluated by keeping in mind the correspondence of the 24 meridians to the 24-hour day. Furthermore, when the moonset curve falls inside the twilight bands at some particular geographic area, an estimate of how low the sun is below the horizon can be immediately visualized by simply knowing that each band represents a six-degree interval.

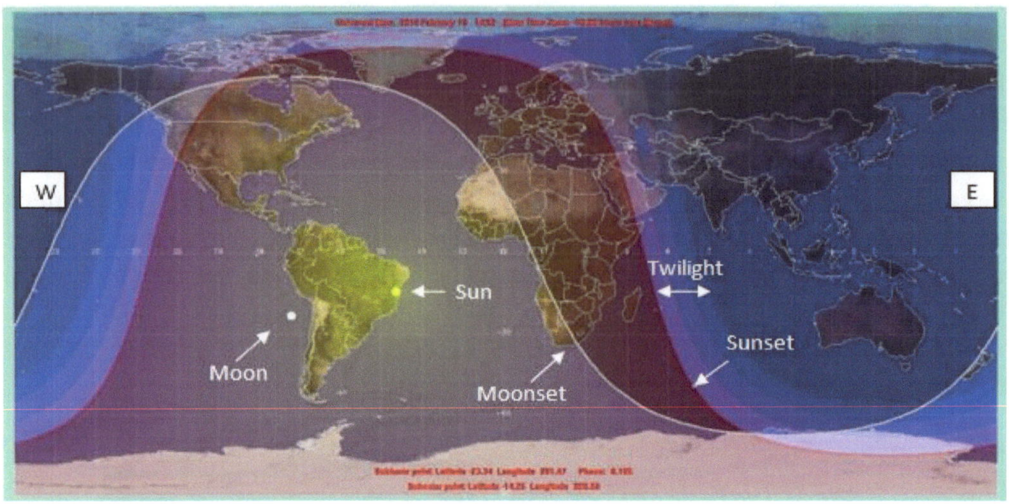

Figure 4.5-1 Global view on February 10 14:52 UT

The new moon of Raby`al-awal 1431 occurs on February 14 at 2:51 UT. Section 11.1 provides the phases of the moon for 2010, a moonrise/moonset table and a sunrise/sunset table for the city of Riyadh, Saudi Arabia, to use as a reference.

Figure 4.5-2 shows the 1-minute old New moon Figure 4.5-2-Figure 4.5-5 show details of the progress of moonset and sunset for the 6, 12, and 16 hr moons. Initially, the curves for sunset and moonset are nearly superposed as conjunction takes place. Then gradually the moonset curve shifts east of the sunset curve in the upper latitudes crossing at some point signaling a post-sunset possibility of new moon sighting in the upper latitudes. Then the two curves start having some width pointing to longer and longer duration between the two settings. And simultaneously, the crossing point drops further and further south. The sun terminator and the moon terminator show inverted bell curves with longer daytime and shorter nights. Initially, the curves for sunset and moonset are nearly superposed as conjunction takes place. Then gradually the moonset curve shifts east of the sunset curve in a near-uniform fashion and the crossing point drops out very early, giving an opportunity for earlier visibility everywhere.

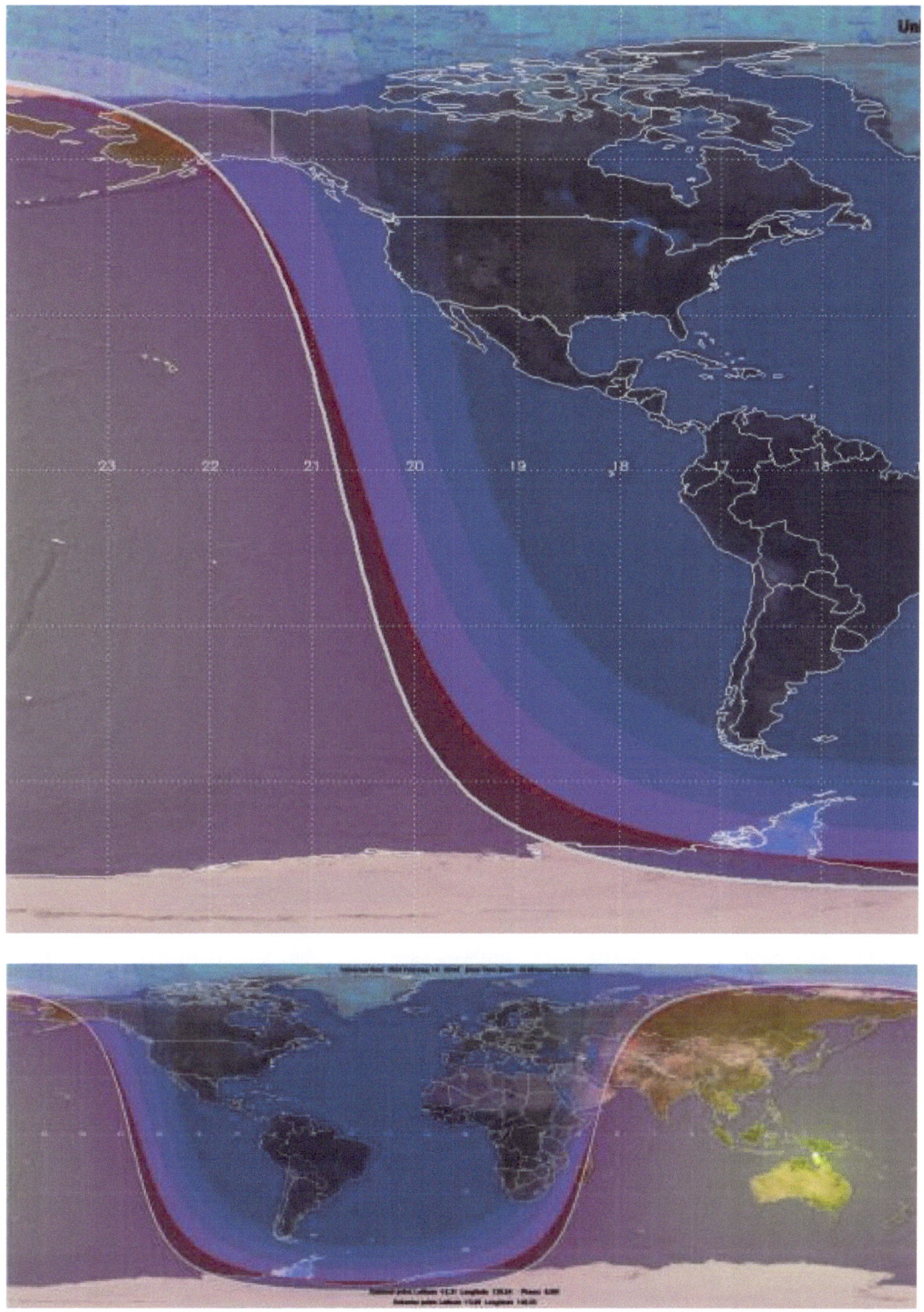

Figure 4.5-2 New moon is 1 minute old, Date is: 2010 February 14, Time is 2:51 UT

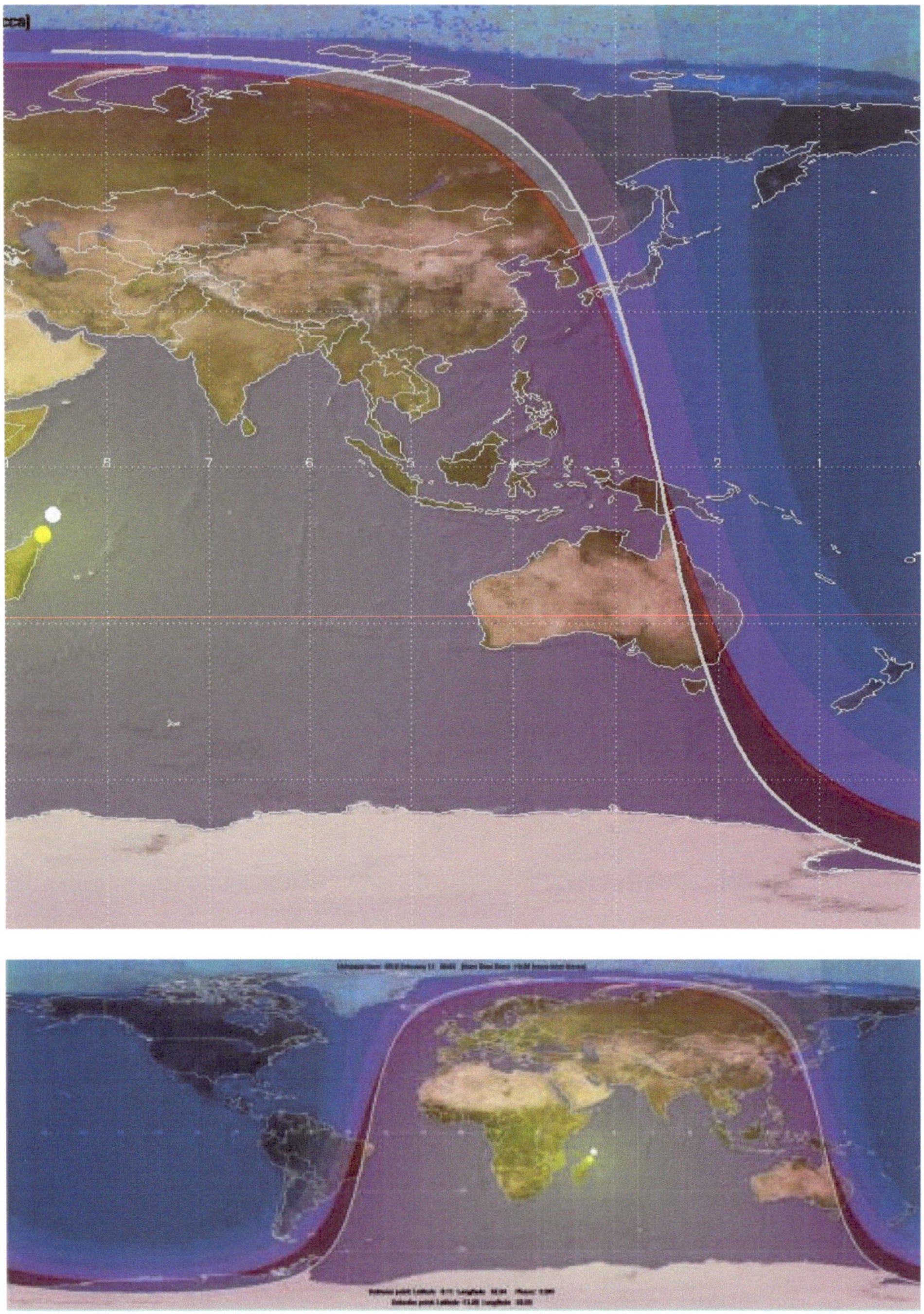

Figure 4.5-3 New moon is 6-hour old; Date: 2010 February 14, Time is 8:51 UT

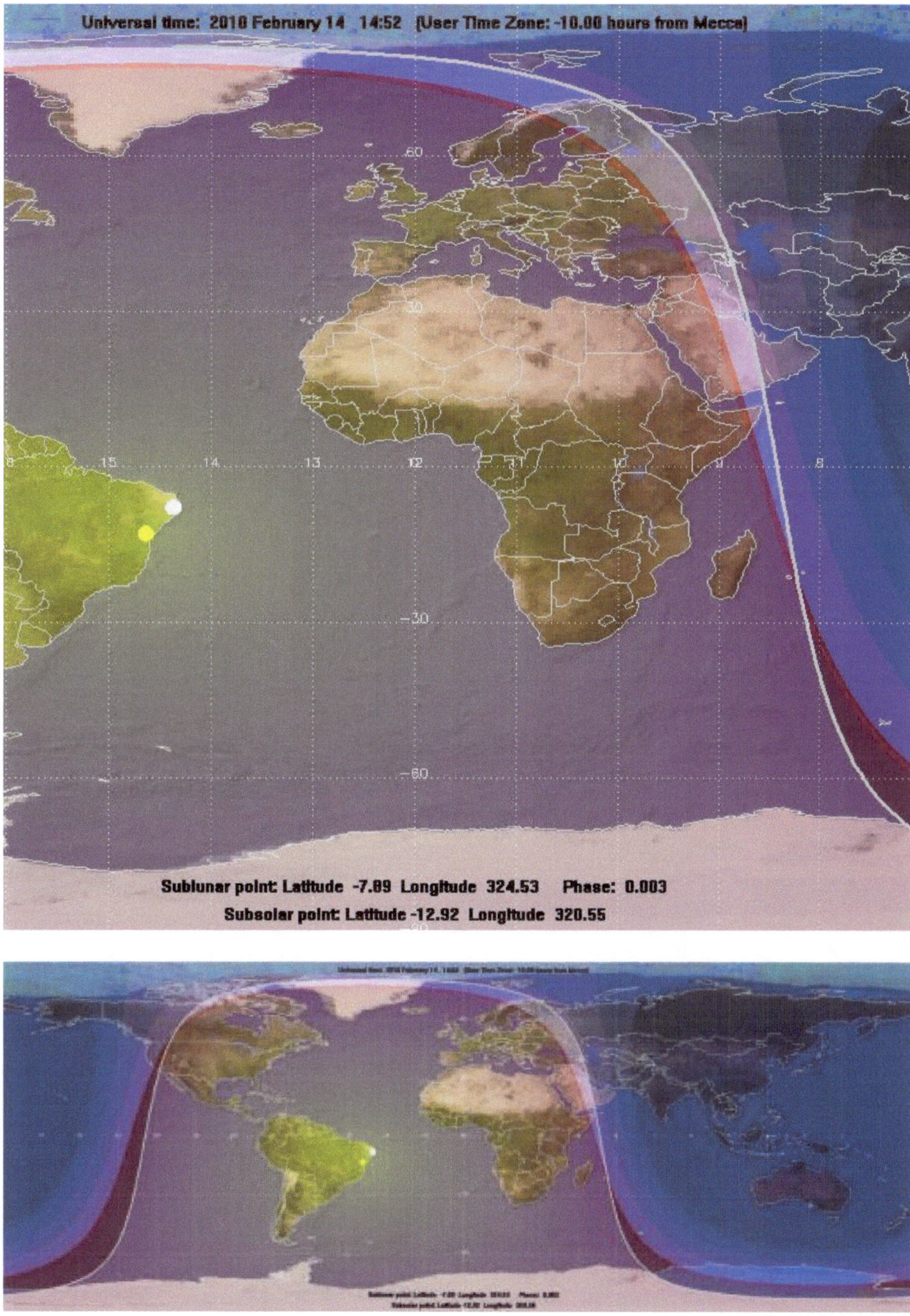

Figure 4.5-4 New moon is 12hr old, Date is: 2010 February 14, Time is 14:52 UT

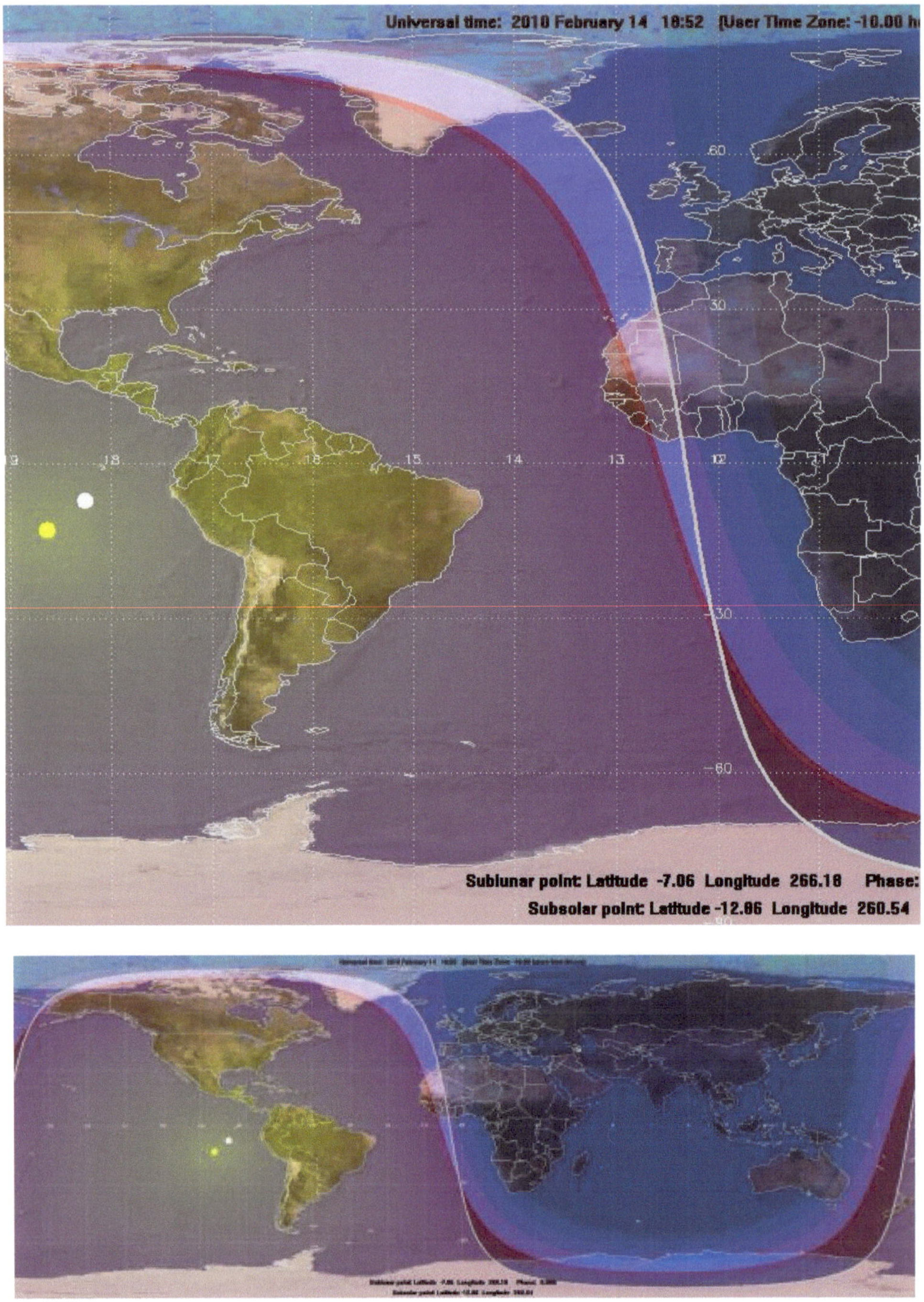

Figure 4.5-5 New moon is 16-hour old; Date: 2010 February 14, Time: 18:52 UT

4.6 Closing statement

The relative timing and position of sunset and moonset at any time and anywhere in the globe are characterized to give an insight into the events in a graphical easy to use format. The twilight bands are added to further illustrate the dynamics of Hilal prediction and sighting. Those who are familiar with new moon tracking knows very well the importance of twilight evolution in new moonlight detection.

The simulation uses a state-of-the-art mapping and computation to aid in the understanding of certain realities. There are quite a few capabilities that were added to the simulation to bring the answers post-sunset one can quickly make sense of the potential for sighting.

The high-resolution imagery can make the display on large projection screens possible to track the events of Hilal observation. One can follow the progress of nightfall and moon setting over particular areas on the globe.

The slopes of the sunset and moonset curves require analysis and characterization from day to day and month to month through the year. The points of inflection and points of crossings point to important physical meanings that need to be looked at in details.

There is much more information and capabilities that can be developed and put at the service of decision-makers to help with the assessment of Hilal sighting. For example, I can see the extension of the material presented herein to include the integration of local and global atmospheric satellite data.

Chapter 5 New moon sighting principles

I often say that when you can measure what you are speaking about and express in numbers, you know something about it; but when you cannot measure it, you cannot express it in numbers, your knowledge is a meager and unsatisfactory kind; it may be the beginning of knowledge, but you have scarcely, in your thoughts, advanced to the stage of science, whatever the matter may be.

William Thomson (Lord Kelvin) 1824-1907

Light from a new moon is observed over a period of about a half-hour, twenty minutes or so after sunset and a few minutes prior to moonset. Given enough time following sunset, the sun falls low enough below the horizon for the earth shadow to rise to a sufficient altitude dampening the effect of the scattering of sunlight caused by the bulk of air molecules, aerosols and other small airborne particles in the lower layers of the atmosphere. At the same time, the new moonlight dips further and further into the sun-heated layers of the atmosphere traveling increasingly longer distances.

In this chapter, we establish the basic geometry of the light rays responsible for the evening twilight sky illuminance in the direction of the setting moon, as the sun dips gradually below the western horizon. The light rays from the setting new moon emerge against a background sky which is illuminated by scattered sunlight rays. Characterization of the lights from the moon and scattered sunlight is carried from the perspective of an observer placed at the surface of the earth. The same derivations are valid for an observer placed at some altitude above the earth that can be simulating High altitude balloon or aircraft observation.

In Section 5.1, we will follow the geometric representation of Rozenberg (Rozenberg, 1966) of the twilight geometry of scattered sunlight. The complex twilight problem solution is then presented to show that the Azimuthal angle between the solar vertical and the line of sight of the moon has only a minor effect when such angle is small, which is the case for the new moon problem. Subsequently, we derive a simple solution to the earth shadow problem using Taylor Series expansions.

5.1 Basic twilight geometry

Figure 5.1-1 shows the geometry of light rays from the sun illuminating one side of the globe. We consider an observer who is located on earth surface at some point *N* with his/her eyesight set on the moon during twilight. The point *N* is chosen deep in the night side for clarity sake and does not invalidate the geometric relations that we are discussing herein. As the earth rotates counterclockwise, the point *N* falls further and further in darkness and will progress to see the moon in turn, setting below the horizon. The location of sunset is at point *M*. The same figure shows the sunlight arriving parallel from infinity at an angle ζ with the zenith at *N,* and ς' with the zenith, at *P*. The line of sight to the moon is defined by its zenith distance *z,* and the azimuth angle *A* relative to the solar vertical. The line of sight intersects the solar tangent ray at point *P,* which has an elevation *h* above the surface. Consider a spherical triangle composed of *N, C* and *S*, where *S* is the intersection of a line parallel the line of sight to the moon and the surface of earth going through point *O*. The scattering angle φ from Point $P(z,A,h)$ is given by the spherical identity

$$\cos\varphi = \cos z \cos\zeta + \sin z \sin\zeta \cos A \,. \qquad 5.1\text{-}1$$

Let ϑ be the angle given by NOP, we have

$$\cos\vartheta = \frac{R + l\cos z}{R+h}, \quad \sin\vartheta = \frac{l\sin z}{R+h}, \qquad 5.1\text{-}2$$

where *l* is the distance between the observer at *N* and point *P* and *R* is earth radius. From the last two relations, we may write

$$(R + l\cos z)^2 = (R+h)^2 - l^2\sin^2 z \,. \qquad 5.1\text{-}3$$

Next, consider the line *OQ* perpendicular to the ray intersecting P, the angle $\chi = QOP$ is given by

$$\frac{R+y}{R+h} = \cos\chi \,. \qquad 5.1\text{-}4$$

Consider the solar ray which would pass through earth center. This ray intersects the surface at point C. Also, consider the point *D* where the zenith at P intersects the surface. Points *D, M, C* form a great circle and the angle $COM = \pi/2$.

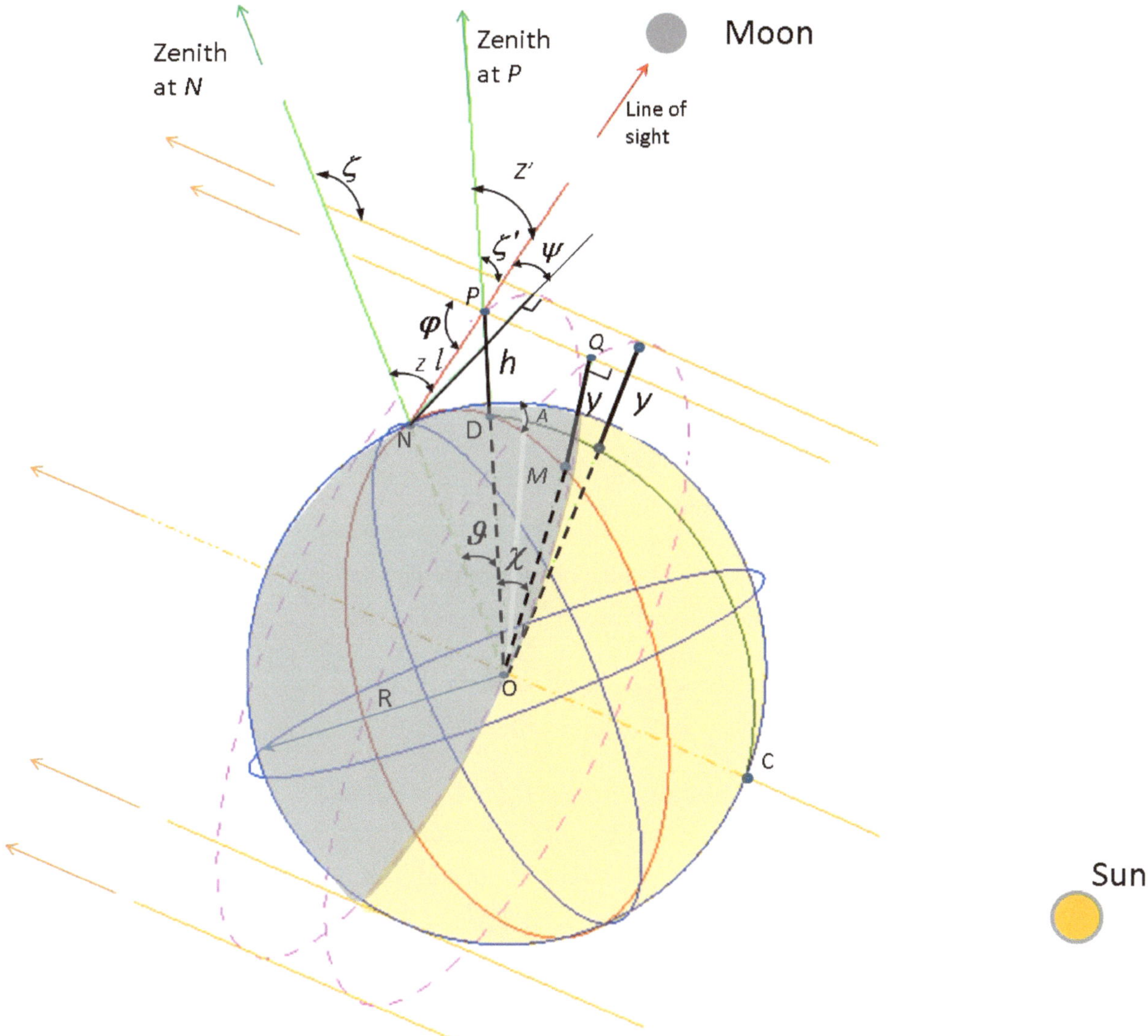

Figure 5.1-1 Geometry of sun and moonlight rays with an observer placed on the earth surface after sunset.

The spherical triangle NDC gives

$$\cos\left(\chi+\frac{\pi}{2}\right)=-\sin\chi=\cos\vartheta\cos\zeta+\sin\vartheta\sin\zeta\cos A\,. \qquad 5.1\text{-}5$$

And taking the square of Equation 5.1.5 and making use of Equations 5.1.2 and 5.1.3, we obtain

$$\left((R+L)\sin\varsigma + L\cos\psi\right)^2 - (R+y)^2 + l^2\sin^2 z\sin^2 A = 0\ . \qquad 5.1\text{-}6$$

Where the angle ψ is between the line of sight to the moon and the segment *NF*, perpendicular to the sunray at elevation y, dropped from point *N* and is given by

$$\cos\psi = \cos z\sin\zeta - \sin z\cos\zeta\cos A,$$

which is the same as Equation 5.1.1 with ψ replacing φ and the angle $\zeta - \pi/2$ is used instead of ζ .

Introducing the dimensionless variables, given by:

$$x = \frac{y}{R},\ u = \frac{l}{R},\ w = \frac{h}{R},$$
$$\varepsilon = \sin^2 z\ \sin^2 A,\ \text{and}\ \upsilon = 1 - \sin\varsigma + x, \qquad 5.1\text{-}7$$

the solution of the following system of equations establish the relation between *h* and *y* as functions of ζ, *z* and *A*

$$(1+u\cos z)^2 = (1+w)^2 - u^2\sin^2 z$$
$$(\sin\varsigma + u\cos\psi)^2 = (1+x)^2 - u^2\varepsilon \qquad 5.1\text{-}8$$

5.2 Earth shadow height

The height of the earth shadow in the direction of the line of sight corresponds to a perigee $y = 0$. Thus, the dimensionless parameters in Equations 5.1.7 reduce to

$$x = 0,\ u = \frac{l}{R},\ w = \frac{h}{R},$$
$$\varepsilon = \sin^2 z\ \sin^2 A,\ \upsilon = 1 - \sin\varsigma\ . \qquad 5.2\text{-}1$$

And the above system of equations is therefore

$$(1+u\cos z)^2 = (1+w)^2 - u^2\sin^2 z,$$
$$(\sin\varsigma + u\cos\psi)^2 = 1 - u^2\varepsilon\ . \qquad 5.2\text{-}2$$

Solving for u_s, the normalized distance traveled by the moonlight photons in earth shadow, from the second equation which is a quadratic function of *u, we* have

$$u_s = \frac{\sin\varsigma\cos\psi + \sqrt{\sin^2\varphi\cos^2\psi + \cos^2\varphi\left(\cos^2\psi + \sin^2 z\sin^2 A\right)}}{\cos^2\varphi\cos^2\psi\sin^2 z\sin^2 A}, \text{ if } A \neq 0 \qquad 5.2\text{-}3$$

$$u_s = (1-\sin\zeta)/\cos\psi, \text{ if } A = 0.$$

Next, we note that the right-hand side of the second equation of 5.2.2 is close to unity for small azimuth angles *of the moon* relative to the solar vertical, *A. This is* the case for a new moon, because $\sin z < 1$, $\sin^2 A << 1$ and $\varepsilon = \sin^2 z \sin^2 A << 1$. This means a good approximation for u_s during a new moon is given by the second equation of 5.2-3. Subsequently, we solve for the normalized earth shadow height w_s from the first equation as a function of u_s and the angle *z*,

$$w_s = h_s / R = \sqrt{u_s^{\,2} + 2\, u_s \cos z + 1} - 1\,, \quad u_s = l_s / R \simeq (1-\sin\zeta)/\cos\psi\,. \qquad 5.2\text{-}4$$

The height of earth shadow and the distance a light ray from the moon travels in earth shadow, and are plotted as a function of the position of the zenith angle of the sun for different elevation angles of the moon in Figures Figure 5.2-1 and Figure 5.2-2 respectively. The angle *A* is assumed to be nil in both plots, as shown in the numerical data of Table 5.2-1.

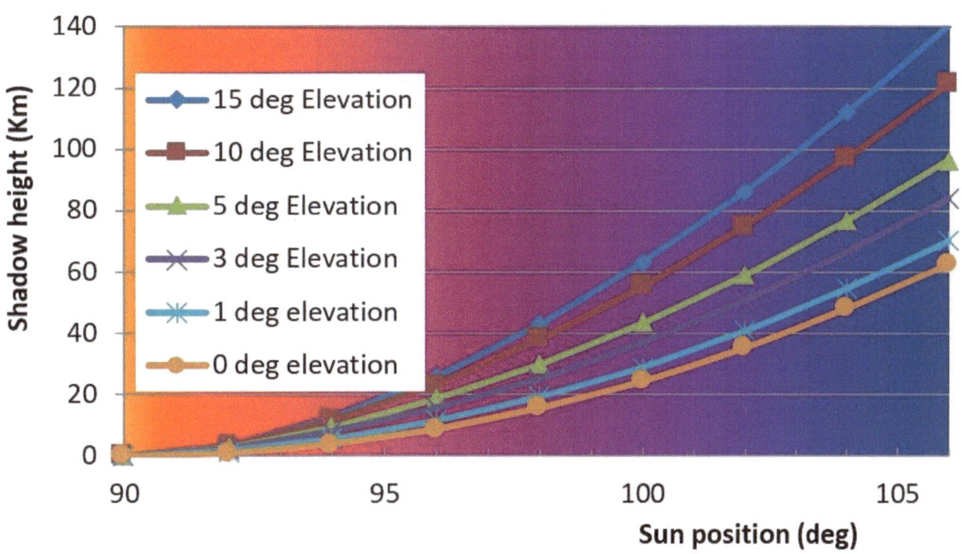

Figure 5.2-1 Shadow height versus sun zenith angle for several line-of-sight angles

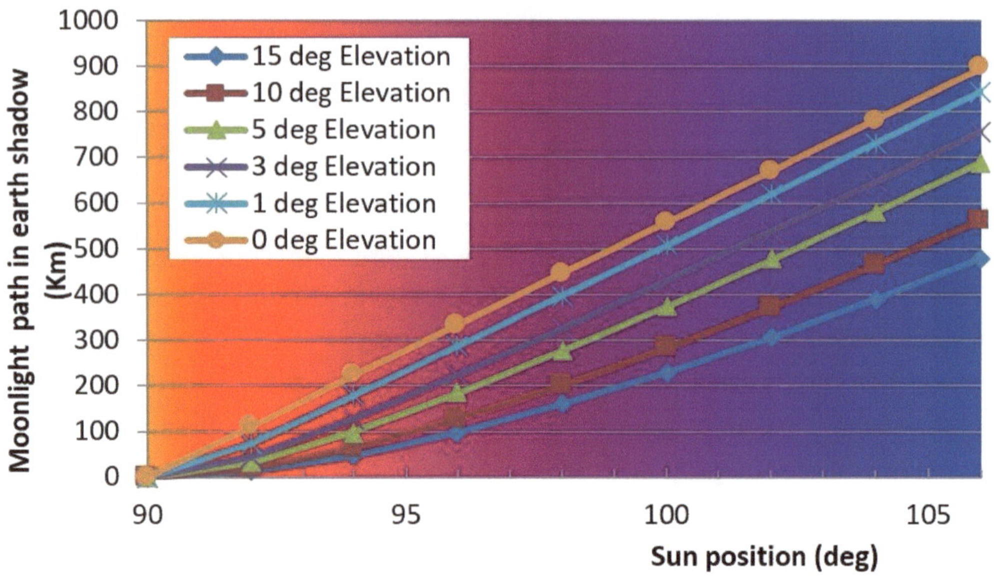

Figure 5.2-2 Distance traveled by moonlight in earth shadow versus sun zenith angle for several line-of-sight angles

As we will see later, the visibility of the new light from the moon depends on two effects: The first is *extinction*. As the moonlight makes its way toward the observer, its intensity diminishes gradually and given a large enough distance traveled, it becomes difficult to perceive. Extinction is independent of whether photons from the moon are going through earth shadow or sunlit layers of the atmosphere; it depends on the density of the layers of the atmosphere being traveled. The longer the path is in lower dense layers, the more pronounced is the extinction, and therefore it helps to have higher elevations of the moon to limit the effect of extinction. The second important and relevant effect is the contrast of the moonlight against the background sky. The higher the earth shadow is, the less scattered sunlight arrives at the observer, thus improving contrast. For a particular sun angle depression below the horizon, the higher the position of the moon in the sky, the less space is covered by the shadow and the higher the number of scattered sunlight photons will reach the observer. The effect of sunlight scatter is, therefore, competing against extinction.

Finally, sunlight scattering is not affected by the azimuth angle of the moon relative to the solar vertical when such angles are small and near the horizon, which is the case for moon sighting. Based on the above, criteria used for moon visibility that include the azimuth angle of the moon relative to the solar vertical as a key parameter are not necessarily wrong. They could be indirectly relating to the main parameters responsible for successful sighting using the parameter *A*. We will revisit this issue in the following chapters. But before proceeding, we should have a close look at the effect of shadow height as a function of the sun and moon positions.

Table 5.2-1 Earth shadow height for different sun & moon positions (units in degrees)

A	z	zeta	cos(psi)	u	w	l	h
0	75	90	0.258819	0	0	0	0
0	75	92	0.292372	0.002084	0.000541	13.27434	3.448547
0	75	94	0.325568	0.007482	0.001963	47.66878	12.50365
0	75	96	0.358368	0.015286	0.004065	97.38874	25.89778
0	75	98	0.390731	0.024907	0.006734	158.6824	42.90172
0	75	100	0.422618	0.035948	0.009901	229.0242	63.08001
0	75	102	0.45399	0.048134	0.013525	306.662	86.1677
0	75	104	0.48481	0.06127	0.01758	390.351	112.0041
0	75	106	0.515038	0.075214	0.022052	479.1912	140.4957
0	80	90	0.173648	0	0	0	0
0	80	92	0.207912	0.00293	0.000513	18.66678	3.26796
0	80	94	0.241922	0.010069	0.001798	64.1506	11.45231
0	80	96	0.275637	0.019874	0.003642	126.6193	23.20319
0	80	98	0.309017	0.031493	0.005947	200.6431	37.8881
0	80	100	0.34202	0.044419	0.008662	282.9945	55.18763
0	80	102	0.374607	0.058334	0.011762	371.6476	74.93513
0	80	104	0.406737	0.073031	0.015232	465.2788	97.04552
0	80	106	0.438371	0.088369	0.019068	562.9972	121.4809
0	85	90	0.087156	0	0	0	0
0	85	92	0.121869	0.004999	0.000448	31.84592	2.854507
0	85	94	0.156434	0.015572	0.001477	99.20727	9.411944
0	85	96	0.190809	0.02871	0.00291	182.9107	18.54041
0	85	98	0.224951	0.043262	0.004695	275.625	29.91418
0	85	100	0.258819	0.058698	0.006815	373.9671	43.42104
0	85	102	0.292372	0.074742	0.009264	476.1803	59.02373
0	85	104	0.325568	0.091238	0.012042	581.279	76.71734
0	85	106	0.358368	0.108096	0.015149	688.6825	96.51366

5.3 Effect of earth shadow height on sighting

Having determined the height of earth shadow as a function of the zenith distance of the moon, we address now the nature of the observation itself. To this end, we consider the quad chart shown in Figure 5.3-1. This chart has a regular pair of axes: the sun angle below the horizon -zenith distance of the sun minus 90° - along the horizontal axis and the moon elevation along the vertical axis. The axes along $\pm 45^o$ correspond sun-moon relative elevation, a critical parameter in a few proposed criteria for new moon visibility, and the approximate time for new moon observation, respectively. The quad chart displays the two civil and nautical Twilight periods with different shades.

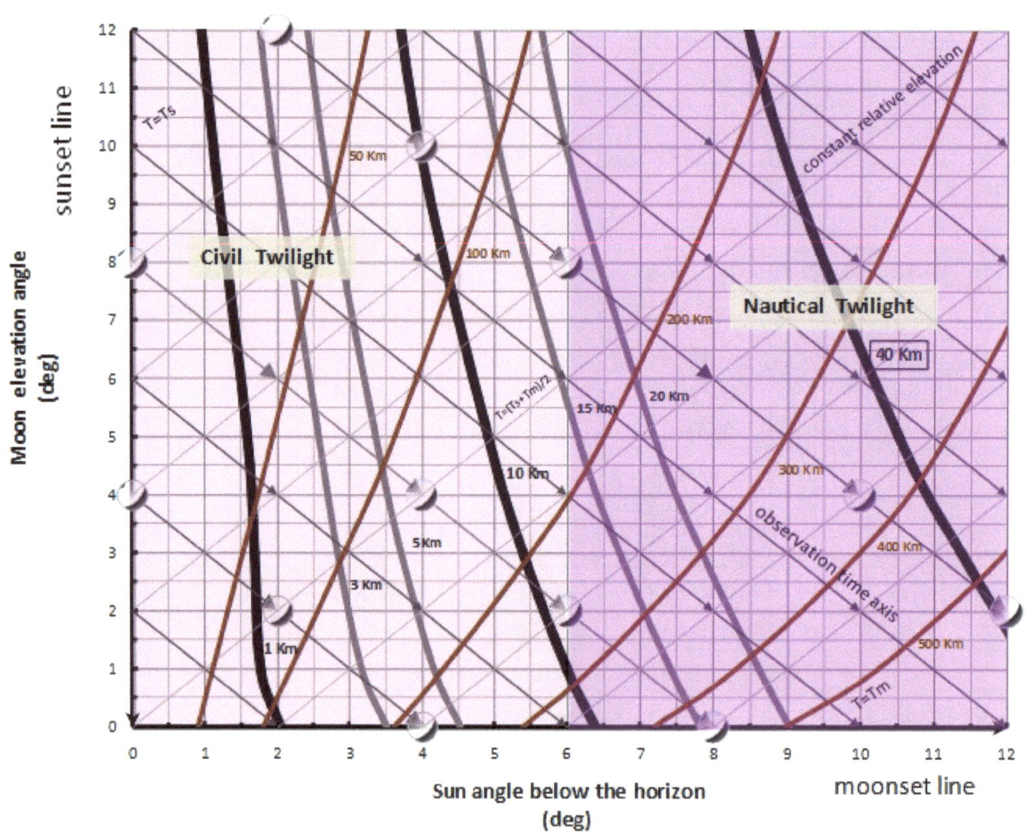

Figure 5.3-1 Quad-chart depicting new crescent observation

From astronomical tables and online software provide the elevation of the moon at the time of sunset following an astronomical new moon. Then at the vertical axis (sunset axis), the origin of the observation is marked down. As time elapses, the moon progresses along a straight line making -45° with the horizontal axis until the moon sets. The length of the segment represents time because 95 % of the motion of the moon in the sky is due to the rotation of the earth and the motion of the moon is small over such a small period of time. It can be verified that the relative elevation remains nearly constant between sunset and moonset. The lines in the orthogonal direction, oriented in the $+45^o$ direction represent the time following the beginning of the observation T_S, time of sunset. The observation is carried all the way until moonset at T_m. Because the moon moves about its own diameter every three minutes and because the sun and the moon are about 1/2 degree each, the scale is about $6\sqrt{2} \simeq 8.5$ minutes per graduation. As time passes the moon descends along the time axis and crosses different black and grey curves with negative slopes depicting the height of earth shadow (1, 3, 5, 10, 15 and 20) *Km* and red curves representing the distance traveled in earth shadow (50, 100, 200, 300 and 500) *km*. It can be seen that the moon elevation needs to be at least 6.35^o before one has a chance to have the entire troposphere at about a 10 *Km* mark in earth shadow, thus avoiding scattering by water droplets, ice crystals, dust, and aerosols. This is the time when the sky enters the nautical twilight.

In conclusion, what really matters for new moon observations is the elevation of the moon at sunset, if scattering is the key factor hampering new moon visibility. The higher the moon elevation at sunset, the higher the shadow is going to be when the observation takes place between sunset and moonset. In addition, a detailed analysis of the photometry will show that the 6.35^o mark is found independently for a perfect atmosphere.

5.4 Approximate solution

The solution to the twilight problem is presented for high zenith angles, where the line of sight is fixed on a setting new moon, and the sun is below the horizon. At the first sunset, following conjunction, the elevation of the moon is usually less than 12 degrees, and by moonset, the sun is at a depression less than about 12 degrees. This is the period when new moon field sighting commonly takes place. That is a span of about 45-60-minute window, which is the time period covering Civil and Nautical twilight.

The twilight problem is quite complex, as stated by Patat and coworkers (Patat, Ugolnikov, & Postyliakov, 2006). The calculation of scattered radiation from the sun entails actual knowledge of the composition of the different layers in the atmosphere and the physical phenomena taking place, a subject of numerous investigations in the past century. The main purpose of this section is to seek a solution for the geometry of the twilight problem for small angles. Let us denote by $v = \pi/2 - z$, and $\rho = \zeta - \pi/2$ the elevation of the moon and the sun declination. For small angles $v + \rho$, Equation 5.2.2 becomes

$$\begin{aligned}&\left(1+u\, sinv\right)^2 - (1+w)^2 + u^2 cos^2 v = 0\\&\left(cos\rho + u(v+\rho)\right)^2 + u^2 A^2 cos^2 v - 1 = 0.\end{aligned} \tag{5.4-1}$$

From the second equation, we solve for u and substitute *A=0* as in Section 5.2, we obtain

$$u = \frac{l}{R} = \frac{1-\cos(\rho)}{(\rho+v)}$$

Then we substitute u back into the first equation and solve for w. By, taking the multivariate Taylor series expansion in terms of ρ and *v*, near 0 results in:

$$u = \frac{l}{R} \approx \frac{1}{2}\frac{\rho^2}{(\rho+v)}, \quad w = \frac{h}{R} \approx \frac{1}{8}\left(\frac{\rho(\rho+2v)}{\rho+v}\right)^2$$

Contour lines of *l* and *h* are shown in the plot below. For a line of sight in the direction of the setting moon, it can be shown that the elevation v can be related to the setting angle of the sun through

$$v + \rho \approx C\ , \tag{5.4-2}$$

where *C* is the elevation of the moon at sunset. Using the above relation, a moon-constrained line of sight we can express the height of the earth shadow and the distance traveled by a photon in the shadow of the earth

$$u = \frac{l}{R} \approx \frac{\rho^2}{2C}, \quad w = \frac{h}{R} \approx \frac{\rho^2(2C-\rho)^2}{8C^2}, \quad \text{for } 0 \le \rho \le C\,. \tag{5.4-3}$$

It is interesting to note that for an observer tracking the moon after sunset, the line of sight intersects the limit of the earth shadow at a point which height depends only on the angle of the sun, and the position of the moon at sunset. This confirms field observations indicating successful sighting seems to depend on how high the moon is at sunset. The reason behind this is the fact that over the short period of time in which sighting takes place, the separation between the moon and the sun is nearly constant as expressed earlier by Equation 5.4.2.

Contour lines of earth shadow height and the distance traveled by a photon from the moon in the shadow are displayed function of the sun angle depression below the horizon and the elevation angle of the moon in Figure 5.4-1. The plot yields similar results to those already obtained from the complete solution shown in Figure 5.3-1.

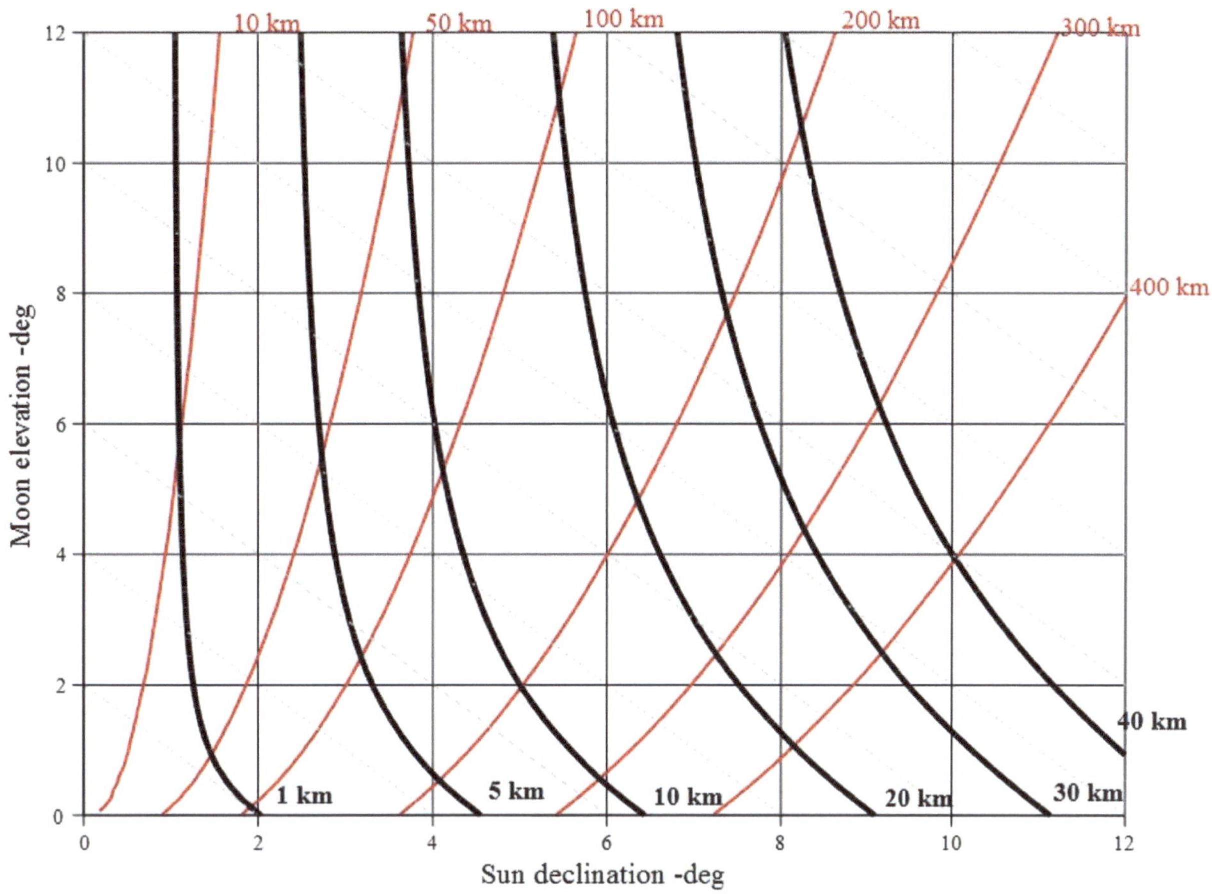

Figure 5.4-1 Contour lines of the shadow height (thick contour lines 1, 5,10,20,30 and 40 Km), and contour lines of distance traveled by a lunar photon in the shadow zone (thin contour lines 10, 50, 100, 200 and 400 Km)

The maximum expected shadow height as a function of the position of the moon at sunset is

$$h_{max} \approx \frac{C^2 R}{8} \qquad 5.4\text{-}4$$

and is shown in Figure 5.4-2. It can be seen that the shadow height is at the 10 Km mark when the sun declination is about 6.4 deg. Depending on the geographic latitude, the minimum elevation of the moon at sunset needs to be between 8 and 9 degrees to guarantee that the entire troposphere gets to be in the shadow by moonset. The troposphere, the lowest section of the atmosphere, contains about 80% of its mass and 99% of its water vapor and aerosols. The average thickness of the troposphere is about 17 km in the middle latitudes (25° N — 55°

N and 25° S — 55° S). It is up to 20 km deep in the tropics (24° S, 24° N) and thinner near the Polar Regions (approximately 7 km in winter). Contour lines providing the shadow height level, as a function of the sun declination ρ and the moon position at sunset, *C*, for the approximation given by Equation **5.4.3** for $0 \leq \rho \leq C$, is shown in Figure 5.4-3. As far as this approximation is concerned, the shadow height reaches a maximum when the sun declination ρ reaches the height of the moon at sunset *C*.

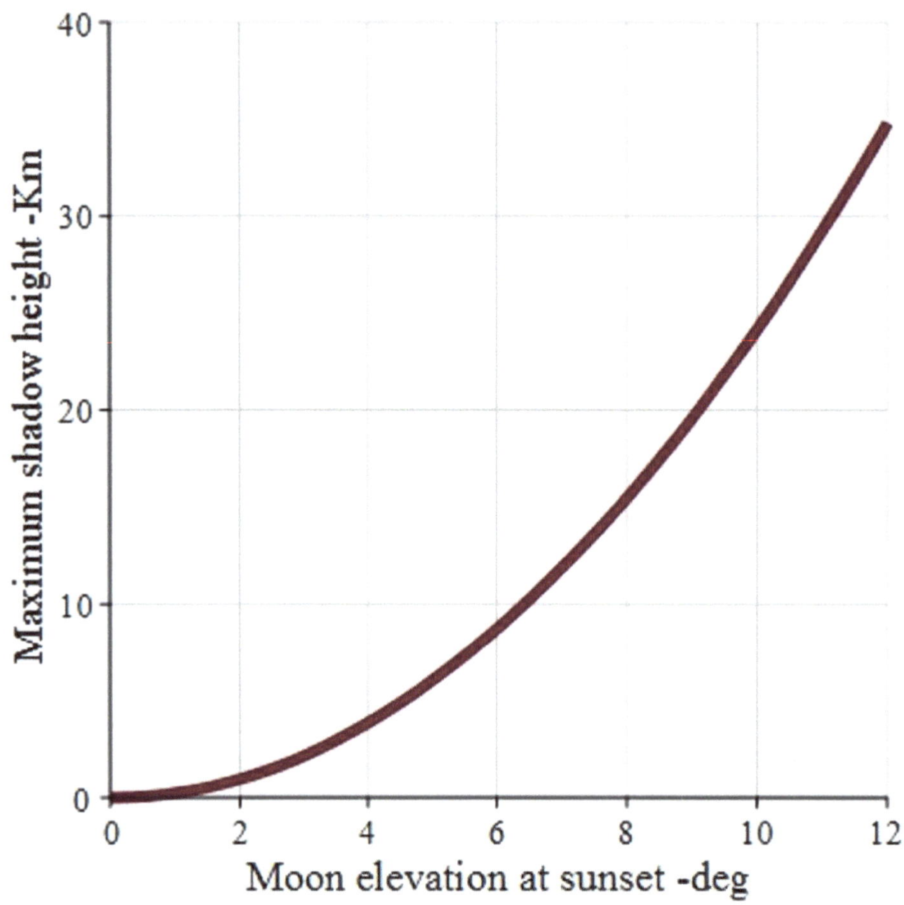

Figure 5.4-2 Maximum expected shadow height as a function of the position of the moon at sunset

In conclusion, we have characterized the geometry of photons from the new moon and from scattered sunlight during evening twilight from the perspective of an observer placed on the on earth surface. We also established the height of the earth shadow and the distance traveled by a photon from the moon through earth shadow for different sun declinations and moon elevation angles. We have shown that the azimuth angle of the moon relative to the sun vertical does not appreciably affect the characteristics of the earth shadow. Through Taylor series expansions, we derived simple expressions for the shadow height and its relationship with the line of sight.

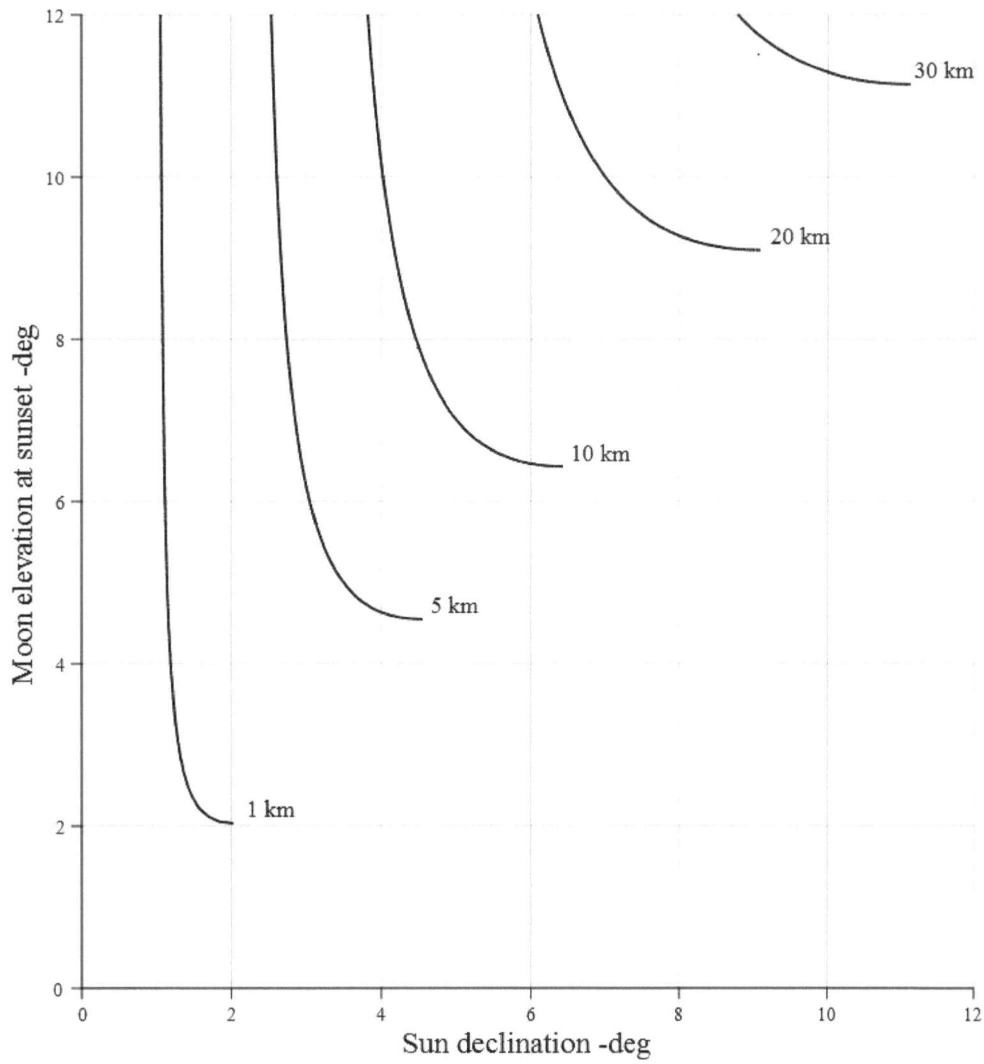

Figure 5.4-3 Contour lines of shadow height as a function sun angle below the horizon and moon elevation at sunset

Chapter 6 Moonlight measurement

Scientific evaluation of celestial bodies relies on the emission or reflection of electromagnetic radiation, and the entire body of knowledge in the field of astronomy relies on the analysis of radiation. In this chapter, we introduce the fundamental parameters used in quantifying and measuring radiation, which will be used in the characterization of the new moonlight problem. Studies of radiant matter started as the field of *photometry*, where the only sensor available is the human eye, and evaluation of visual perception is measured in relative quantities. Photoelectric sensors, film, charge-coupled devices, and radiofrequency equipment propelled the new field of radiometry to make it reign supreme. By now, Radiometry encompasses multiple disciplines involving electromagnetic radiation, and of which photometry has become small -yet important- subset. Radiometric parameters characterizing *radiant* matter are measured in absolute quantities, such as joules and watts. The adjective *luminous* is often used to qualify corresponding photometric quantities. The nomenclature which has evolved over the past century can be quite confusing for an outsider to the fields of radiometry and photometry because of the variation in the use of terminology from one reference to another.

There are two sources of dynamic lights at play during the period of twilight in which new moon sighting takes place. In addition to the transient light from the new moon, we must deal with the rapidly changing background light of the sky. We shall focus on how the light from the thin crescent moon can be characterized by what is known in literature and add some insight to the topic. We shall defer addressing the interaction of these lights with the atmosphere before reaching the human eye to Chapter 7.

6.1 Brightness, Flux Density, and Magnitude

Consider some arbitrary electromagnetic energy of a ***specific intensity*** I_ν at a frequency ν going through a planar surface dA as shown in Figure 6.1-1. The part of this energy which enters the solid angle $d\omega$, oriented along a direction defined by an angle θ from the normal to the surface, over an infinitesimal time dt is given by $I_\nu \cos\theta \, dA \, d\nu \, d\omega \, dt$. The ***total intensity*** or ***brightness*** of the electromagnetic energy I (the two terms are used interchangeably) is obtained by integrating I_v over the frequency content of the radiation and therefore we have

$$I = \int_0^\infty I_v dv .$$

Some of the most important parameters in radiometry are the ***flux density*** F_v and *the* ***total flux density*** F**.** They measure the power of radiation per unit area, in units of $W\ m^{-2}\ Hz^{-1}$ and $W\ m^{-2}$ respectively. They are given by

$$F_v = \int_S I_v \cos\theta\ d\omega \ \ and \ \ F = \int_S I \cos\theta\ d\omega ,$$

where the integration takes place over all directions. In spherical coordinates, the solid angle $d\ \omega$ is given by the area $\sin\theta\ d\varphi\ d\theta$ on a unitary sphere with coordinates (φ,θ).

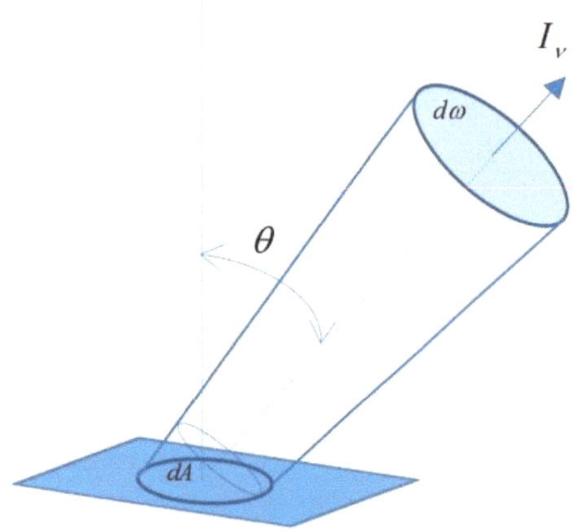

Figure 6.1-1 Radiation Intensity I_v passing through an elemental surface dA from a solid angle $d\omega$ oriented along a direction θ from the surface normal

For uniform radiation emanating from an object, integrating over one hemisphere gives the radiation leaving the unitary surface, i.e., $0 \le \theta \le \pi/2$ and $0 \le \varphi \le 2\pi$ yields $F = \pi I$. The total flux of a star with isotropic radiation is called ***luminosity*** and is given by

$$L = 4\pi r^2 F$$

The flux density varies as a function of $1/r^2$ which makes the luminosity constant. The ***Intensity*** is defined as the luminous flux per unit solid angle crossing unit area normal to the direction of the light. The corresponding ***Illuminance E*** is given by

$$E = I \cos\theta\ d\omega .$$

The ***brightness class*** or ***apparent magnitude*** of a star with a flux density F is defined as

$$m = -2.5 \log_{10} \frac{F}{F_0}, \quad 6.1\text{-}1$$

where some preselected flux density F_0 is chosen to have zero magnitude. The "magnitude" classification of brightness is logarithmic and has no units. Traditionally, the abbreviation "mag" is used in front of the *m* value to remind the reader that reference is made to the magnitude scale. The brightness of the sun is -26.8 mag, the full moon is -12.5 mag and the faintest object in the sky is +25 mag. For details about the origin of the definition and different magnitude systems, the reader is directed to classical astronomy textbooks –see, e.g. reference (Kartttunen, Kröger, Oja, & Donner, 1984).

For extended objects that are not point stars such as comets, galaxies, and nebulae that have a surface *A* usually measured in [arc-sec^2] unit, we define the surface brightness as

$$S = F / \omega$$

S is independent of the distance *r* ($\omega = A / r^2$ and *F* varies as $\omega = 1 / r^2$). The surface brightness ***S***, expressed in units of **magnitude / arc-sec^2,** of a light source of magnitude ***m*** spread over an area ***A***, given in **arc-sec^2**, is not computed by a simple division of magnitude by the surface area. Instead, because of the logarithmic nature of the magnitude scale, the surface brightness is computed using the following expression:

$$S = m + 2.5 \log_{10} A \quad 6.1\text{-}2$$

The calculated surface brightness of a full moon, based on a diameter of about 31 arc-min and a magnitude m= -12.5 is calculated to be about +3.6 mag / arc-sec^2. In Section 6.2, we discuss several reported lunar brightness measurements for larger phase angles, and in Section 6.3, we calculate a rough approximation of the surface area of a crescent moon and establish a preliminary approximation of its surface brightness as a function of the phase angle.

6.2 Moon brightness

Like in any other case, when it comes to the crescent moon, the human eye's resolving ability depends on the contrast between the background surface brightness and the crescent moon surface brightness. Before addressing the topic of contrast in Chapter 7, we seek to develop an understanding of the parameters involved in the calculation of a crescent surface brightness. We first start by deriving approximations for the magnitude of a crescent moon and subsequently derive the surface area of the partially illuminated moon. Both quantities are derived as functions of the phase angle during a lunation and make use of Equation 6.1.2 to derive the surface brightness.

The magnitude of the moon can be approximated using (Allen, 1976) and (Krisciunas & Schaefer, 1991):

$$m = -12.73 + 0.026|\alpha| + 4\times10^{-9}\alpha^4,$$

where α is the phase angle given in degrees. Figure 6.2-1 shows this approximation against other data compiled in reference (Lane & Irvine, 1973). No measured data is available beyond 120 deg, and therefore, the data was extrapolated to 180 degrees. Also, shown is data from Zombec (Zombeck, 2006), adapted from Wertz (Wertz, 1980), down to 20-degree separation angle, based on the expression

$$m = 0.23 + 5\log R - 2.5\log P(\alpha)$$

In which R is the observer-moon distance in AU (mean value of 0.00257 AU) and $P(\alpha)$ is the moon phase law with $\alpha-$deg the phase angle between the sun and earth, measured at the moon.

Table 6.2-1 The phase angle between the sun and earth, measured at the moon.

$\alpha-$deg	0	40	80	120	160
$P(\alpha)$	1	0.377	0.127	0.027	0.001

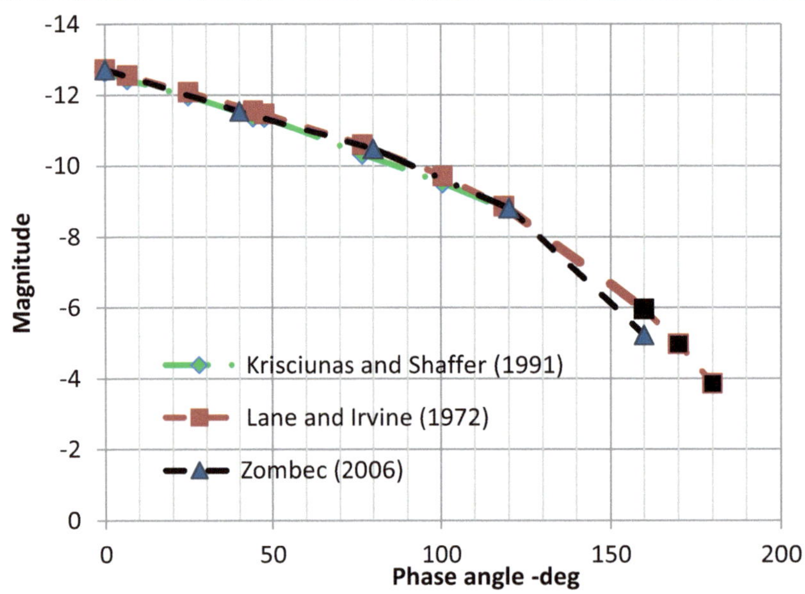

Figure 6.2-1 Magnitude of the moon as a function of the phase angle

6.3 The surface brightness of the crescent moon

In this section, we derive an approximation of the surface brightness of a crescent as a function of the phase angle. We first calculate an approximation of the surface area and the crescent thickness for phase angles slightly less than 180°, as seen from earth. We subsequently use Equation 6.1-2 to obtain an estimate of the surface brightness for a given magnitude of the moon. The brightness magnitude of the moon is available from a handful of investigations but unfortunately for relatively larger phase angles as discussed earlier in Section 6.2. Understandably, these investigations were not addressing smaller phase angles, and interest had been in cases where more light from the moon was available. It is obvious that measurements are needed for more accurate evaluations. Due to this limitation, we shall extrapolate the reported data to lower phase angles using curve fitting.

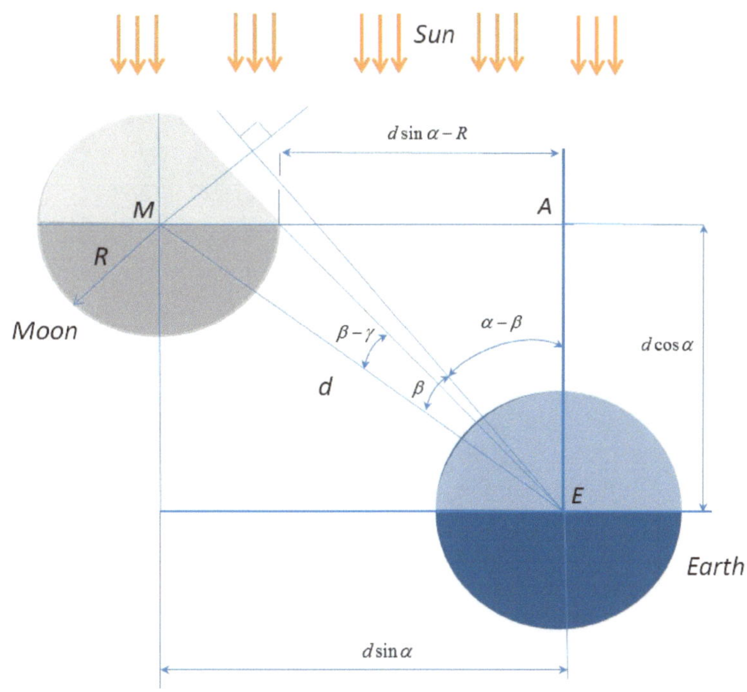

Figure 6.3-1 Visible portion of the illuminated moon when the phase angle is $\pi-\alpha$

Consider the simplified two-dimensional view of the earth-moon system with a separation angle α from the sun rays assumed to arrive uniform and parallel due to the long earth-sun distance, as shown in Figure 6.3-1. We further assume that the moon, earth, and sun are in a single plane going through a slice of the moon of radius *R*, where $0 \le R \le Rm$ with *Rm* is the radius of the moon. The moon is assumed to be located at a distance *d* from the earth. The angular size of the moon is therefore given by the angle $\phi_m = 2R_m / d$ and the size of the illuminated part of the disk is denoted by the angle γ, which is adjacent to the angle $\alpha-\beta$ from the solar vertical. Consider the right-angle triangle *MAE* in Figure 6.3-1, we have

$$\tan(\alpha-\beta+\gamma)=\frac{d\sin\alpha-R}{d\cos\alpha}. \qquad \textbf{6.3-1}$$

Substituting $\sin^{-1}(R/d)$ for the angle β, we have

$$\gamma=\tan^{-1}\left(\tan\alpha-\frac{R}{d}\frac{1}{\cos\alpha}\right)-\left(\alpha-\sin^{-1}\frac{R}{d}\right). \qquad 6.3\text{-}2$$

Consider the elevation angle θ of the moon slice. We have $R=R_m\sin\theta$. Expressing Equation 6.3.2 in terms of ϕ_m, α and θ, we have

$$\gamma=\tan^{-1}\left(\tan\alpha-\frac{\phi_m\sin\theta}{2}\frac{1}{\cos\alpha}\right)-\left(\alpha-\sin^{-1}\frac{\phi_m\sin\theta}{2}\right). \qquad 6.3\text{-}3$$

Bivariate Taylor series expansion of the angle γ near $(\alpha,\phi_m)=(0.0)$ gives

$$\gamma\simeq\frac{\phi_m}{4}\sin\theta\left(\alpha-\frac{\phi_m}{2}\sin\theta\right)^2. \qquad 6.3\text{-}4$$

And the maximum thickness is given by

$$\gamma_{\max}\simeq\frac{1}{4}\phi_m\left(\alpha-\phi_m/2\right)^2. \qquad \textbf{6.3-5}$$

Because the angle ϕ_m is smaller than the angle α by almost two orders of magnitude, the last two Equations can be further reduced to

$$\gamma\simeq\frac{\phi_m}{4}\sin\theta\,(\alpha^2-\alpha\phi_m\sin\theta),\text{ and }\gamma_{\max}\simeq\frac{1}{4}\phi_m(\alpha^2-\alpha\phi_m). \qquad 6.3\text{-}6$$

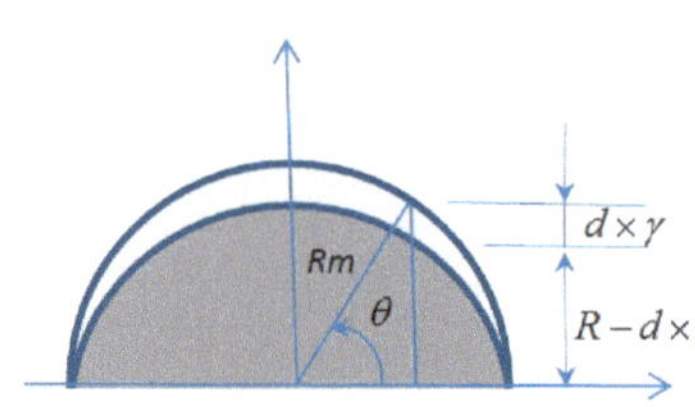

Figure 6.3-2 Crescent moon geometry

Finally, we note that the dimension of the illuminated area is given by

$$R-d\sin(\beta-\gamma)\simeq R-d(\beta-\gamma)\simeq d\gamma, \qquad \textbf{6.3-7}$$

and an infinitesimal length along the diameter of the moon is $R_m \sin\theta \, d\theta$. Integrating $d\,\gamma$ with respect to θ, we obtain the area of the crescent moon

$$A_c \simeq \int_0^{\pi} d\left(\frac{1}{2}\frac{R_m \sin\theta}{d}\alpha^2\right)(R_m \sin\theta)\, d\theta = \frac{\pi}{4}R_m{}^2\alpha^2 = \frac{\pi}{16}d^2\phi_m{}^2\alpha^2 . \qquad 6.3\text{-}8$$

In conclusion, the illuminated area can be written in angular units as

$$A_c \simeq \frac{\pi}{16}\phi_m{}^2\alpha^2 , \qquad 6.3\text{-}9$$

where α is in radians and the angular diameter of the moon ϕ_m and A_c are given in consistent angular unit. The normalized area of the illuminated crescent moon is approximated using

$$A_c \,/\, A_m \simeq \frac{1}{4}\alpha^2 , \qquad 6.3\text{-}10$$

where A_m is the area of the full moon. The above expression is valid up to a separation angle of about 40° and about a quarter radius illuminated moon, as illustrated in Figure 6.3-3. Also shown in the Figure is the illuminated area as a function of the phase angle. Furthermore, in order to properly plan for detectability and imaging, the maximum thickness of the moon is provided for different separation angles in Table 6.3-1. The mean value of the moon radius of 1737.1 Km and the mean earth-moon distance of 384,402 Km were used to generate the numerical values of the Table. Figure 6.3-4 shows the thickness of the crescent at different elevation-radii, from $R=R_m/20$ to $R=R_m$ and for separation angles from new moon to a waxing quarter moon.

In planetary astronomy, the phase F of a planet is defined as the fraction of the visible illuminated disk and is given in terms of the solar elongation of the earth as measured at the planet. In the case of the moon and the configuration of Figure 6.3-1, this angle is $\pi-\alpha$, and the expression for the phase of the crescent moon is given by

$$F = \frac{1}{2}(1+\cos(\pi-\alpha)). \qquad \textbf{6.3-}11$$

Taylor series expansion of the phase near conjunction, that is near $\alpha = 0$, leads to

$$F = \frac{1}{4}\alpha^2 - \frac{1}{48}\alpha^4 + O(\alpha^6)$$

This shows that the phase F has the same leading term as the one given by Equation 6.3.10. It is a third-order accurate which explains why the results are good all the way to a phase of about 40 degrees, as shown in Figure 6.3-3. Also, the development leading to the approximation given by Equation 6.3.10 yields the phase Equation given by Equation 6.3.11.

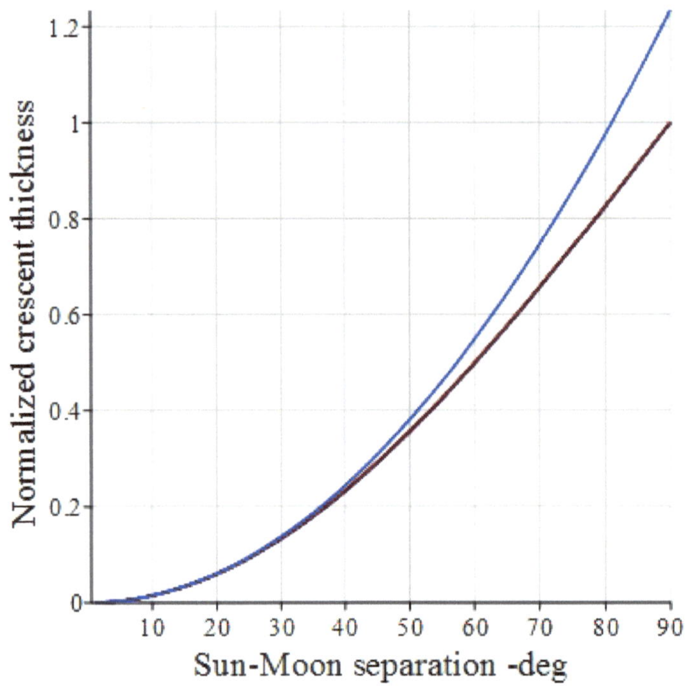

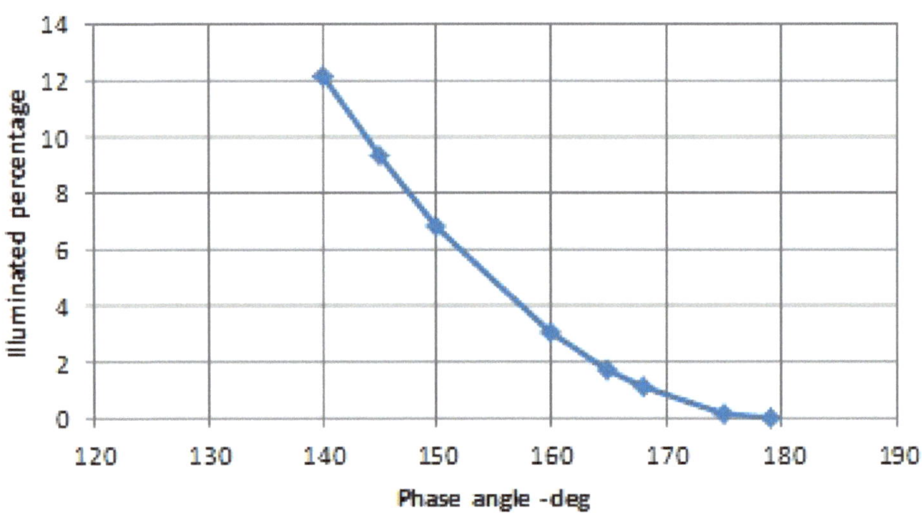

Figure 6.3-3 Normalized crescent thickness approximation versus exact expression (top) and Percent area illuminated moon (bottom)

Table 6.3-1 Maximum crescent thickness function of moon separation angle

Separation angle -deg	Max. crescent thickness -arcmin
1	0.001
3	0.018
4	0.033
5	0.053
6	0.078
7	0.108
8	0.142
9	0.181
10	0.225
11	0.273
12	0.326

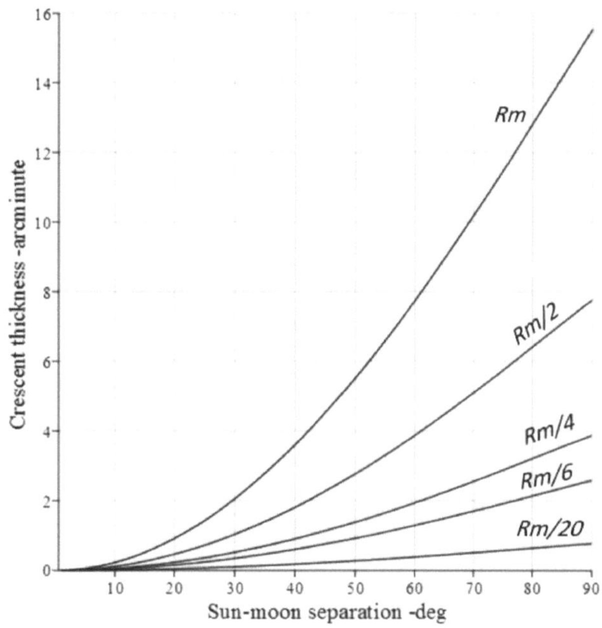

Figure 6.3-4 Crescent moon thickness for separation angles 0 to 90 deg

Back to the matter of brightness, the crescent surface brightness shown in Table 6.3-2 does not vary substantially over a good size separation interval. This means that if there is no moisture or other particulates in the horizon, the crescent is bright enough irrespective of its size. Experienced new moon watchers agree that there is "plenty of light" in a crescent moon. The brightness in mag per arcsec^2 is converted to the photometric unit of candela per m^2 using

$$S = 108000 * 10^{-0.4m} \quad [cd/m^2].$$

Table 6.3-2 Surface brightness of the moon

Phase angle	Separation angle	Brightness	Illumination	Area	Surface brightness	
deg	deg	mag	%	$arcsec^2$	$mag/arcsec^2$	cd/m^2
140	40	-7.6	12	332543	6.3	341
145	35	-7.2	9	254603	6.3	319
150	30	-6.8	7	187055	6.4	304
160	20	-5.9	3	83136	6.4	311
165	15	-5.5	2	46764	6.2	358
168	12	-5.2	1.1	29929	6.0	424
171	9	-4.9	0.62	16835	5.7	566
174	6	-4.5	0.27	7482	5.1	944

Our goal is to determine the magnitude of a crescent moon. Obviously, such data is hard to measure for low sun-moon-earth phase angles because of the scatter. Reference (Pilar Montanes-Rodriguez, 2007) used the photometric observation of the moon at Big Bear Solar Observatory to determine the surface brightness of earthshine and its variation for the calibration of lunar flashes. The Data which was taken between 1998 and 2005 had an excellent straight-line fit for the moonshine from 150 deg all the way to 30 deg separation for both waxing and waning moons. The computed data shows excellent agreement above 30 deg but deviates from the straight line near a new moon, as shown in Figure 6.3-5.

The linear relationship between the crescent brightness and lunar phase (in degrees) could be expressed as

$$B = 0.019167\alpha + 3.533 \ (\mathrm{mag/arcsec^2})$$

which extrapolates to an initial magnitude of about 7 $mag/arcsec^2$ at an astronomical new moon, a crescent magnitude of 6.9 $mag/arcsec^2$ at a phase angle of 6.35 deg and 6.75 $mag/arcsec^2$ at 12 deg.

In the 1960s geologists worked extensively to model lunar brightness, which happens not to track with the illuminated surface area as illustrated in Figure 6.3-6. The normalized brightness curve, which is adapted from (Much, 1970), closely follows the formula from (Allen, 1976) and (Krisciunas & Schaefer, 1991). The same formula is also used in the brightness calculation of Table 6.3-2. The annotated "part of the sunlit surface" curve is the phase function given by Equation 6.3.11.

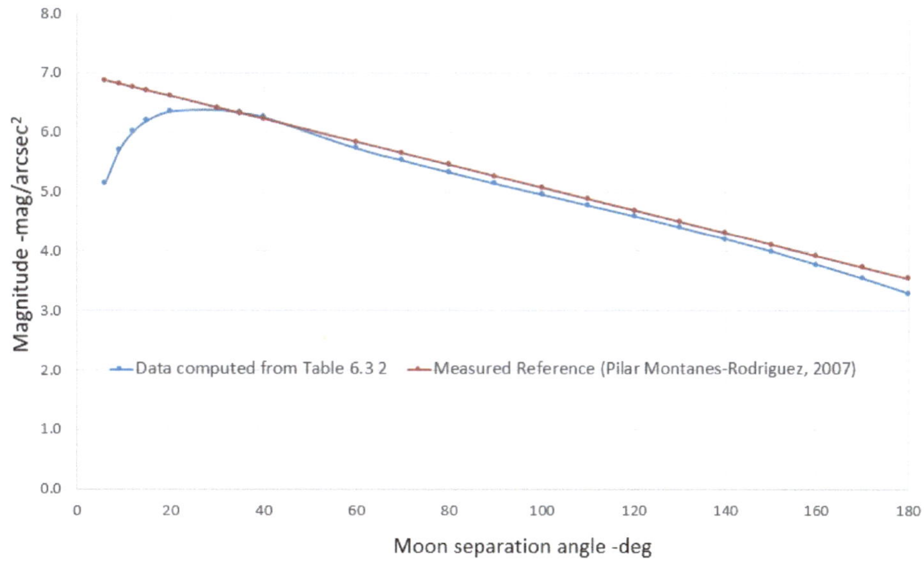

Figure 6.3-5 Surface brightness of the crescent moon

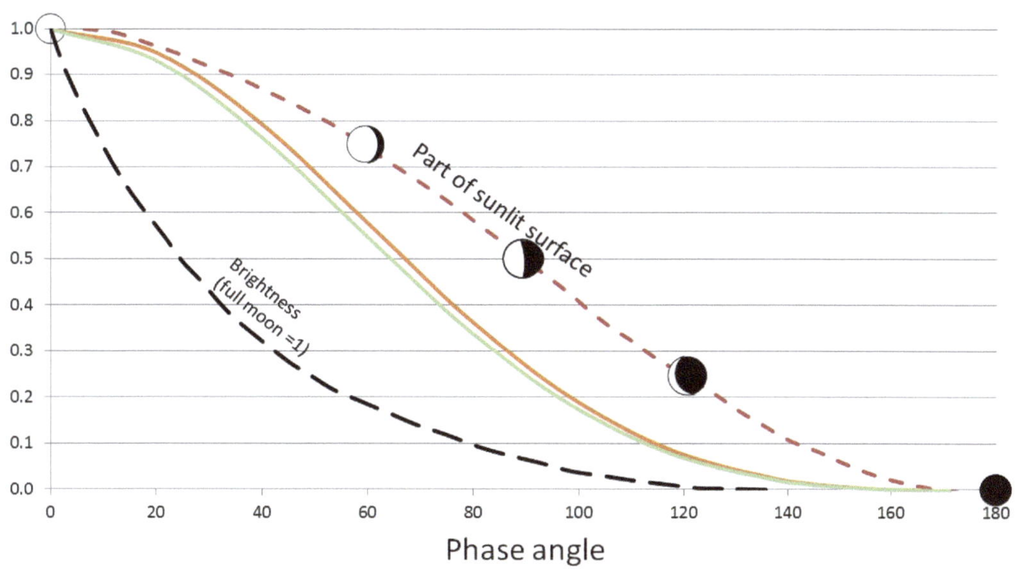

Figure 6.3-6 Normalized lunar Brightness decreases faster from full moon than the partially illuminated side of the moon (Much, 1970). The middle two curves are due to the view factor effect as discussed in Section 6.4

Except near a full moon, the surface brightness of the partially sunlit moon does not correlate with the illuminated surface area. The concept of surface brightness is used in astronomical objects with large luminous surface areas such as galaxies. However, the crescent moon is an elongated object which does not fit such two-dimensional representation.

Adjustment to account for the effect of the geometry of the crescent moon on the perceived radiation will be the topic of Section 6.4. The perceived brightness of the moon from the earth's surface is not uniform, not only due to the difference in reflectance of its geological formations but also because of the view factor of the sunlit surface. The latter is more prominent at small phase angles as the three-dimensional shape of the moon affects the emitted radiation in the direction of an observer on the earth surface.

6.4 Radiation view factor

Let us denote by dS_1 an infinitesimal surface with an outward unit normal vector n_1 at a point P_1 of spherical coordinates (r,θ,φ), located on the hemisphere of the moon surface which faces the sun as shown in Figure 6.4-1. This surface is assumed to reflect light in a diffuse and uniform manner. Some of this reflected radiation may or may not reach another infinitesimal surface dS_2 with an outward unit normal vector n_2 at a point P_2 on the earth surface, characterized by its unit outward vector n_2. The unit vectors n_1 and n_2 are tilted by the respective angles θ_1 and θ_2 from the direction P_1P_2.

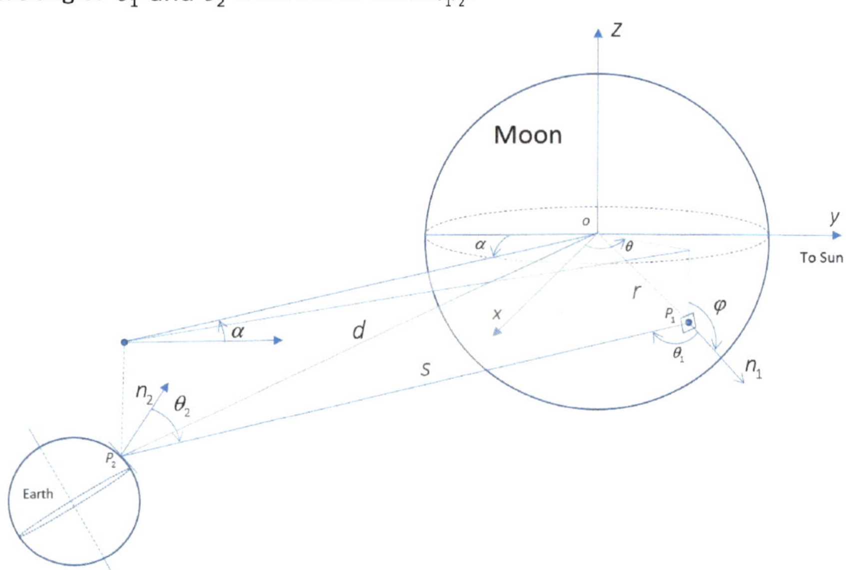

Figure 6.4-1 Geometric View Factor parameters for lunar emittance (shown not to scale)

From an observer's perspective, the direction of the moon is assumed at an angle α from that of the sun in the ecliptic plane and at inclination angle i_0 in the earth-moon plane perpendicular to the ecliptic.

Let us denote by dF_{12} the fraction of energy leaving point P_1 and reaching point P_2 separated by a distance s. The fraction dF_{12} is referred to as the infinitesimal view factor. The radiation power intercepted by dS_2 is given by the product

$$L_1 dS_1 \cos\theta_1 d\Omega_{12} ,$$

where is the radiance, the product $dS_1 \cos\theta_1$ is the normal area in the direction of point P_2 and $d\Omega_{12}$ is the solid angle subtended by dS_2. The energy reaching point P_2 is given by

$$L_1 dS_1 \pi s^2 / \cos\theta_2 dS_2 ,$$

and the fraction of energy arriving at surface dS_2 is given by

$$dF_{12} = \frac{L_1 dS_1 \cos\theta_1 d\Omega_{12}}{L_1 \pi s^2 / \cos\theta_2 dS_2} = \frac{\cos\theta_1 \cos\theta_2}{\pi s^2} dS_2 dS_1 \ .$$

For finite surfaces A_1 and A_2, the view factor is given by

$$F_{12} = \frac{1}{A_1} \int_{A_1} \int_{A_2} \frac{\cos\theta_1 \cos\theta_2}{\pi s^2} dS_2 dS_1 \ .$$

The vector n_2 is restricted to the line of sight of an observer or an instrument placed at P_2, which makes $\theta_2 = 0$. Also for a ray to emanate from point P_1, we have to have $0 \le \theta_1 \le \pi / 2$. The surface A_1 is the area of the sunlit hemisphere of the moon, and A_2 is the area of a detector which is the human eye in the case of a naked-eye observer. The unit vectors at points P_1 and P_2 are given in the moon-centered spherical coordinate system shown in Figure 6.4-1 by

$$n_1 = [\sin\varphi\cos\theta,\ \sin\varphi\sin\theta,\ \cos\varphi]^T ,$$

$$n_2 = -\frac{P_1P_2}{\|P_1P_2\|} = -\frac{P_1O + OP_2}{\|P_1P_2\|} = -\frac{-rn_1 + OP_2}{s} = \frac{r}{s} n_1 - \frac{d}{s}[\sin\alpha, -\cos\alpha, -\sin i_0].$$

In the above, the angle i_0 is the inclination of the moon and can be approximated by the mean inclination of the lunar orbit to the ecliptic plane at small phase angles. The distance s between points P_1 and P_2 is given by

$$s = \left(d^2(2 - \cos^2 i_0) + r^2 + 2dr(\sin\varphi\sin(\theta - \alpha) + \cos\varphi\sin i_0)\right)^{1/2} .$$

The cosine of the angle θ_1 is computed by taking the dot product of $-n_1$ and n_2. That is

$$\cos\theta_1 = -n_1 n_2 = -\frac{r}{s} + \frac{d}{s}(\sin\varphi\sin(\alpha - \theta) - \cos\varphi\sin i_0). \qquad 6.4\text{-}1$$

Because the radiation arrives parallel and uniform at the point P_2, the function inside the integral is constant with respect to surface A_1, and because the surface area of the hemisphere of the moon facing the sun is $A_1 = 2\pi r^2$, we have

$$F_{12} = \frac{A_2}{2d^2\pi^2}\int_0^{\pi}\int_0^{\pi}\frac{\langle -\rho + (\sin\varphi\sin(\alpha-\theta) - \cos\varphi\sin i_0)\rangle \sin\varphi}{(\rho^2 + 2\rho(\sin\varphi\sin(\theta-\alpha) + \cos\varphi\sin i_0) + (2-\cos^2 i_0))^{3/2}}\,d\theta\, d\varphi \,. \qquad 6.4\text{-}2$$

Note that we used $dS_1 = r^2\sin\varphi d\theta d\varphi$ and $\rho = r/d$ in the final form of the double integral. The term inside the angle brackets is zero when negative to ensure summation occurs only when $0 \le \theta_1 \le \pi/2$. The view factor can be written as

$$\frac{F_{12}}{A_2} = \frac{1}{d^2}G_{12} \,. \qquad 6.4\text{-}3$$

The fraction of energy received by an observer is roughly proportional to the inverse of the square of the earth-moon distance, the area of the sensing element and a geometric G_{12} which depends on the phase angle as shown in Figure 6.4-2. Even during a full moon, we have about 18 % of the radiation from the illuminated hemisphere arrives to earth due to the view factor. This value drops to 10 % at quarter moon and to about 2 % at 12 degrees. These approximate values were established using numerical integration, assuming a linear approximation of the inclination angle.

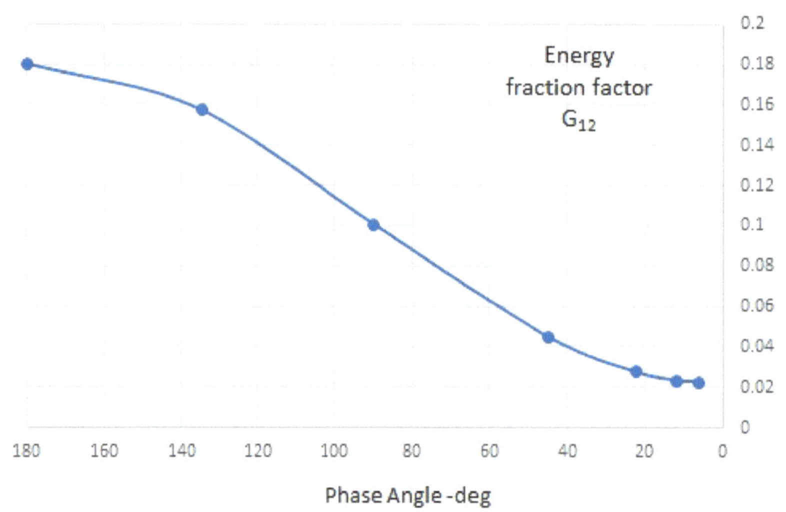

Figure 6.4-2 Perceived fraction of emitted Energy due to view factor

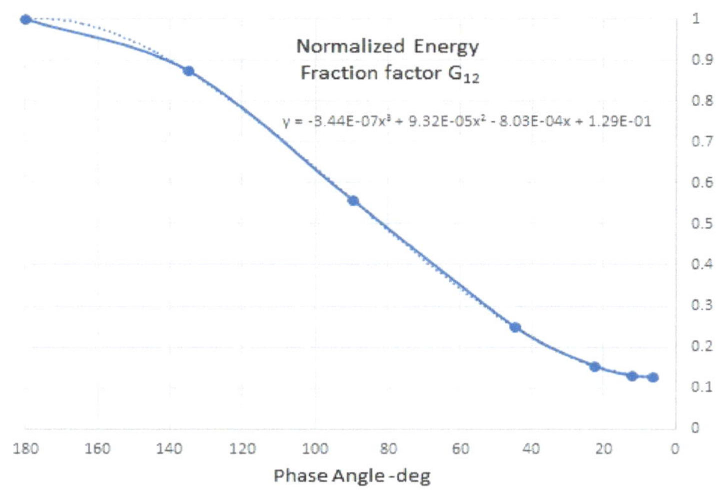

Figure 6.4-3 Normalized perceived fraction of emitted radiation energy

Figure 6.4-3 shows a normalized version of the geometric factor, which we denote by $\bar{G}_{12}$. At small phase angles less than about 12 degrees, the fraction of emitted energy perceived by an earth observer is about 13 percent of the ratio obtained for a full moon. That is, the smaller the phase angle is, the smaller the perceived fraction of emitted energy.

Brightness data is available for a full moon, near full moon and mid-range phase angles, but is unavailable for small angles. The geometric view factor contributes to the lack of correlation between the surface brightness of the partially sunlit moon and the illuminated surface area. Because good correlation exists near full moon, we use the geometric view factor to extrapolate to data near new moon phase angles. Recall Equation 6.1.1 giving the apparent magnitude of an object in terms of a flux density *F* and a preselected flux density F_0. We have $m=-2.5\log_{10}F/F_o$. We also have Equation 6.1.2 giving the surface brightness of an object of surface area *A*, as $S=m+2.5\log_{10}A$. We write the flux density from the moon at a certain phase is related to the flux density during a full moon using $\bar{G}_{12}F$, and therefore we may write

$$\begin{aligned} s(\alpha) &= m(\alpha)+2.5\log_{10}A(\alpha)=-2.5\log_{10}\frac{\bar{G}_{12}F}{F_0}+2.5\log_{10}A(\alpha) \\ &=-2.5\log_{10}\bar{G}_{12}-2.5\log_{10}\frac{F}{F_0}+2.5\log_{10}A(\alpha) \\ &=-2.5\log_{10}\frac{F}{F_0}+2.5\log_{10}\frac{A(\alpha)}{\bar{G}_{12}}\quad . \end{aligned} \tag{6.4-4}$$

The last two equalities show that for the moon, the terms involving $\bar{G}_{12}$ is embedded in the measurements and explain the dependence of the surface magnitude on the view factor.

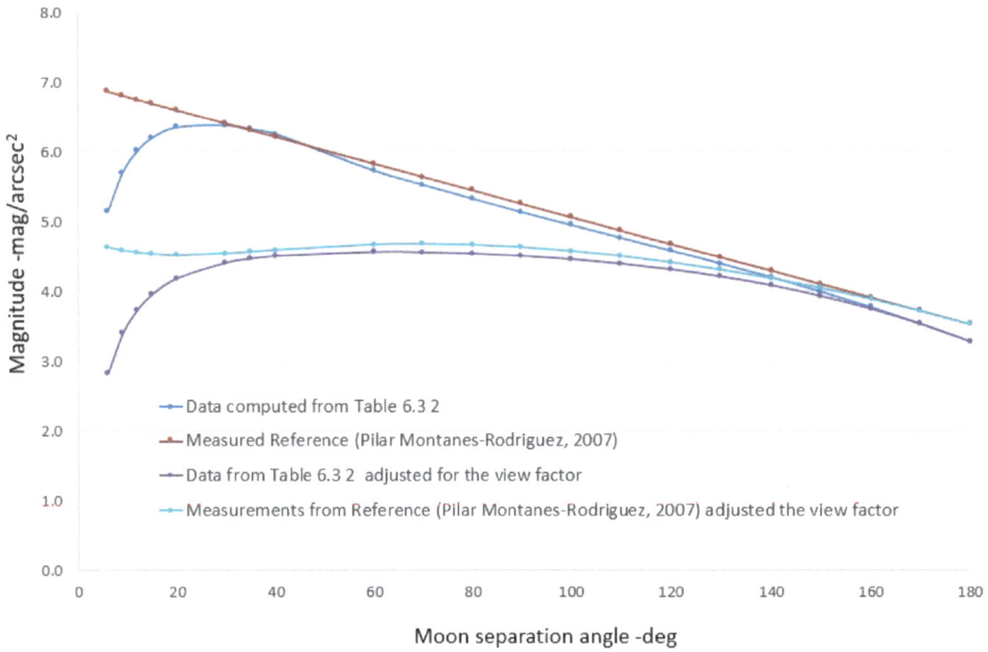

Figure 6.4-4 Magnitude data adjusted for view factor

Figure 6.3-6 shows improvement in correlation with an adjusted surface area represented by the two inner curves. Note that one of the two curves accounts for the earth-moon distance variation with phase angle. Certainly, the profile of the adjusted curves near a full moon does not show the rapid decay near a full moon. However, the data when the two curves are adjusted to eliminate the effect of the view factor a near-constant magnitude over a wide range of the phase angle, as shown in Figure 6.4-4.

Chapter 7 Optics of the atmosphere

Light from the crescent moon travels hundreds of kilometers through multiple layers of the atmosphere during twilight. One can think of these layers as a large optical filter with complex dynamic characteristics that need to be well understood. We shall draw from the field of atmospheric optics to formulate first crescent moonlight visibility. Major efforts to understand and characterize the optical properties of the atmosphere were initiated during the second half of the 20th century. They had applications in multiple fields including remote sensing, civilian and military aerial navigation, atmospheric sciences and astronomy. These efforts have provided a wealth of information from which the physics of new moon sighting can be developed.

Electromagnetic radiation is practically the only source of information that we exploit to learn about celestial objects. This radiation is affected by several atmospheric properties, including molecular composition, temperature, pressure, density, water vapor, aerosols, airspeed, and turbulence. And as a result, scientific evaluations can be difficult due to the large number of parameters which affect measurements. Fortunately, unlike nighttime astronomy, new crescent moonlight confirmation does not require science extraction, and therefore does not have the same limitations on the quality of the collected light and instrumentation. All that is needed in our case is a confirmation of sighting. To this end, we have two topics of relevance to our objective. The first topic entails the study of visible moonlight transmission in the atmosphere, while the second topic pertains to the nature of the background sky. Because the new moon is near the sun, and sighting occurs during twilight, we need to fully understand the mechanism of scattered sunlight which defines the background sky in which the new crescent moonlight is manifested.

Depending on its frequency, incident radiation is partially retained by gases, partially scattered by these gases and suspended particulates, and the rest is transmitted through the layers of the atmosphere. Our atmosphere is mainly transparent in the visible light spectrum. Outside the visible spectrum, the ozone in the atmosphere blocks UV wavelengths, which are harmful to biological life on earth. Also, water vapor and carbon dioxide absorb radiations mainly outside the visible spectrum. Our atmosphere is transparent in the visible spectrum; this is the reason the human eye (and those of other animals) is specialized to sense radiation in the visible spectrum.

The first section of this chapter addresses the optical transmittance of the atmosphere in different wavelengths. While our focus is on the visible spectrum, we venture beyond the human range of perception to have an idea on how an extended capability may help with sighting verification.

Scattering of sunlight is the result of redirected photons by suspended particles and molecules in the atmosphere. We are all familiar with a partially illuminated moon being visible in broad daylight, when it is dozens of degrees away from the sun. The light from the moon is visible because of two factors: the first and most important one is the reduced sunlight scatter in the line of sight. The second, and less important factor, is the reduced thickness of the atmosphere in a direction close to the zenith. The size of the bright area of the moon is irrelevant to the potential for visibility. Observers often say, "there is plenty of light from the moon."

Scattering is classified into three types: Rayleigh, Mie, and Non-selective. Rayleigh scattering is caused by molecules in the atmosphere and happens to be responsible for the blue color of the sky. Mie scatter is due to particles in the lower atmosphere of size equal to incident solar wavelengths. As a result, moisture, aerosols, and smoke are the sources of the grey color appearance in the sky. Finally, the larger particles, also in the lower section of the atmosphere, produce the non-selective scattering responsible for the hazy properties of certain skies. Scattering is the most dominant effect impacting new moon sighting. Forward scattering of light can be quite large, often resulting in images which are several times the size of the target when observed near the horizon.

Another issue affecting new crescent moon sighting, which classical nighttime astronomers struggle with as well, is thermal turbulence. The "seeing" aspects of astronomical observations are separate from scatter-causing diffusion of light during twilight. The former is similar to what an object would look like when observed from a distance over a sun-heated road, resulting in image distortion of the order of 2 to 20 arc-seconds, while the latter is responsible for the enlargement of vehicle light beams in a snowy or dusty night. Because this area is not too relevant to our subject, we still need to have an idea on what it entails as it removes questions from our topic of interest and affects some of the hardware used for observation. Section 7.2 briefly discusses thermal turbulence and gives a classification of the different types of seeing. Section 7.3 introduces a geometric model describing sunlight scatter, which is responsible for illuminating the background sky. The standard model for light attenuation through scattering and absorption used in the field of atmospheric optics is subsequently introduced in Section 7.4 where the reader will learn about the concepts of atmospheric Transparency, and Optical Thicknesshickness.

The brightness contrast of a new crescent moon light against a background sky during twilight is proposed as a basis for new moon crescent visibility. Section 7.5 formulates a logarithmic measure of contrast relating scattered sunlight to new crescent moon light attenuation. By integrating over the visible spectrum and making use of the standard human eye efficiency in perceiving different wavelengths, a bulk contrast measure of new moon brightness is subsequently derived. This measure happens to depend mainly on one parameter: the scatter angle.

We intend to present two methodologies for quantifying the onset of new moon sighting. We shall differentiate between photometric and radiometric crescent moon visibility. While both methodologies seek the same result, the former is used to quantify the perceived light brightness by the human eye and compare it to that of the background sky, the latter deals with absolute energy power and is best used for in instrument measurements. We shall present a criteria for crescent moon visibility using photometry in Section 7.6 and a more sound criteria for visibility using radiometry in Section 7.7. In Section 7.8, we use the performance of the human eye to establish the critical angle of moon elevation for which first detection occurs. Finally, background sky measurements reveal in Section 7.9 and show that first-order scattering dominates the illumination of the sky during twilight when new moon sighting takes place.

7.1 Atmospheric transmittance

The extent of the electromagnetic spectrum from the short gamma-ray waves all the way to radio waves is shown in Figure 7.1-1. The visible spectrum which the human eye can detect happens to be between 0.4 µm and 0.7 µm, which is the range where the sun radiates at its maximum level. This range holds only a small part of the electromagnetic spectrum. Adjacent to the visible spectrum is the infrared spectrum which extends from 0.7 µm to 1.0 µm. The electromagnetic waves outside the visible spectrum are manifested in a variety of forms and require specialized sensors to interpret. For example, satellites use these sensors to peer in the realm of the invisible to gather information from the ground. Depending on the wavelength, these waves are not fully transmitted in their journey through the layers of the atmosphere.

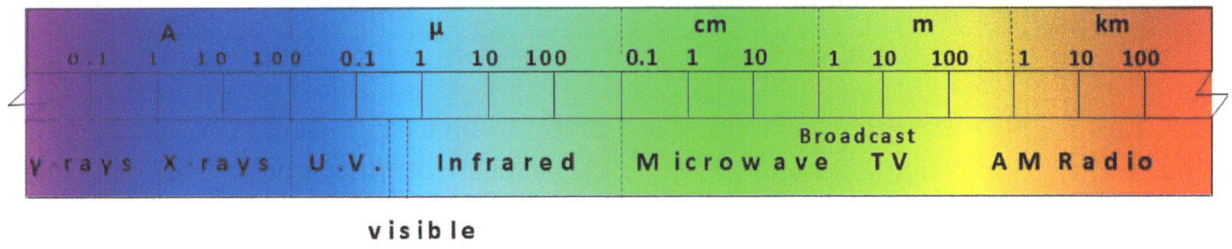

Figure 7.1-1 Wavelength spectrum bands

Atmospheric Transmittance is the ratio of transmitted radiation to the total radiation incident upon the layers of the atmosphere. The individual layers of the atmosphere exhibit different

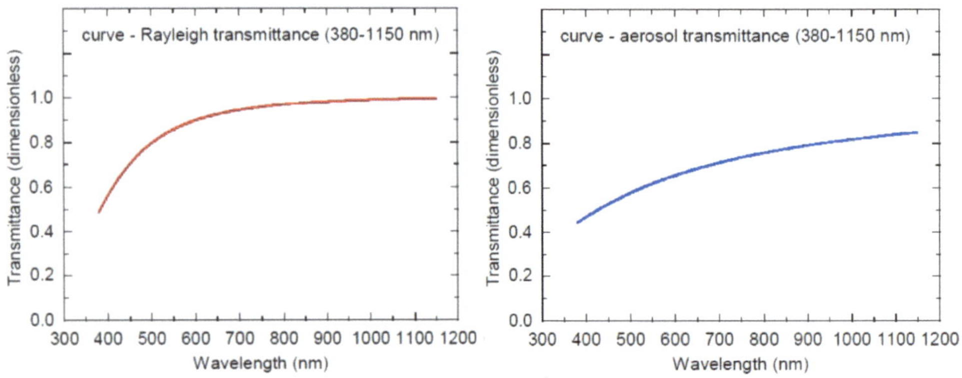

Figure 7.1-2 Rayleigh and aerosol transmittances - Courtesy of the NASA SeaWiFS Program

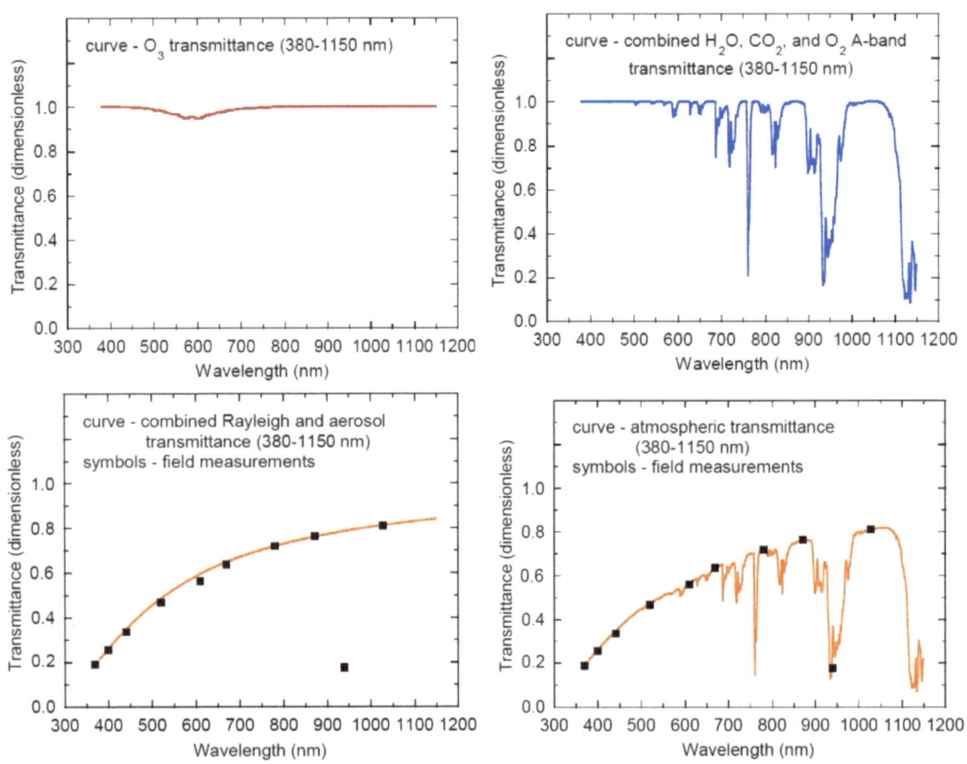

Figure 7.1-3 MODTRAN data for Ozone, water vapor, carbon dioxide and Oxygen gases (top), Combined Rayleigh and aerosol transmittance and overall atmospheric transmittances against field measurements (bottom) - Courtesy of the NASA SeaWiFS Program.

transmittance characteristics depending on the wavelength. Transmittance also varies whether the direction of the light is horizontal, vertical, or at an oblique angle. There are two main processes behind the transmittance characteristics of the atmosphere. They are *scattering* and *absorption*. Part of the electromagnetic energy is scattered by the molecules and aerosols in the atmosphere. Another part is absorbed by the different constituents in its way to the observer. Considerable data was produced to characterize transmittance. This characterization includes both field measurements and analytical validations. For example, during the SeaWiFS NASA solar Irradiance calibration program, atmospheric transmittance was evaluated between wavelengths 380 and 1150 nm.

Transmittances due to Rayleigh and aerosol are shown in Figure 7.1-2, while transmittances through atmospheric gases are shown in the two top plots of Figure 7.1-3. In addition, measurements from a 10-band sun photometer are shown against a MODTRAN[13]-simulated overall atmosphere which accounts for the proper mix of atmospheric gases in the bottom of the same figure. Figure 7.1-4 shows the scattering losses and absorption losses below the scattering line as a function of wavelength. Loss through scattering dominates transmission in the visible spectrum, while most of the absorptions occur in the infrared regime and is due to water, Carbon monoxide, carbon dioxide, methane, and ozone.

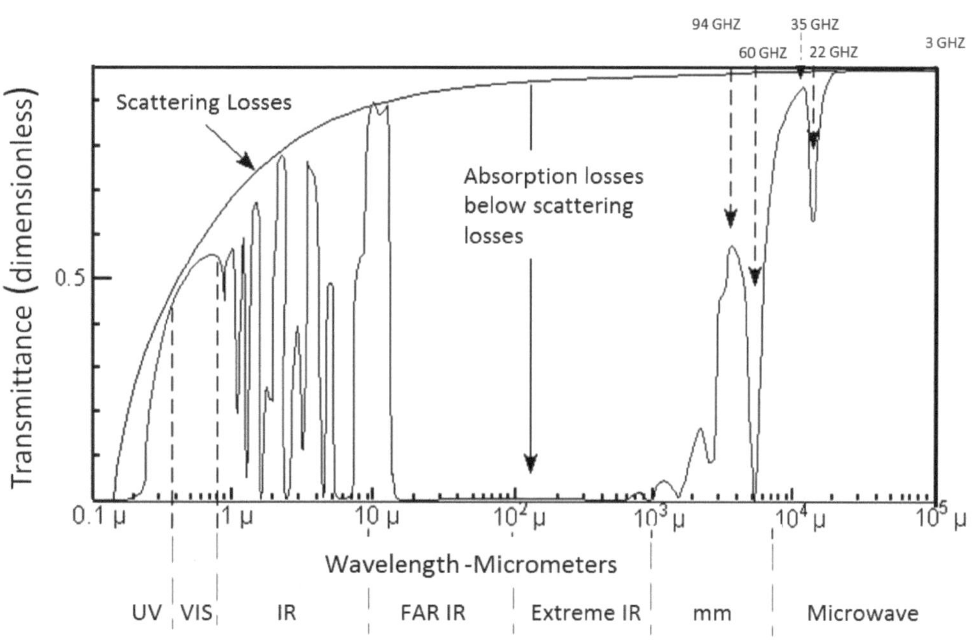

Figure 7.1-4 Atmospheric transmission along a 1 NM sea level path, US Navy –public domain

Obviously, an observation system designed to detect the new moon ought to leverage the higher reflectance which occurs at longer wavelengths. However, except for specific "bands" of wavelengths that transmit well, the atmosphere absorbs most infrared light and therefore cannot be seen from the ground as shown from Figure 7.1-4. An observation system for new

[13]Computer program designed to model atmospheric transmission in from middle ultraviolet to far infrared

moon detection should operate somewhere in such bands. The choice of the wavelengths over which new moon sensing takes place depends on several additional parameters. Not only does the atmosphere need to be transparent to the chosen wavelengths, but the optics of the instrumentation used also has to be transparent to those wavelengths. Furthermore, the sensor itself must have enough sensitivity to detect the first light from the moon. Specialized photodetectors are selected to target the right frequency range. For example, by utilizing CdTe substrates, thin HgCdTe layers can be grown to maximize peak sensitivity to the moon terminator at specific wavelengths. Overall sensitivity requires that the telescope be large enough to collect as much light as possible, yet not too large for the noise to overwhelm the signal.

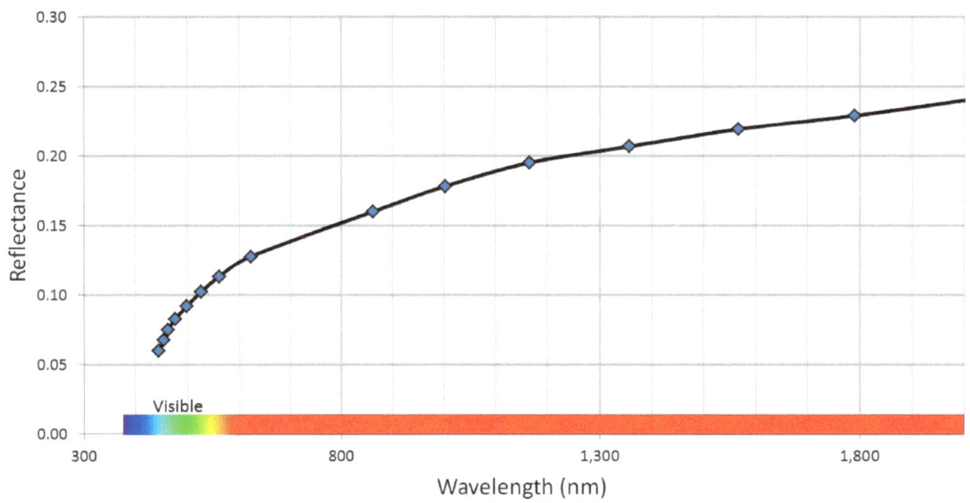

Figure 7.1-5 Moon surface reflectance as a function of wavelength (Kieffer & Stone, 2005)

The moon reflects sunlight differently depending on the wavelength. Figure 7.1-5 shows higher reflection with increasing wavelength, all the way into the infrared regime, i.e. (>1000nm). Scaled data was obtained from the Apollo program soil was used to corroborate the USGS ROLO[14] moon observation campaign and derive the lunar spectrum model of Figure 7.1-5.

When considering new moon detection outside the visible spectrum, the choice of wavelengths over which sensing takes place depends on the transparency of the atmosphere, transparency of the optical system, and detector performance at those wavelengths. Optical glass and specialized photodetectors are selected to target the right frequency range. For example, by utilizing CdTe substrates, thin HgCdTe layers can be grown to maximize peak

[14] USGS's Robotic Lunar Observatory used for space –based remote sensing calibration, see website: www.moon-cal.org

sensitivity to the moon terminator at 8-9 μm, a band in which the atmosphere is transmissible. The main advantage in considering the use of longer wavelengths is in the avoidance of diffraction of light by aerosols. Even though non-visible spectrum observations reveal other physical manifestations and properties, e.g., temperature below the surface of the target, such manifestations can, in certain cases, be correlated to new moon detection.

Before getting too excited about the nonvisible spectrum, there is a fundamental difference between radiation from the visible spectrum and the longer wavelengths. While at optical wavelengths the moon is seen essentially in the sun's reflected light, detection in longer wavelengths picks different physical manifestations that are inherently different from those of reflected solar radiation. For example, thermal blackbody emission is dominant in comparison with the reflected radiation from the sun. The moon, as a "black body," absorbs radiation emitted by the sun at all frequencies, and then re-emits this radiation (see the illustration in Figure 7.1-6).

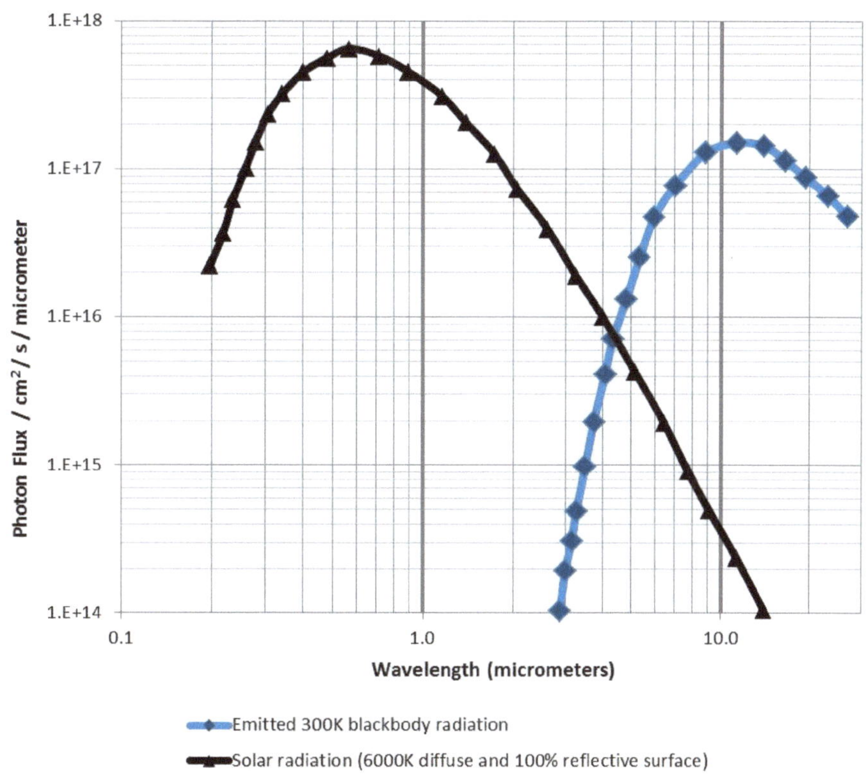

Figure 7.1-6 Solar and black body emittance under no atmospheric absorption

Radio detection can be used to reveal the moon's temperature at different depths from the surface. For example, the energy picked up by the 140-Foot Radio Telescope NRAO in Green Bank, WV in the US generates images from about one meter below the moon's surface. It is possible to relate the temperature of specific locations on the moon to crescent brightness and extract the location of the moon terminator. This would provide an independent verification method for new moon detection, providing that reliable calibration for phase lag is achieved, which in turn depends on the geology of the observed locations on the moon.

The moon in microwave exhibits a time-lag of several days and an asymmetry with respect to the maximum temperature, as shown in Figure 7.1-7. But infrared temperature is symmetrical with respect to the onset of a full moon; it reaches a maximum at a full moon and a minimum at a new moon, making it the area of choice for immediate interest for non-visible sighting. It remains, however, that Infrared wavelengths are stopped by water vapor.

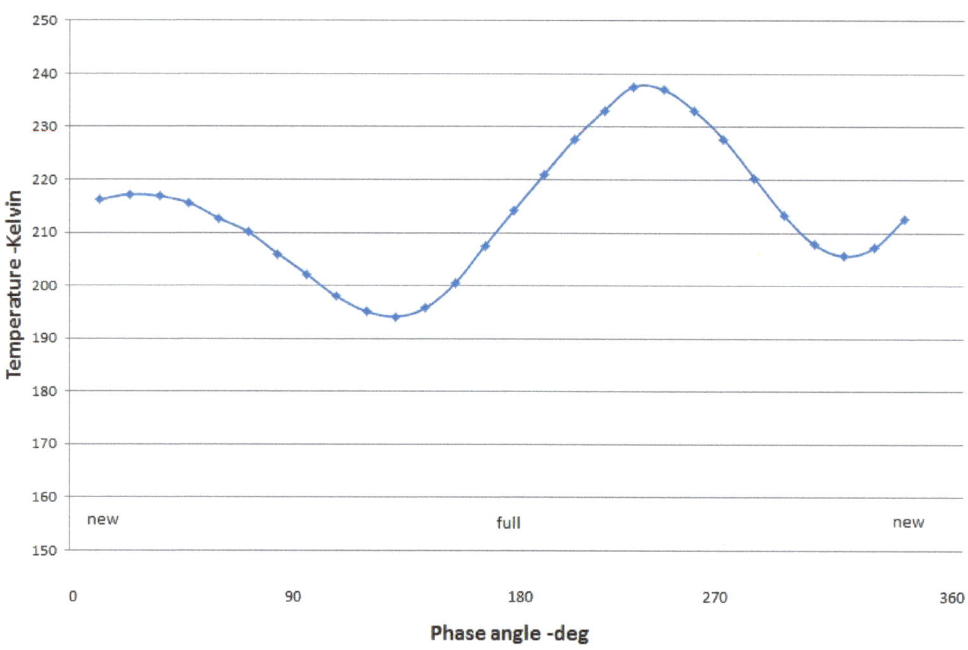

Figure 7.1-7 Moon Surface Temperature at 2.77 cm (Monstein, 2001)

Finally, sub-millimeter wavelength astronomy (0.3 — 1.0 mm) is the newly studied electromagnetic band. This is a technologically difficult area because of the complexity of the detectors used and the lack of transmissibility of the atmosphere in microwave. The radio image shown in Figure 7.1-8 represents the temperature distribution of the moon, illustrating the performance of the antenna, whereas the image in visible light shows the sunlight reflected by the moon surface. The sub-millimeter frequency band lies between the regions transparent to radio astronomy and those open in visible light. Therefore, sub-millimeter detectors borrow from both optical and radio astronomy. The sub-millimeter telescope looks like a radio telescope, although it is housed in an assembly similar to an optical telescope. Observations are done using specialized detectors called bolometers, which are devices for measuring the energy of incident radiation. They include an "absorber" connected to a heat sink (volume at constant temperature) via an insulating link. Any radiation received by the absorber raises its temperature above that of the heat sink. For sub-millimeter wavelengths, bolometers are considered the most sensitive detectors. To achieve the best sensitivity, they are cooled down to a 50 to 300 milli-kelvins, i.e., a fraction of a unit deg K.

At lower elevations, the atmosphere is not transparent at sub-millimeter wavelengths. This is mainly due to water vapor absorbance of such radiation. The situation is much better at higher elevations where the atmosphere is thin and dry, which limits the choices for such observatories to a very few locations on earth.

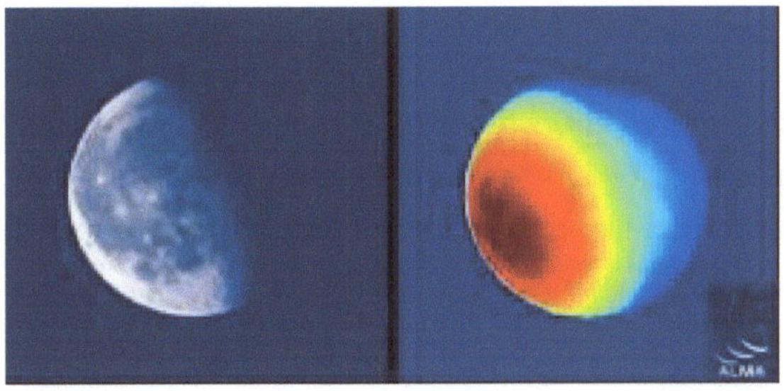

Figure 7.1-8 First Image of the moon taken with an ACA Antenna (Left: Optical image taken with a digital camera; Right: Sub-millimeter Radio image taken with an ACA antenna) at the Atacama Large Millimeter/submillimeter Array (ALMA) interferometer. Credit: National Radio Astronomy Observatory.

7.2 Thermal turbulence and astronomical seeing

Ground-based observers are often restricted by poor viewing, known as the *"seeing"* condition, which refers to the optical quality of the air column in the line of sight. Air, being a refracting medium, is part of the optical system in which changes in density or movement in the air column affects observation.

Experience showed that optical quality is mainly due to thermal turbulence caused by convective currents in the layers of the atmosphere. Observers often have to put up with poor to average seeing conditions for a long time to get a brief moment of steady, sharp images, even if they use high-speed photography to get rid of twinkling stars and other objects. While this issue is encountered in classical astronomy, during nighttime observation, daytime and twilight astronomy have even more difficult challenges due to heat dissipation from the ground and higher air temperature. There are three types of seeing which can occur independently or concurrently.

In poor seeing of *Type I*, the image changes rapidly and has blurred edges. This is usually attributed to air currents near or within the telescope. Non-homogeneities of about 10% to 50% the diameter of the telescope prevent rays from reaching the image focus simultaneously. When such conditions take place, images are not expected to yield an adequate resolution.

Seeing of *Type II* is associated with steady but blurred images and is due to low-lying rapid air movement away from the telescope tube. This is caused by non-homogeneities that are smaller or equal to 50% the diameter of the telescope. This condition may occur near artificial structures with a different cooling/heating rate. Weather fronts can produce serious *Type II* image degradation. Lunar photography will be hampered by these effects, but they will not eliminate chances for detection.

Finally, in seeing *Type III*, sharp yet unsteady images are obtained. This is caused by high-level atmospheric non-homogeneities larger than the aperture of the telescope. In this seeing condition, the image of the crescent would ripple but would be sharply defined.

Fortunately, because the need is to make a positive identification of a new moon, the quality of a new moon image is not relevant because there is no need for science extraction. It remains however that in extreme conditions, a new moon observation system has to be carefully designed to overcome image degradation due to non-homogeneities in the optical path, especially we have no control over site selection, nor we have the luxury of waiting for the air to stabilize. Positive sighting is limited to a small window of time.

7.3 The geometry of light scattering

Much of the visible light and shorter wavelengths that hit the atmosphere is scattered as illustrated in Figure 7.1-4. Scattering results in a diffuse light that brightens the background sky and reduces contrast, making the new moon difficult to observe. The scattered light is also polarized. Therefore, polarizing filters designed for the operating bands of a moon-observing system help eliminate the glare of the background sky surrounding of a New moon. In this section, we characterize sunlight scattering in the new moon line of sight, when observed from the earth surface. We shall neglect multiple scattering because the twilight sky brightness is dominated by single scattering during sighting time. This will be verified through measurements in Section 7.7. We will also neglect the effects of atmospheric refraction and altitude. The addition of these effects is simple, should there is a need to account for them.

Consider the simplified planar geometry of twilight solar illumination shown in Figure 7.3-1. The plane shown is that of the large circle going through the earth center C_E, the center of the sun C_S, assumed to be at infinity and an observer, placed at point *O* on the earth surface. We mentioned in Section 5.2 that the height of earth shadow is not affected by the azimuth angle *A*, separating the earth-moon great circle with that of the earth-sun great circle. Therefore, at small phase angles of the moon, it is assumed that a line of sight locked on the setting moon is in the same plane as that of sunlight scattered to the observer. We further consider a pencil of light from the sun which is attenuated in its journey across the atmosphere until it reaches the line of sight *(O t)* where it gets scattered towards the observer at point *O*. The scattered light gets further attenuated along the line of sight in the direction of the moon before reaching point *O*.

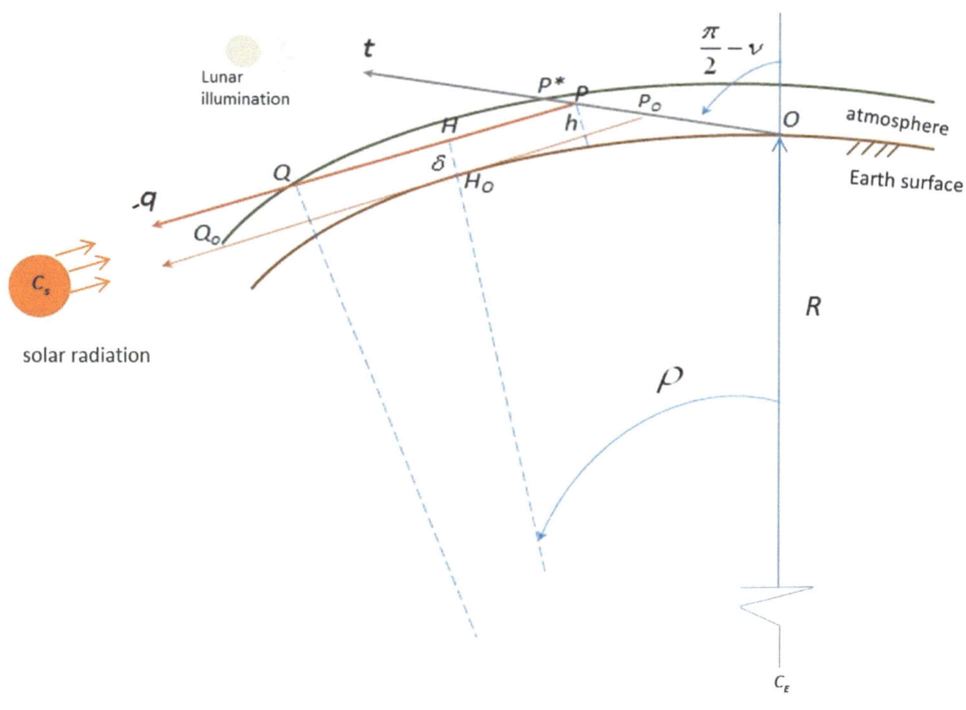

Figure 7.3-1 Simplified geometry of the optical solar illumination of the atmosphere

We denote by *t* a non-dimensional coordinate along the line of sight and define a second non-dimensional coordinate *q* along the sunray *QP* with origins at *O* and *Q* respectively. We recognize the sunray tangent to earth surface at point H_o from which the radial line to the center of earth makes an angle with the Zenith at the Observer equals to the declination angle of the sun denoted by ρ. Furthermore, we denote by ν the elevation angle of the moon above the horizon. It can be easily verified from geometry, that for a moon-constrained line of sight, the *scattering angle* by which the sunray *QP* gets deviated at *P* towards the observer is the sum of angles ρ and υ. It can be shown that the scattering angle remains nearly constant during the time of a sighting, and is equal to the elevation of the moon at sunset, *C*. Therefore, we may write

$$\theta = \rho + \upsilon \approx C . \qquad 7.3\text{-}1$$

The sum is nearly constant over the time of sighting because the motion of the moon relative to that of the sun is negligible, and from an earth observer's perspective, most of the position change is due to the rotation of the earth.

Let us first focus our attention on the attenuation of solar radiation in its path through the atmosphere. We recall the approximation of the height of earth shadow *h* and the distance traveled by a photon from the moon inside earth shadow *l* given by the normalized Equations 5.4.3, which we restate here as following:

$$u = \frac{l}{R} \approx \frac{\rho^2}{2C}, \quad w = \frac{h}{R} \approx \frac{\rho^2 (2C - \rho)^2}{8C^2}, \quad \text{for } 0 \le \rho \le C.$$

At any time following sunset, and for any point *P* along the line of sight in the direction of the setting moon, an approximation of the coordinates *t*, normalized with respect to earth radius *R*, is given by *u*. The three angles of the triangle OPC_E, are given by $\nu + \pi/2$, $\pi/2 - \nu - \gamma$, and γ (not shown in the Figure) . Applying the law of sines to the triangle OPC_E, we have

$$\frac{R}{\sin(\pi/2 - \nu - \gamma)} = \frac{R + h}{\sin(\pi/2 + \nu)} = \frac{R\,\gamma}{\sin\gamma}.$$

Bivariate Taylor series expansion of the left equality for small angles ν and γ gives, after dropping third-order terms,

$$\frac{1}{2}\gamma^2 + \nu\gamma - h/R = 0 .$$

Solving the above quadratic equation, we have

$$\gamma = \sqrt{\nu^2 + 2h/R} - \nu \approx \sqrt{2h/R} - \nu .$$

An acceptable estimate for $R\gamma$ *is* Rt, which gives after replacing the variable *h* by its non-dimensional equivalent *h/R*:

$$t \simeq \begin{cases} 0 & at\ O, \\ t_o = \dfrac{\rho^2}{2C} & \text{at } P_o, \\ \sqrt{2h} - v = \sqrt{2h} - (C - \rho) & \text{at } P. \end{cases} \qquad 7.3\text{-}2$$

Then, using the same argument, we also have

$$h \simeq \begin{cases} 0 & at\ O, \\ \dfrac{\rho^2 (2C - \rho)^2}{8C^2} & \text{at } Po, \\ \dfrac{1}{2}t^2 + tv = \dfrac{1}{2}t^2 + t\,(C - \rho) & \text{at } P. \end{cases} \qquad 7.3\text{-}3$$

In addition, the normalized approximation of the distance H_oP_o is given by

$$HoPo = \tan\rho - \frac{\cos\nu}{\cos\rho} t \simeq \rho - \frac{1}{2}\frac{\rho^2}{C}. \qquad 7.3\text{-}4$$

And the normalized perigee height, minimum height of an arbitrary sunray, is given by

$$\delta = (t - t_0)\sin(v + \rho) \simeq tC - \frac{\rho^2}{2}. \qquad 7.3\text{-}5$$

If the atmosphere extends to $R_l = r_l R$, the normalized distances *QH* and *HP* are given by

$$QH = \sqrt{(1 + r_l)^2 - (1 + \delta)^2} \simeq \sqrt{2(r_l - \delta)} \simeq \sqrt{2\left(r_l - tC + \frac{\rho^2}{2}\right)}. \qquad 7.3\text{-}6$$

$$HP = H_oP_o - (t - t_o)\cos(\rho + v) \simeq \rho - \frac{1}{2}\frac{\rho^2}{C} - \left(t - \frac{1}{2}\frac{\rho^2}{C}\right)\left(1 - \frac{C^2}{2}\right) \simeq \rho - t. \qquad 7.3\text{-}7$$

Before intersecting a line of sight, a sunray undergoes a path of normalized length

$$QP = QH + HP = \sqrt{2\left(r_l - tC + \frac{\rho^2}{2}\right)} + (\rho - t), \text{ where } t > t_o. \qquad 7.3\text{-}8$$

We also have the normalized height, at a position q given by

$$h(q,t,\rho,C)=\sqrt{(QH-q)^2+(1+\delta)^2}-1\simeq\sqrt{(\rho-t-q)^2+\left(1+tC-\frac{\rho^2}{2}\right)^2}-1$$

$$\text{for } 0\le q\le q^*,\ \frac{\rho^2}{2C}\le t\le t^*, q^*=\sqrt{2\left(r_l-tC+\frac{\rho^2}{2}\right)}+(\rho-t) \text{ and } t^*\simeq-(C-\rho)+\sqrt{2r_l}.$$

7.3-9

The expression for t^* is obtained by satisfying $h(q^*,t^*,\rho,C)=r_l$ and dropping the higher-order terms in the third Equation of 7.3.4.

Having established the geometric parameters for solar and lunar light pencils during twilight, we proceed to discuss how both lights are affected by the atmosphere prior to reaching an observer, placed on the earth surface facing the horizon in a westerly direction.

7.4 Light attenuation and atmospheric transparency

The illuminance of an object by a light source inside the atmosphere is related to the corresponding exo-atmospheric illuminance using

$$T=\frac{I}{I^*}=e^{-\tau(\lambda)} \qquad 7.4\text{-}1$$

where T is referred to as the *Transparency*[15] of the medium and $\tau(\lambda)$ is referred to as the *optical thickness* of the layer traveled by the light pencil, given by

$$\tau=\int_0^l k(\lambda,x,y,z)dl,$$

in which k is the *attenuation or extinction coefficient,* which depends on the wavelength and the position. Because part of the energy is scattered, and another part is absorbed by the medium. Therefore, we have

$$k=\alpha+\sigma,$$

Where α is called the *absorption coefficient*, a scalar quantity for an isotropic atmosphere, and σ is the *scattering coefficient* of the medium. Furthermore, for an incoming ray at an angle z to the zenith, the elementary optical thickness of air along dh is given by

[15] Complementary to transmittance introduced in Section 7.1

$$d\tau = k(h,\lambda)\ \sec z\ dh\ .$$

For a homogenous air and neglecting refraction, the optical thickness between two heights h_1 and h_2 is given by

$$\tau(h_1,h_2) = \sec z \int_{h_1}^{h_2} k(h,\lambda)\ dh = \sec z\ \tau_v(h_1,h_2)\ ,$$

Where $\tau_v(h_1,h_2)$ is the optical thickness in the vertical direction. The variation of the angle z with height and with earth's curvature is accounted for by assuming an integrated mean value for $\sec z$, denoted by m, called *air mass* (not to be confused with the magnitude which is also indicated by the variable m). Therefore, for small angles, we have

$$m = \overline{\sec z} \simeq \overline{1/\nu}\ .$$

And assuming a homogenous atmosphere, we have $\overline{1/\nu} = 1/\nu$ and $\tau_v = \tau$ which leads to

$$T = \frac{I}{I^*} = e^{-m\ \tau(\lambda)} \simeq e^{-\frac{\tau(\lambda)}{\nu}}\ . \qquad 7.4\text{-}2$$

Most investigators consider the optical thickness to be proportional to a normalized density distribution. Therefore, for an arbitrary path between two distinct points traveled by a pencil of light, the optical thickness is given by

$$\tau(\lambda,\Gamma) = C_e(\lambda) \int_\Gamma \frac{\rho(\gamma)}{\rho_o} d\gamma\ , \qquad \textbf{7.4-3}$$

where Γ is the path traveled, and λ is the wavelength. The integration variable γ is a coordinate along the path, ρ is the air density at a position γ which depends on the height h, ρ_o is a characteristic normalizing density, and $C_e(\lambda)$ is some proportionality constant to be determined from optical thickness data for a unit airmass at sea level.

The MSISE model[16] from the Community Coordinated Modeling Center (CCMC) at Goddard Space Flight Center gives the neutral temperature and densities of the atmosphere from the surface to thermospheric heights (altitude > 90 km). Below 72.5 km, the model is primarily based on reference (Labitzke, Barnett, & Edwards, 1985). Below 20 km, the data is padded

---

[16] considered as the standard for international space research, see e.g. (Hedin, 1991).

with averages from the US National Meteorological Center. And above 72.5 km, we supplemented data given in (Rozenberg, 1966), up to 600 Km.

Figure 7.4-1 shows the normalized atmospheric density distribution for 2013/3/14, 17.00 UTC, at 40 deg N latitude and 255 deg longitude. The optical thickness of Table 7.4-1 was derived by Pendorf (Penndorf, 1957) for a molecular atmosphere and is used herein in computing moonlight attenuation.

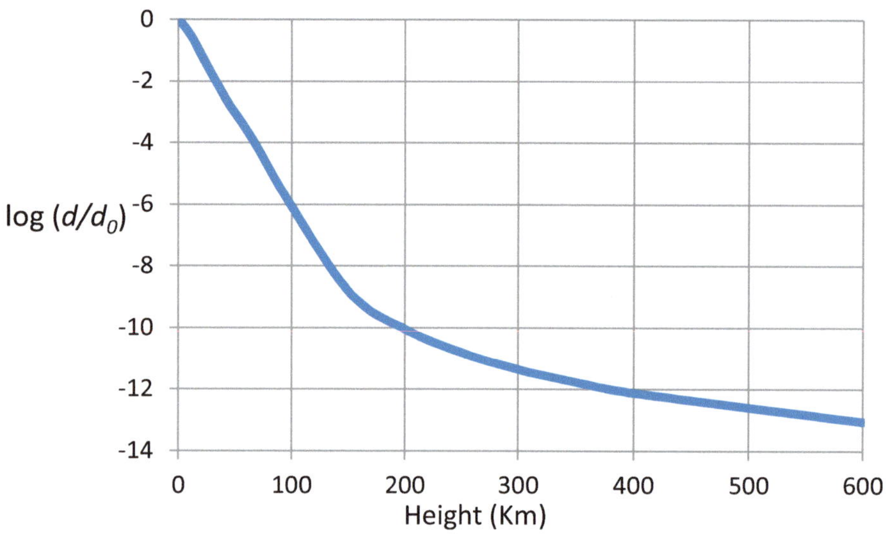

Figure 7.4-1 Normalized atmospheric density distribution

Table 7.4-1 Optical thickness (Penndorf, 1957 -see Rozenberg, 1966)

λ, μ	τ^{*mol}	λ, μ	τ^{*mol}	λ, μ	τ^{*mol}
0.40	0.3630	0.51	0.1335	0.61	0.06433
0.41	0.3277	0.52	0.1233	0.62	0.06022
0.42	0.2966	0.53	0.114	0.63	0.05643
0.43	0.2691	0.54	0.1057	0.64	0.05294
0.44	0.2447	0.55	0.09805	0.65	0.04971
0.45	0.2231	0.56	0.0911	0.66	0.04673
0.46	0.2038	0.57	0.08477	0.67	0.04397
0.47	0.1865	0.58	0.07897	0.68	0.04141
0.48	0.1711	0.59	0.07367	0.69	0.03903
0.49	0.1572	0.60	0.0688	0.70	0.03682
0.50	0.1447				

The background solar irradiance attenuation along PQ for a wavelength λ can be expressed using

$$I_{\odot}(\lambda)/I_{\odot}^{*}(\lambda) = e^{-\tau(\lambda,\Gamma_q)}, \quad 7.4\text{-}4$$

where $I_{\odot}(\lambda)$ is the solar irradiance at the end of the optical path and $I_{\odot}^{*}(\lambda)$ is the solar irradiance just outside the atmosphere. The term $\tau(\lambda,\Gamma_q)$ is the optical thickness along the path $\Gamma_q = PQ$. If this solar energy is scattered at point *P* towards the observer at point *O*, the perceived irradiance is then given by:

$$I_{\odot}(\lambda)/I_{\odot}^{*}(\lambda) = \varphi(\theta)\, e^{-\tau(\lambda,\Gamma_q)}\, e^{-\tau(\lambda,\Gamma s)}, \quad 7.4\text{-}5$$

where $\Gamma_s = PO$ and $\varphi(\theta)$ is the canonical scattering phase function for air molecules derived in (McCartney, 1976), adopted in (Patat, Ugolnikov, & Postyliakov, 2006), and given by

$$\varphi(\theta) = \frac{3}{16\pi}\left(1 + 0.9324\cos^2\theta\right), \quad 7.4\text{-}6$$

where θ is the scattering angle. Because the scattering angle is small during new moon sighting, about equal to the moon elevation at sunset *C* and less than 10 degrees, the phase function can be approximated by the constant $\varphi_C \simeq 0.115$. Figure 7.4-2 shows a plot of the phase function and Table 7.4-2 shows its numerical values for angles between 0 and 10 deg.

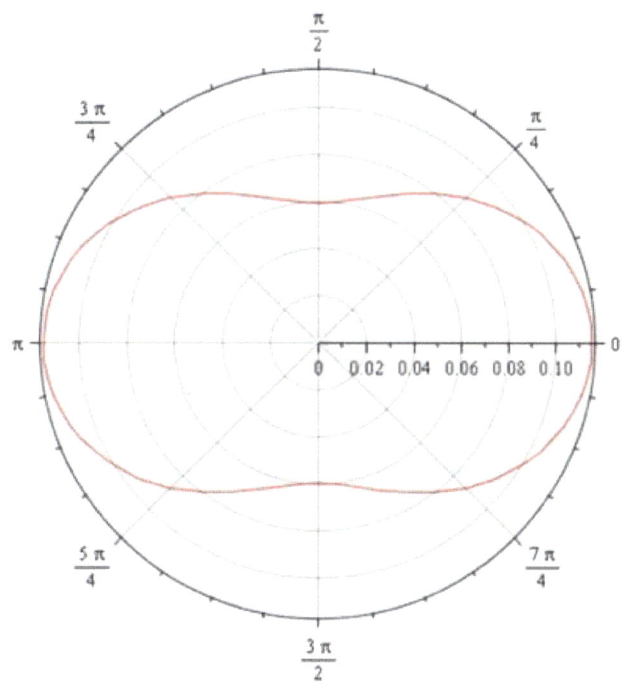

Figure 7.4-2 Scattering phase function for air molecules

Finally, to account for the variation of the attenuation among the parallel solar rays, we introduce $I_{\odot}^{b}$, the bulk solar illumination of the sky in the direction of the line of sight given by

$$I_{\odot}^{b} / I_{\odot}^{*} = \frac{1}{l_{P_0P^*}} \int_{\Gamma s^*} \varphi_C \, e^{-\tau(\lambda,\Gamma_q)} e^{-\tau(\lambda,\Gamma s)} ds \,, \qquad 7.4\text{-}7$$

where Γs^* is the path along P_0P^* and $l_{P_0P^*}$ is the distance between P_0 and P^*.

Table 7.4-2 Scattering phase function for air molecules for small angles

Scattering angle -deg	10	8	6	5	4	3	2	1	0
Scattering Phase function	0.1137	0.1143	0.1147	0.1149	0.1151	0.1152	0.1153	0.1153	0.1153

7.5 Contrast measure

Objects are perceived because they differ from their background, and contrast is the fundamental parameter for object detectability. While addressing the brightness of the moon through measurements, Krisciunas and Schaefer (Krisciunas & Schaefer, 1991) wrote "With the eye, video system, or CCD camera the detection of an object depends on the contrast of the brightness of the object in question versus the brightness of the sky. We can think of this as achieving a signal to noise ratio. Whatever brightens the sky degrades the limiting magnitude of the eye or video system by approximately the same amount." In the same subject, Clark wrote that visual astronomical observation depends on contrast discrimination, and not just on detecting faint light (Clark, 1990). McCartney (McCartney, 1976) also wrote, "We distinguish an object from its surroundings because of differences in luminance and chromaticity between various parts of the viewed field. Such differences are usefully expressed as *contrasts*." He further clarifies that for visibility along an atmospheric path, luminance contrasts are more important than chromaticity.

In Photometry, the *contrast* between an object of luminance L_o and a background of luminance L_b is measured using the ratio C, which is defined by

$$C = \frac{L_o - L_b}{L_b} = \frac{L_o}{L_b} - 1 \,. \qquad 7.5\text{-}1$$

The *contrast-threshold* is a measure of the smallest perceived difference in luminance between an object and its background, defined by

$$\varepsilon = \frac{(L + \Delta L) - L}{L} = \frac{\Delta L}{L} \,, \qquad 7.5\text{-}2$$

where the quantity $L+\Delta L$ is the luminosity of the object and L is the luminance of the background. We shall adopt the concept of contrast-threshold to the photometry of the new crescent moon visibility in Section 7.6.

Fechner, in 1859 investigated visual perception to stimulus. He noted that when the field luminance changes from L_1 to L_2, the change in sensation is proportional to $\text{Log } L_2 / L_1$. Let us denote by λ the *luminance ratio* L_o / L_b. We have $\lambda = C+1$, and

$$\text{Log}(\lambda) = \text{Log}\left(\frac{L_o}{L_b}\right) = \text{Log}(C+1) = C - \frac{C^2}{2} + \frac{C^3}{3} - \cdots \; . \qquad 7.5\text{-}3$$

The logarithm of the luminance ratio λ not only approximates the contrast measure C when $L_o - L_b \ll L_b$, at which first visibility of the crescent moon occurs but also provides the correct measure of the progression of contrast when variations of the radiation from the crescent moon and scatter of sunlight change substantially and or depart from linearity. There is another advantage in the adoption of this measure, which stems from the use of the logarithm. By treating all multiplication factors that do not depend on the varying parameters during new crescent observation as a bias, we do not have to determine several physical unknowns as it will become clear next when we use radiometric quantities.
Near horizon transmittance

7.6 Near-horizon transmittance

Transmission of exo-atmospheric radiation in the line of sight of a setting moon at a low elevation ν, over a frequency band $\Delta\lambda$, can be defined by the *Near-Horizon Mean Transparency*, also referred to *as the Near-Horizon Mean Transmittance*,

$$T_m(\nu) \simeq \frac{1}{\Delta\lambda} \int_{\Delta\lambda} e^{-\frac{\tau(\lambda)}{\nu}} d\lambda \, . \qquad 7.6\text{-}1$$

This expression provides a gross simplification of the optical transmittance of a medium over the visible frequency range and neglects absorption losses due to gases. Gas absorptions do not account for much of the transmittance drop in the visible spectrum, shown in Figure 7.1-4. This mean transparency shall provide us with a mean to characterize visibility in a bulk fashion over the visible band and independently of a particular wavelength.

Using the optical thickness of Table 7.4-1, the Near-Horizon Mean Transmittance in a molecular atmosphere is calculated for the spectrum between 400 and 700 nanometers. This range covers most of the visible spectrum, which is between 380 and 740 nanometers. For an atmosphere with aerosols, we use the evaluate the optical thickness using the procedure discussed in Section 7.4. The results are plotted in Figure 7.6-1 for low elevation angles of a setting moon below 12 degrees, for a molecular atmosphere condition , and for a typical atmosphere with aerosols. We will use the latter to evaluate crescent visibility in conjunction with field measurements of the background sky brightness in Section 7.10.

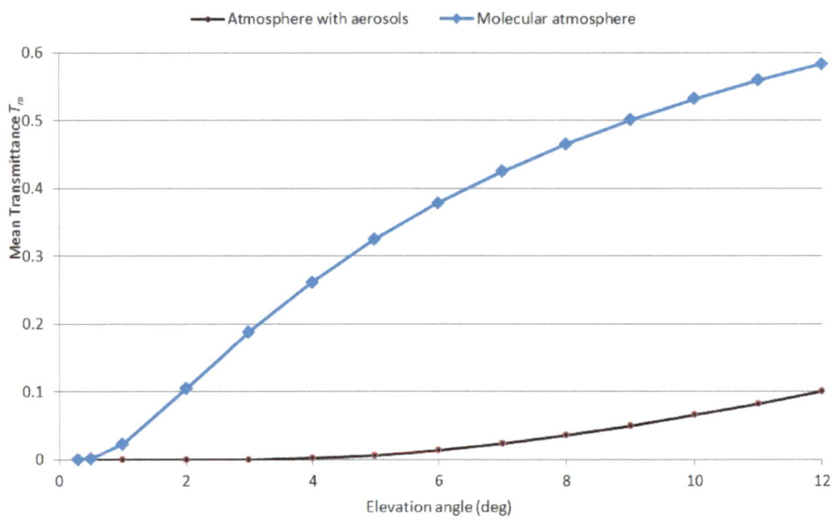

Figure 7.6-1 Mean Transmittance in the visible frequency band in a molecular atmosphere free of aerosols and in a typical an atmosphere with aerosols

7.7 Photometry of crescent visibility

We introduced the concept of the contrast threshold in Section 7.5. Separately, data from The Tiffany Foundation experiment (Blackwell, 1946) established measurements of the contrast threshold of the human eye under multiple observation conditions. Figure 7.7-1 shows the logarithmic luminance-contrast threshold plotted as a function of the background luminance for objects of various sizes.

The curve corresponding to an object size of 3.6 minutes of arc is chosen to derive the visibility threshold curve produced in Figure 7.7-2. Note that we would have used data with smaller object sizes if it was available. The threshold curve is generated for a sky background luminance having a moon elevation of 12 degrees at the time of sunset, shown in a solid line with markers. Also, the plot shows the moon luminance in a molecular atmosphere and in an atmosphere with aerosols, derived from data discussed in Section 7.1. The luminance of the crescent is shown at a phase angle of 12 degrees and corresponds to an exo-atmospheric luminance of 215 cd/m2 (6.75 mag /arcsec^2 from the data referenced in (Pilar Montanes-Rodriguez, 2007) and discussed in Section 6.3. An additional curve for a hypothetical luminous source crossing a clear atmosphere is displayed in intermittent lines.

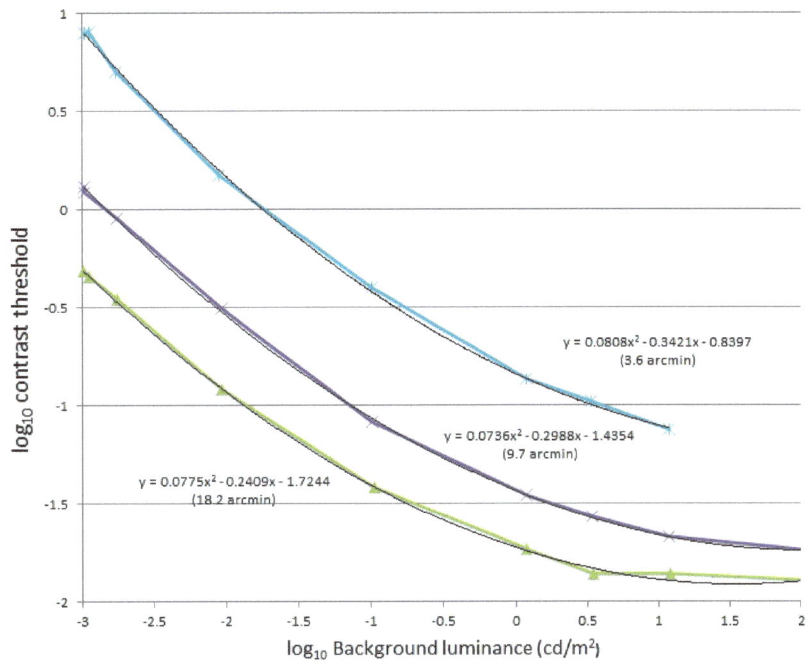

Figure 7.7-1 Luminance-contrast threshold for different circular object sizes

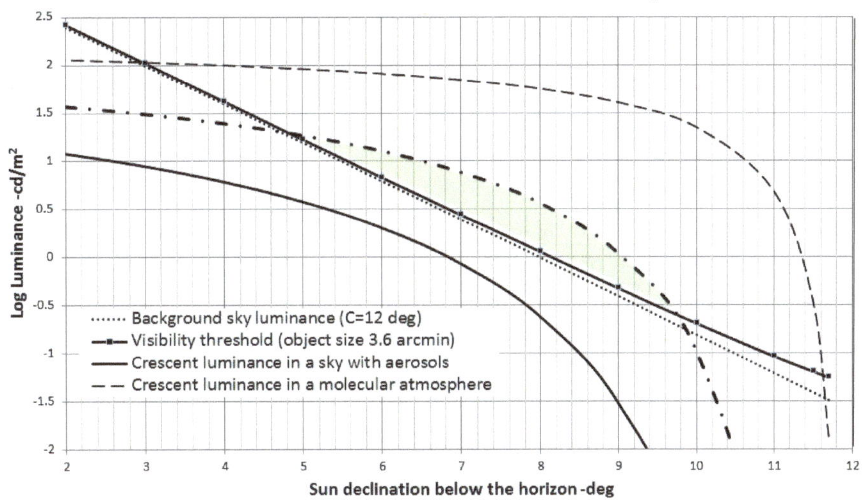

Figure 7.7-2 Visibility for crescent luminance for moon elevation of 12 deg at sunset assuming molecular atmosphere, a clear and an aerosol- filled atmosphere.

Any luminance curve above the threshold curve presents a condition for positive crescent sighting. See, for example, the highlighted region between the curve of crescent moon through a clear sky and the threshold line. Finally, it may be stated that photometrically, the visibility of the crescent moon depends weakly on the visibility threshold and depends mainly on (a) the elevation of the moon at sunset, (b) the corresponding zero air-mass crescent moon luminance and (c) the optical depth and the aerosol content in the atmosphere.

7.8 Radiometric measure of crescent visibility

Let us denote by C_t the *contrast* between a new crescent moon and its background sky can be measured using the logarithmic ratio of the apparent irradiance of the crescent moon, I, to the irradiance of the background sky in the direction of the line of sight which moves with a setting moon during twilight $I_{\odot}^{b}$. Making use of Equations 7.4-1 and 7.4-7, we can write

$$C_t = \log\left(I / I_{\odot}^{b}\right) = \log\left(I^{*} e^{\frac{\tau(\lambda)}{v}} / \frac{I_{\odot}^{*}}{I_{P_oP^{*}}} \int_{\Gamma s^{*}} \varphi_C \, e^{-\tau(\lambda,\Gamma_q)} e^{-\tau(\lambda,\Gamma s)} ds \right), \text{ or}$$

$$C_t = \log\left(I^{*} / I_{\odot}^{*}\right) + \log\left(e^{\frac{\tau(\lambda)}{(C-\rho)}} / \frac{\varphi_C}{I_{P_oP^{*}}} \int_{\Gamma s^{*}} e^{-\tau(\lambda,\Gamma_q)} e^{-\tau(\lambda,\Gamma s)} ds \right). \qquad 7.8\text{-}1$$

The first term in the last equality is constant over the short observation time and can be considered as a bias value to ignore, as it would shift the measure up or down. Our interest is in the relative contrast from one condition to another and not in absolute magnitudes, and the contrast is then given by

$$C_t = \text{Log}\left(e^{\frac{\tau(\lambda)}{(C-\rho)}} / \frac{\varphi_C}{I_{P_oP^{*}}} \int_{\Gamma s^{*}} e^{-\tau(\lambda,\Gamma_q)} e^{-\tau(\lambda,\Gamma s)} ds \right). \qquad 7.8\text{-}2$$

Note that we switched to the natural logarithm from its base-10 equivalent, as they differ by a multiplication constant, and because C_t is a measure, we can adopt either one. Next, we notice that we can drop the scattering constant φ_C because it also contributes to the total bias , and after a few simplifications, we express the contrast as

$$C_t = \text{Log}\, I_{P_oP^{*}} - \frac{\tau(\lambda)}{(C-\rho)} - \text{Log}\left(\int_{\Gamma s^{*}} e^{-\tau(\lambda,\Gamma_q)} e^{-\tau(\lambda,\Gamma s)} ds \right). \qquad 7.8\text{-}3$$

Or writing $I_{P_oP^{*}} = t^{*}$ for clarity and, we have

$$C_t = \text{Log} \frac{t^{*} e^{\frac{\tau(\lambda,\Gamma_t)}{(C-\rho)}}}{\int_{\Gamma s^{*}} e^{-\tau(\lambda,\Gamma_q)} e^{-\tau(\lambda,\Gamma s)} ds}. \qquad 7.8\text{-}4$$

Let us turn our attention to the extinctions along the lines PQ and QO of Figure 7.3-1. We have

$$\tau(\lambda,\Gamma_q)=C_e(\lambda)\int_0^{q^*}\frac{\rho h}{\rho_o}\,dq\,,\ \text{and}\ \tau(\lambda,\Gamma_s)=C_e(\lambda)\int_0^{t}\frac{\rho h}{\rho_o}\,ds\,.$$

Assume that the entire atmosphere is below 150 Km, which is a reasonable assumption, we may use only one branch of the bilinear density distribution as it can be immediately seen in Figure 7.4-1. Therefore, we may use the approximation $\rho/\rho_0=e^{-a\,h}$ in the expression of the contrast and obtain

$$C_t=\text{Log}\,\frac{t^*\ e^{-\frac{\tau(\lambda)}{(C-\rho)}}}{\int_{t_0}^{t^*}\int_0^{t}e^{-ah(s)}ds\,.\int_0^{t^*}e^{-ah(q)}dq\ \ dt}\,, \qquad 7.8\text{-}5$$

where a is a constant. The limits of the integrals and h are given by Equations 7.3-9. Note that the constant term $C_e^{\,2}(\lambda)$ has been dropped as it can be regarded as a bias term. The contrast is shown as a function of the sun declination for several values of the moon elevations at sunset in Figure 7.8-1 .

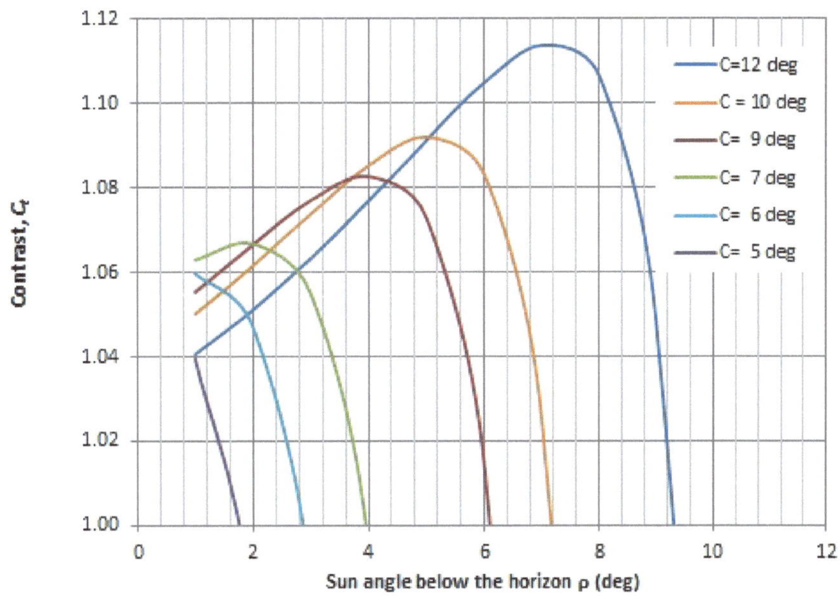

Figure 7.8-1 Contrast versus sun declination for various moon elevations at sunset, and $\tau^{*mol}(\lambda=0.5\mu)=0.1447$

For C, less or equal to about 6 degrees, there is an immediate decrease in contrast, and therefore it is impossible to have a positive sighting. The increase, in contrast, starts with curve C =7 degrees. Upon sunset, the background sky brightness gradually decreases, which increases the contrast. At the same time, the brightness of the setting new moon crescent decreases as the airmass increases. We have therefore a race: does the background sky brightness decrease fast enough to overcome the loss of the crescent brightness due to airmass increase, resulting in an increase of contrast? The elevation angle of the moon at sunset for which the contrast starts increasing is somewhere between 6 and 7 degrees as we will see later. Obviously, accounting for atmospheric refraction would lead to a higher value of *C* by about a fraction of a degree which corresponds to more immersion of radiation in earth's shadow where the layers of the atmosphere are denser.

For longer wavelengths, we see from Figure 7.8-2 that we have an increase in contrast at lower moon elevations at sunset, and the increase is extended another degree or so at higher angles. However, one would not expect that we have an improvement in contrast. The advantage in going to longer wavelengths is in the opportunity to have a positive sighting when *C* is in the lower range, which means the potential for sighting starts occurring earlier, and in the extended opportunity for sighting before the contrast dives down.

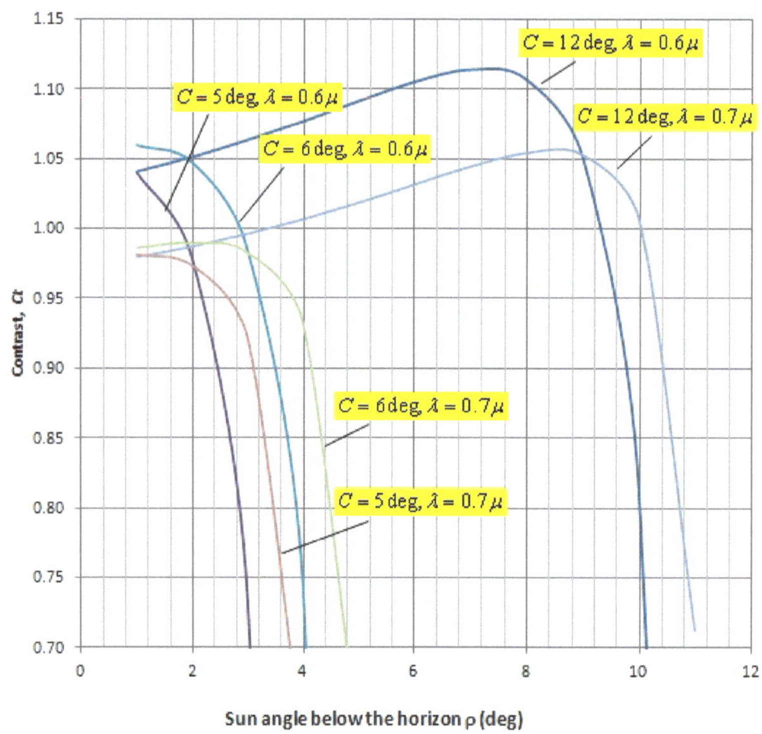

Figure 7.8-2 Contrast versus sun declination for various moon elevations at sunset for $\tau^{*mol}(\lambda = 0.6\mu) = 0.0688$ and $\tau^{*mol}(\lambda = 0.7\mu) = 0.03682$

The central criterion for crescent visibility is found to be a function of how large is the scatter angle C, which is the elevation of the moon at sunset. From experience, we learned that the higher the moon elevation at sunset, the longer the observation window is and the better the odds are to locate the new crescent in the horizon. The rise, in contrast, is about 5 % for larger *C* angles and drops at lower angles when considering longer wavelengths. The level of contrast needs field measurements and additional analytical work but will not be addressed further here. Finally, given perfect atmospheric conditions, free of aerosols and moisture, when is the earliest time positive sighting occurs by a common observer? Because we seek a lower bound for first visibility, the effect of aerosols on transmittance was eliminated in the development of the contrast discussed above. This may be an ideal condition that can be achieved only in a remote dry condition.

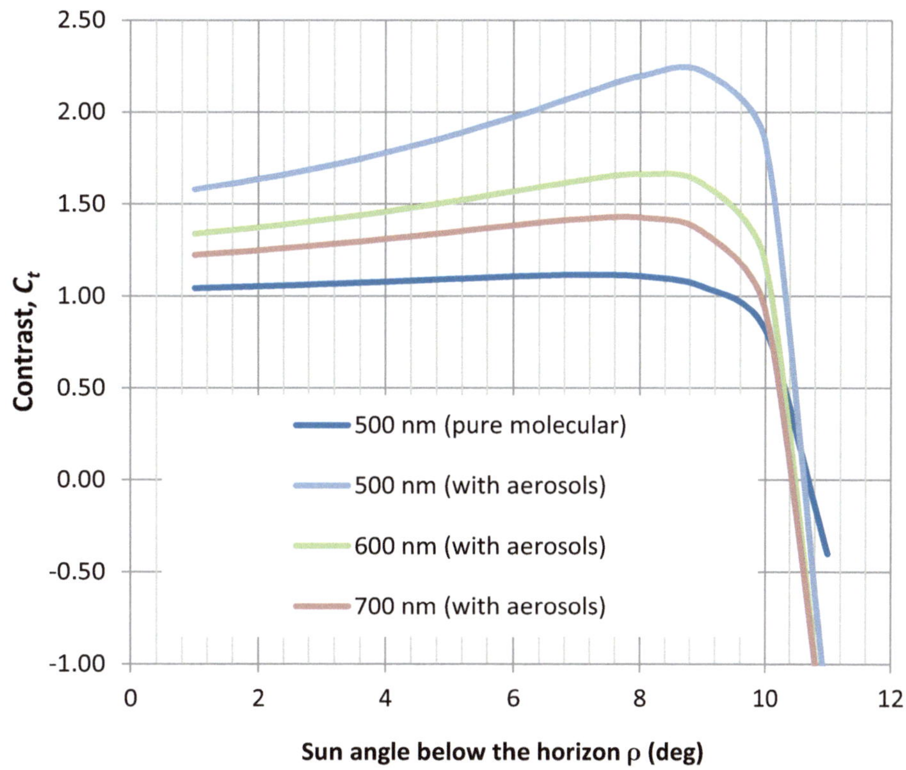

Figure 7.8-3 Contrast versus sun declination for moon elevation of 12 degrees s at sunset

The atmosphere comes with aerosols which further attenuate radiation as illustrated in Figure 7.1-2. Using data from the NASA SeaWiFS program, the absorption coefficient due to aerosols is added to that of molecular atmosphere to produce the contrast for a case of moon elevation at sunset of 12 deg. The results of the simulation are shown in Figure 7.8-3 for 500, 600, and

700 nm. For comparison, data for pure molecular attenuation at 500 nm is also shown in the same plot. It may be concluded that similar results to those obtained for a molecular atmosphere are obtained when considering the effect of aerosols. In addition, a larger rate of change is obtained in contrast. Note that the contrast magnitude of the pure molecular atmosphere case cannot be compared directly to that of an atmosphere with aerosols because of the bias elimination.

7.9 Sighting in the visible spectrum

The environment in which traditional telescopes are exploited is a night sky when the human eye is well adapted to darkness. Telescope research and development has therefore focused on a user who is well accommodated to darkness, and not concerned with the limitation of the human eye in detecting light out of a background of sunlight scatter.

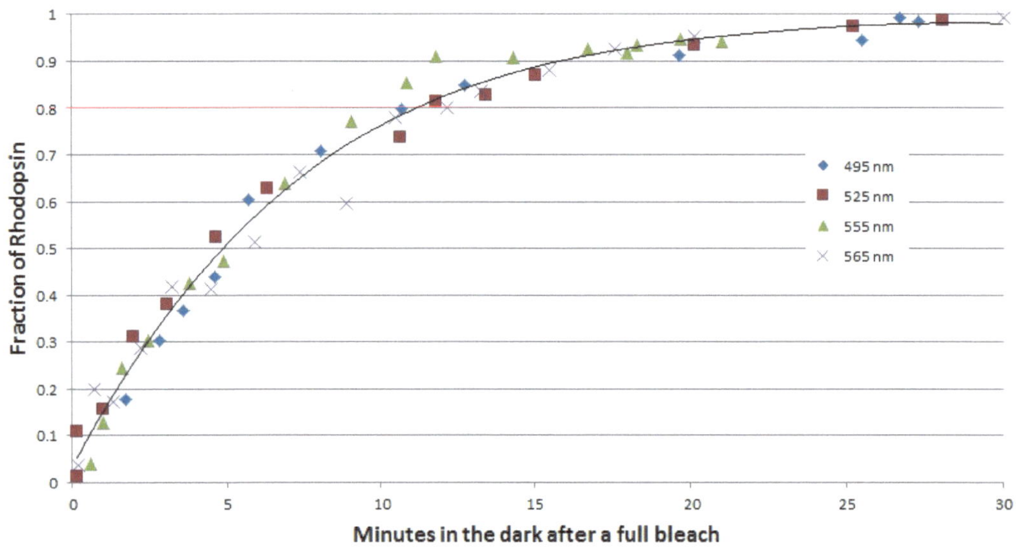

Figure 7.9-1 Rhodopsin kinetics in the human eye **(Alpern, 1971)**

New crescent moonlight detection occurs during twilight, a time of transition between daylight and nighttime in which the visual acuity of the human eye can vary by a factor of 1000. This property makes it a specialized sensor in which night vision is nearly three orders of magnitude more sensitive than daylight-adapted vision. However, the transition between the two modes of detection does not happen instantaneously. Studies have shown that it takes the eye nearly thirty minutes to fully transition from daylight to nighttime viewing. Figure 7.9-1 illustrates the ability of the eye to adapt to dark viewing conditions. Rhodopsin is a necessary chemical for the retina night vision. It is depleted by bright light and is restored at the rate shown in the graph when the eye is exposed to darkness. The response of the eye is an important factor unique to sighting the new crescent. Even though thehe observation time of the new crescent is short and takes place during twilight, the slow response of the eye to adapt could not cause the observer to miss sighting. While this matter needs to be investigated, experience shows that the new moon is bright enough and adaptation to darkness is not critical to sighting.

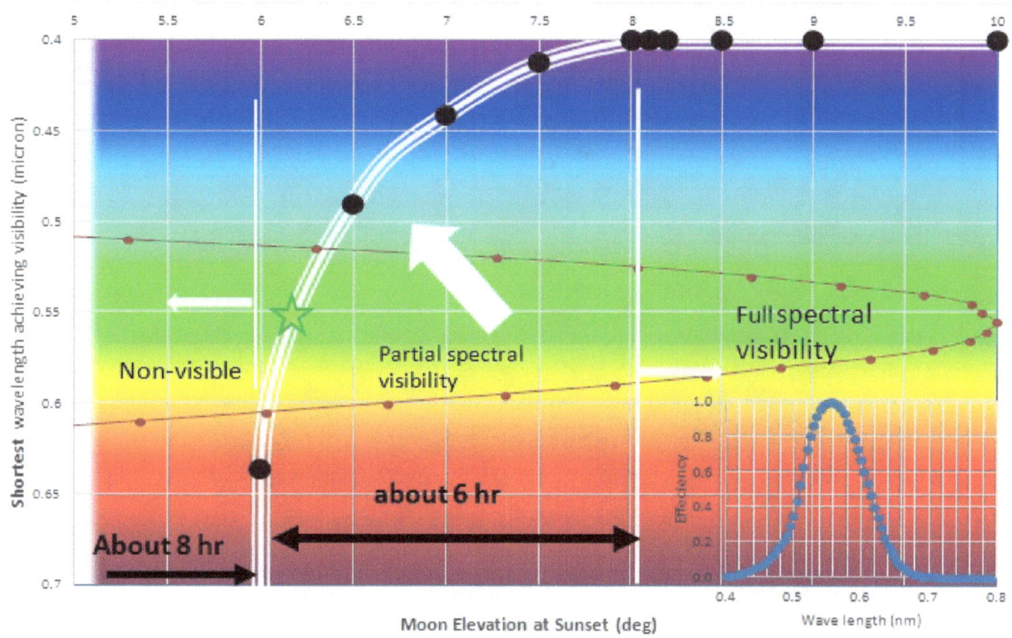

Figure 7.9-2 Optimal sighting occurs when C=6.35 deg at a wavelength of 558 nm. Also, shown is the approximate time of 8 hours for first visibility, at the longest wavelength of the visible spectrum, following conjunction of the lunation of June 27, 2014. Detection over full-spectrum occurred 14 hrs following conjunction.

While adaptation to darkness may not be relevant to new moon sighting, the efficiency of the human eye as a function of wavelength is important. We established earlier that contrast increases first at longer wavelengths. But then the human eye is better at detecting certain wavelengths than others. The luminosity function or luminous efficiency function, established by the *Commission Internationale de l'eclairage,* describes the average spectral sensitivity of the human perception of brightness. The function is shown in the lower right corner of Figure 7.9-2. The Figure displays the shortest wavelength to achieve sighting using the contrast expression of Section 7.8 for different values of moon elevation at sunset. Also, highlighted is the ***mean wavelength*** at which first visibility occurs is **558 nm. It happens that this wavelength corresponds to a scatter angle or moon height at sunset of C=6.35 deg**. For the new moon of June 27, 2014 (conjunction 8:08 UT), first visibility occurred about 8 hrs following conjunction. Full-spectrum visibility occurred 14 hrs following conjunction. Again, it must be reminded that these values are for a molecular atmosphere, free of aerosols. We shall adopt the value of $C=Ccr=6.35$ as our criteria for first visibility as we move forward in this text.

7.10 Background sky measurement

To calibrate the contrast measure and verify the optical model, which is based on the relevant and early twilight sky brightness being dominated by single scattering, field measurements were needed. The brightness of the moon and atmospheric transmittance were heavily investigated for satellite calibration, and a large body of data is available from NASA and from the USGS. However, data on dynamic twilight sky brightness in a line of sight locked on a setting moon is not directly available. To perform such measurements, a special-purpose instrument was developed.

Figure 7.10-1 Background sky measurement setup includes a celestial tracking system with a light meter, a leveling tool, and a compass. The system shown is aligned with earth axis (at 40 deg inclination in Boulder, Colorado.

The instrument, which consists of a light sensor and optics, measures the twilight sky brightness in magnitudes per square arcsecond (mpsas) at regular intervals. The sensor is mounted on an accurate uniaxial tracking system which moves with the setting sky. The system has a microcontroller which communicates with a PC through a USB port. Readings from 2 mpsas to 26 mpsas are achievable. Figure 7.10-1 depicts the setup, oriented towards the western twilight horizon in the outskirts of Boulder, Colorado. Figure 7.9-2 shows the progress of the horizon brightness during Nautical twilight in a clear summer evening, and the data presented in Figure 7.9-3 shows the background sky brightness in the two clear summer nights of August 18 and 21, 2014, just about a mile north of the 40th parallel in Boulder, Colorado.

The horizon sky was monitored at the speed of earth rotation to track the background sky. The background sky brightness is shown to be dominated by a linear drop in surface magnitude during nautical twilight, which is attributed to single scattering events. The brightness becomes subsequently nonlinear during astronomical twilight when the sun dips more than about 12 degrees below the horizon. Therefore, during the time of new moon sighting, which corresponds to a sun depression of 6 to 10 degrees, the background sky brightness is linear. In addition, the drop in the background sky brightness is about 1 mag/arcsec^2/deg of sun depression. Extrapolating to the onset of sunset, the background sky has an intrinsic brightness of about 4.6 mag/arcsec^2.

Figure 7.10-2 Ttwilight sky in the outskirts of Boulder, Colorado, (Aug. 18, 2014)

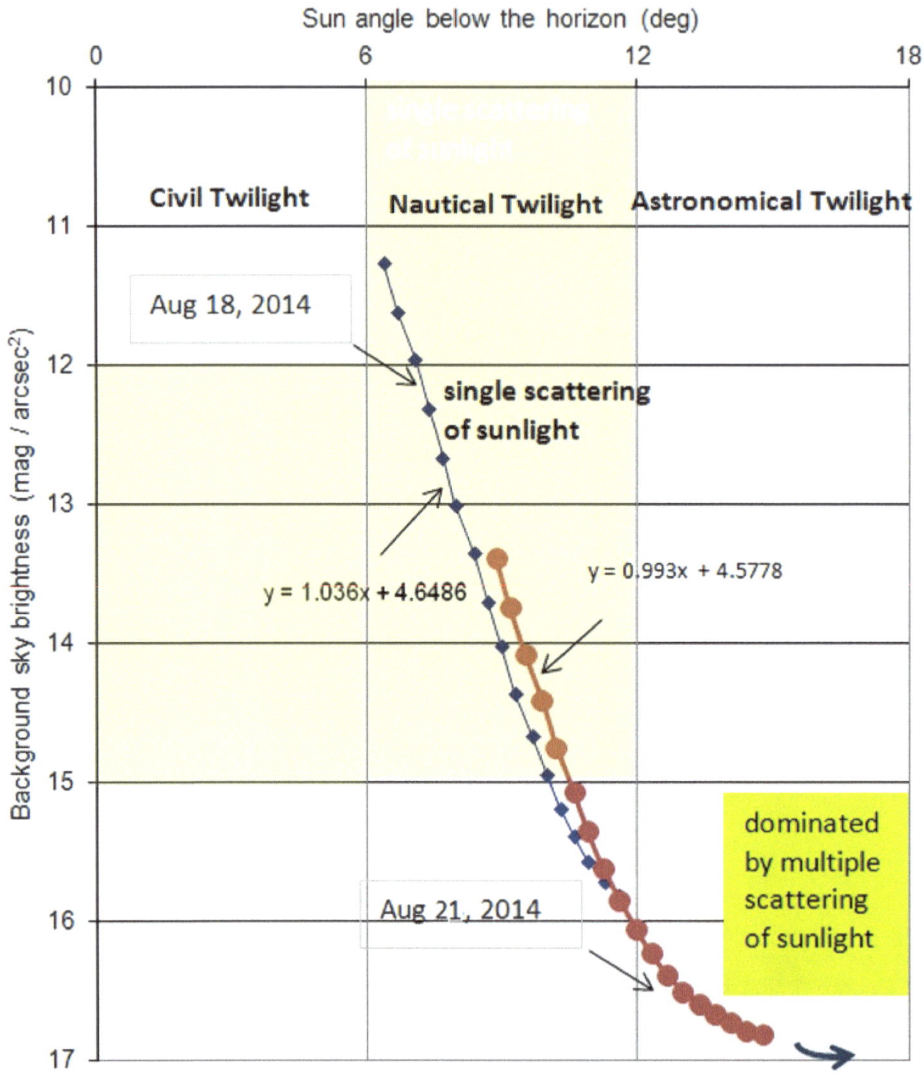

Figure 7.10-3 Background sky brightness measured while tracking a twilight summer sky in the outskirts of Boulder, Colorado (measurements about 15 and 7 degrees southwest of the position of the sun in two separate summer evenings)

The twilight background sky magnitude is about 13 and 15 mag/arc-sec^2, at 15 and 30 minutes after sunset, respectively, e.g. (Clark, 1990). This compares well with the data obtained from current measurements and shown in Figure 7.10-3.

For reference, Table 7.10-1 shows published approximate values of horizon sky luminance under various conditions.

Table 7.10-1 Approximate values of horizon-sky luminance under various conditions -Data is adapted from (Middleton, 1958)

	Approx. luminance	
Sky condition	**mag/arc-sec^2**	**cd/m^2**
Clear sky	3	10000
Overcast day	5	1000
Heavily overcast day	8	100
Sunset, overcast day	10	10
1/4 hr after sunset, clear	13	1
1/2 hr after sunset, clear	15	0.1
Fairly bright moonlight	18	0.01
Moonless, clear night sky	20	0.001
Moonless, overcats night sky	23	0.0001

As discussed earlier, starting from a certain value of moon elevation at sunset, the contrast between the new moonlight and the background sky starts increasing. Such condition is a necessary one but not sufficient for positive sighting. It does not give the critical contrast, but as a minimum, we need to have $C \geq C_m$, where C_m is minimum for contrast to start increasing, and therefore a chance for visibility. The contrast can be obtained using the approximation

$$C_t = \text{Log}\,\frac{I * e^{-\frac{\tau}{C-\rho}}}{I_b}. \quad 7.10\text{-}1$$

Using the measured linear relationship between the background sky brightness, m, given in surface magnitude and the sun declination ρ , given by

$$m = \alpha\rho + m_o,$$

where the measured slope $\alpha \approx 1$ mag / arcsec^2 / deg, and $m_0 \approx 4.6$ mag/arcsec^2, and adopting the conversion expression to candela units,

$$I_b = 108000 \times 10^{-0.4m} = 108000 \times 10^{-0.4(\alpha\rho + m_o)}, \quad 7.10\text{-}2$$

Note that the numerator is written in terms of the near-horizon mean transmittance presented in Section 7.6 and the data shown in Figure 7.6-1. The contrast is plotted for different elevations of the moon at sunset as a function of the sun declination. The contrast increases at 6.35 deg. Below this value the contrast decreases immediately from sunset. This result is similar to the one obtained analytically in the case of a molecular atmosphere.

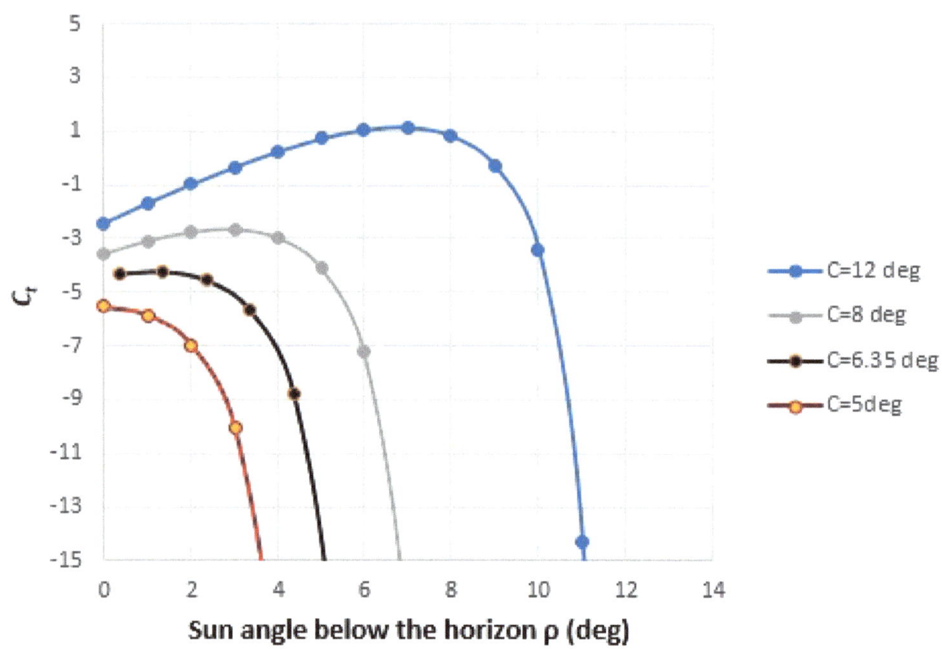

Figure 7.10-4 Contrast from measured background sky brightness versus solar depression angle ρ shown for the visible band and a typical atmosphere with aerosols . The contrast starts increasing at about 6.35 deg.

Chapter 8 Elements of astronomy

8.1 Introduction

Crescent visibility needs to be interpreted from an astronomical perspective, which will be the topic of Chapter 9. But first, relevant results from elementary spherical astronomy need to be covered so that we set the stage for such interpretation. Readers who are familiar with the fundamentals of astronomy should at least glance through the sections of this chapter to get some familiarity with the notation that we will be using as we move forward.

The proximity of the moon to the sun near conjunction permits the use of approximation theory in developing when and where new moon visibility occurs. The position of the moon is approximated using multivariate expansions, also known as Taylor series approximation, around the position of the sun. The following section discusses the mathematical topic of series expansion, which will be heavily relied upon.

The exploitation of expansion and linearization takes place in the transformation between the different coordinate systems, where either one or more coordinates are either nil or do not vary substantially. We shall introduce the coordinate systems which are commonly used in classical planetary astronomy. These are the *Horizon coordinate system*, the *Equatorial coordinate system,* and the *Ecliptic coordinate system*. Coordinate system transformations are used to move or relate information from one system to another. We will make use of such transformations because some concepts are either defined or easily described in one coordinate system, but the needed results are sought in a different coordinate system. For example, the setting of celestial objects is defined in the Horizon coordinate system. However, such information is needed in the equatorial coordinate system when expressing the moon-sun relative setting while deriving crescent first and limit visibility.

We shall also cover fundamentals related to angular separation between celestial bodies. This is a frame-invariant key parameter in first-crescent visibility. Its accurate computation is best carried using approximations due to the small angular separation near conjunction. Finally, the chapter covers the mathematically defined circumstance of moon conjunction, which depends on orbital information and does not relate directly to lunar photometry. The conjunction is, however, the origin of the lunar cycle and bounds the coordinates of the photometric cycle.

The chapter ends with the discussion of the position of the moon at sunset during conjunction and the celestial objects that never set or rise. These two topics provide limiting parameters for new moon visibility.

8.2 Taylor series expansions

A Taylor series expansion of a continuous real function f of a single variable *x,* near a point x=*a*, can be written as an infinite summation of the form:

$$f(x)=f(a)+f'(a)(x-a)+\frac{f''(a)}{2!}(x-a)^2+\frac{f^{(3)}(a)}{3!}(x-a)^3+\ldots+\frac{f^{(n)}(a)}{n!}(x-a)^n+\ldots.\ ,$$

where $f^{(n)}(a)$ is the *nth* derivative of the function at point *a,* which is assumed to exist. One can immediately see that if the function f is a simple straight line, all the higher derivative terms are zero and the expansion is nothing but an expression for the slope of the line. The same concept is extended to functions of more than one variable. For example, a function of two variables $x=(x_1,x_2)$ can be expanded about a point $a=(a_1,a_2)$ as following:

$$f(x_1,x_2)=f(a_1,a_2)+\frac{\partial f}{\partial x_1}(a_1,a_2)(x_1-a_1)+\frac{\partial f}{\partial x_2}(a_1,a_2)(x_2-a_2)+\frac{\partial^2 f}{\partial^2 x_1}(a_1,a_2)\frac{(x_1-a_1)^2}{2!}+$$
$$\frac{\partial^2 f}{\partial^2 x_2}(a_1,a_2)\frac{(x_2-a_2)^2}{2!}+\frac{\partial^2 f}{\partial x_1\partial x_2}(a_1,a_2)\frac{(x_1-a_1)(x_2-a_2)}{2!}+\ldots,$$

and similar expressions can be derived for functions of three, four variables and so on. It is customary to retain several terms up to and including a certain degree of the polynomial. This is called the "truncation order" of the series beyond which the terms are dropped and are assumed to be negligible. Given the fact that we are developing the function about some point *a*, expanding a single variable function up to a *first order* can be written as

$$f(x)=f(a)+f'(a)(x-a)+O(x^2)\,.$$

The term $O(x^2)$ indicates that the error is of second error if the series is approximated linearly and the function f is then assumed to be

$$f(x)\simeq f(a)+f'(a)(x-a).$$

In later developments, we shall expand to a first order whenever the behavior is known to be almost linear over the interval of interest. For example, the path of a setting celestial object is linear over a good interval of time and limiting our expressions to such form will provide much-needed simplifications. We will keep track of how much error we introduce and ensure that we are getting adequate resolution from our approximations.

8.3 Coordinate systems

Multiple coordinate systems are often used in the field of astronomy. There are three of them that we shall adopt throughout this work. In the case of the new moon, we shall use the Horizon coordinate system to obtain an observer's perspective. The Equatorial coordinate system is used to relate to non-local quantities for fundamental expressions that should be valid everywhere on the surface of the globe and any time. And finally, the Ecliptic coordinate system which is used to address the origin of the lunar cycle, which is the onset of conjunction. We shall use a common notation in pertinent literature. The reader is referred to a complete, yet concise and simple, description of these coordinate systems in (Duffet-Smith, 1996). Advanced technical treatment of the subject can be found in (GREEN, 1985). We shall also present a summary of the conversion equations between the different coordinates systems.

8.3.1 The celestial sphere and cardinal points

To reference objects in the sky, astronomers devised a fixed imaginary sphere, centered at the observer with an infinite radius and over which surface all celestial bodies are projected. Even though Even though such a sphere does not consider how far the celestial bodies are, it is very useful in setting up coordinate systems to track motions and explain some fundamental concepts of visibility such as what we have in the case of the moon. Because the positions of celestial bodies vary when the observer is situated at various locations on earth and also vary under the influence of cyclic changes and perturbations, standard observer positions at earth center and at the center of the sun are also considered. Reference is then made to *geocentric* and *heliocentric* celestial spheres respectively. When the observer is situated on the earth's surface or at certain altitudes, we refer to the sphere as *topocentric*. Similar nomenclature is used for associated coordinate systems discussed in the following sections.

Consider the celestial sphere shown in Figure 8.3-1. The observer placed at point *O* is situated at an arbitrary location on earth surface with the *Zenith* direction intersecting the sphere at point *Z* making gravity oriented in the direction *ZO.* The opposite point from Z on the celestial sphere is called the *Nadir.* The earth's axis of rotation is parallel to the axis going through point O intersects the celestial sphere at the two poles *P and Q.* The plane perpendicular to *ZO* is called the *horizon plane* or simply *horizon.* By extending the arc *ZP,* we intersect the horizon plane at point *N,* which is closest to the axis of rotation. This is the North point of the horizon plane. The point diametrically opposite to N from point *O* is *S* or the South point of the horizon. And the points at 90 degrees from the North and the South are *W* and *E* (the latter is omitted from the figure for simplicity). With the observer facing north, *W* is chosen such that it is on the left of the observer. *N, W, S,* and *E* are called the *cardinal points* of the horizon. The diurnal rotations of earth from West to East make celestial bodies which are closest to the axis of rotation in the visible hemisphere to the observer remain always visible. In the northern hemisphere, the North Star Polaris, historically used for guidance, is about one degree off the axis of earth's rotation and is always in the line of sight even during the day when it is not visible. The southern hemisphere does not possess a similar star.

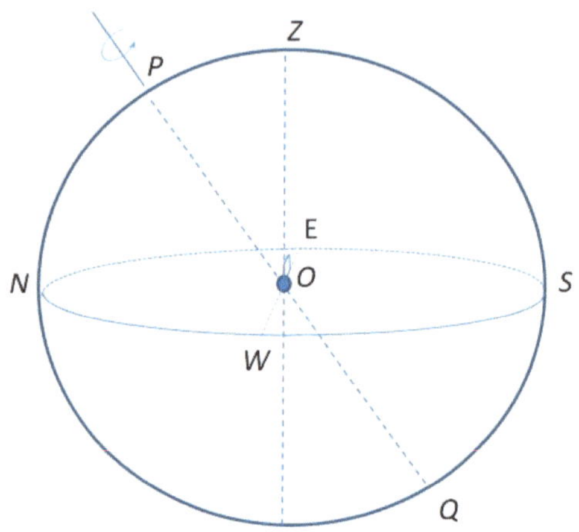

Figure 8.3-1 Celestial Sphere with an observer at center point *O*

We have just established the proper definition for what constitutes reference directions for an observer. Multiple definitions can be found in different fields for what constitutes North, South, East, and West. These definitions are either inaccurate or of little use for what we are set to do. One definition that used to be part of Google's dictionary now incorporated with the online translator states that the word West is a noun which means "the direction toward the point of the horizon where the sun sets at the equinoxes", while the word East is defined as "the direction toward the point of the horizon where the sun rises at the equinoxes". Furthermore, the North is defined as "the direction in which a compass needle normally points, toward the horizon on the left side of a person facing east, or the part of the horizon lying in this direction."

8.3.2 Georeferencing System

Geographic Information Systems (GIS) is a framework for gathering, managing, and processing data related to our surrounding. This discipline, rooted in the field of geography, has evolved into a science over the past few decades. It connects multiple information sets to provide useful data for decision making. Principles, techniques, and practices can be found in multiple references, e.g., (Paul A. Lomgley, 2001) .

Georeferencing, a term commonly used in GIS, is the process of associating data sets with physical space. For people to communicate and plan across geographic locations and time, there is a need for standards to be uniformly used by everyone. As discussed earlier, the use of a uniform calendar was a major step forward for mankind to interact in a global socio-economic world. Georeferencing also needed standardization for the people of the world to interact across boundaries as well. Separate georeferencing systems are hard to use and inevitably lead to errors. For example, the position of an airplane handed over to successive control towers, require the use of standards so that paths are chartered across boundaries and over international land, water, and space. In our application, computer-generated information of solar and lunar photometry, subsolar and sublunar locations, visibility parameters, historical cloud fraction data, western horizon elevation visibility angles are stacked over global geographic maps to produce information on new moon visibility at any time. The georeferencing system of longitude and latitude is universally used in multiple disciplines, including the field of astronomy. This system provides fine resolution of position knowledge and accurate distance measurement between arbitrary locations on the earth's surface. It is often called the geographic system of coordinates. Each point on the earth surface first-crescent by two coordinates: the longitude and the latitude which are defined as follows.

First, the earth center of mass is located on its axis of rotation. The plane perpendicular to the axis of rotation and going through its center of mass is called the Equatorial plane. If we consider the wedge between two planes, each defined by earth's axis of rotation and a radial line emanating from the center of mass on the equatorial plane, each half-plane of such wedge constitutes a constant longitude, called meridian. The zero-longitude is taken as a line drawn on the ground of the Royal Observatory in Greenwich, England, with increasing positive angles going in the direction of the East to +180 degrees and to -180 degrees to the West for a total of 360 degrees around the earth. The symbol used for the longitude is commonly denoted as λ (lambda).

The shape of the earth is not a perfect sphere. It is best described by an oblate spheroid (ellipsoid) with a reduced diameter along its axis of rotation. Such flattening is measured by the parameter $f=(a-b)/a$, where a and b are the radii along the major and minor axes, respectively. The flattening is about one part in 300. Several proposals were put forward to standardize the shape of the earth over the past two centuries. This was initially important for mapping, but when more and more precision was needed for ballistic missiles, a continuous model refinement has resulted over the years until greater precision was finally provided by satellites.

WGS84 (The World Geodetic System of 1984, with the latest revision in 2004 based on earth gravitational model of 1996) is a widely accepted standard that provides a model of the earth's size, shape, gravity, and geomagnetic fields. Part of the model is an earth-fixed terrestrial reference system and geodetic datum. The coordinates origin is at the center of mass position which is known today to about 0.02 m. The reference meridian is 5.31 arc seconds or 102.5 m East of the original George Biddell Airy's 1851 meridian at the Royal Observatory, Greenwich. It is known as the International Reference Meridian, or the prime meridian, i.e., 0° longitude.

The ellipsoid WGS84 describes the sea level surface and was derived using spherical harmonics. It is defined by its semi-major axis of a = 6378137 m and its flattening number given by f = 1/298.257223563.

The ellipsoid permits the accurate development of the needed second coordinate, the latitude, usually denoted by the symbol φ (phi). It is the angle between the perpendicular to the tangent to the ellipsoid at a geographic location and the Equator. For the northern hemisphere, this angle varies between 0 and + 90 degrees and between 0 and -90 degrees in the southern hemisphere as illustrated in Figure 8.3-2. A line of constant latitude is called a parallel.

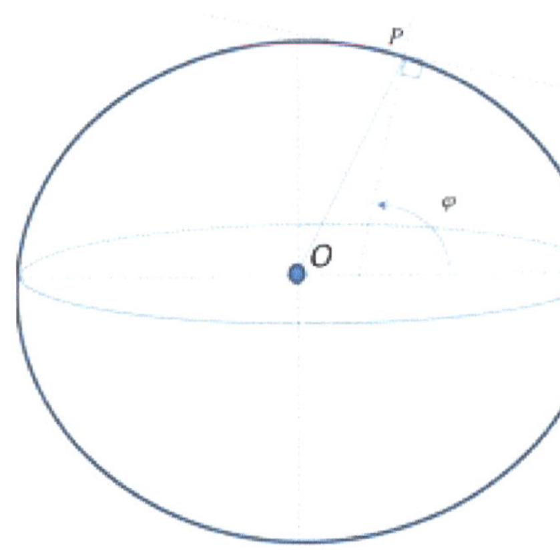

Figure 8.3-2. Definition of the latitude at point P as the angle between the perpendicular to the ellipsoid and the Equator

There is a problem with precise and accurate determination of a position on the earth surface. earth itself is constantly changing shape. earth's crust motion needed to be accounted for. *The International Earth Rotation and Reference Systems Service* was created in 1988 to address this issue. Two dynamic reference frames were established and maintained: a Celestial Reference Frame, tied to celestial objects, and a Terrestrial Reference Frame. Earth Orientation Parameters which connect these two frames together are supplied on a regular basis. These frames provide a common reference to compare observations from various locations on the Globe. The system and data collection have evolved over the past two decades, and today, several geodetic methods are used to compute accurate coordinates.

These are the Global Positioning System (GPS), the Very Long Baseline Interferometry (VLBI), Satellite Laser Ranging (SLR), and Doppler Orbitography and Radiopositioning Integrated by Satellite (DORIS). This service is maintained by *The Institut Géographique National,* based in France.

8.3.3 Horizon coordinates

The *Horizon coordinate system* is also known as the *alt-azimuth* coordinate frame. The Horizon coordinates of an object in the sky are given by two variables: the **azimuth** A and **altitude** a, as illustrated in Figure 8.3-3. Let us denote by *N, E, S,* and *W* the North, East, South, and West respective directions. The North is defined as the direction of the North Pole. An observer placed at the intersection point of segments *NS* and *EW*, looking at the horizon in a sweeping view would position an arbitrary object in the sky by how far its projection is on the horizon plane *NESW* from the North via the azimuth angle *A.* This angle takes origin at the North *N* and increases clockwise from the North in the direction ESW. The azimuth angle *A* is given in degrees and varies between 0 and 360 deg. The second coordinate is given by altitude of the object, which takes origin at the horizon plane and measures the angular height of the object in the sky in degrees. The altitude reaches 90° at the zenith *Z* of the observer and is negative when the object is below the horizon. At any time, an object in the sky is fully characterized by the coordinates A and a.

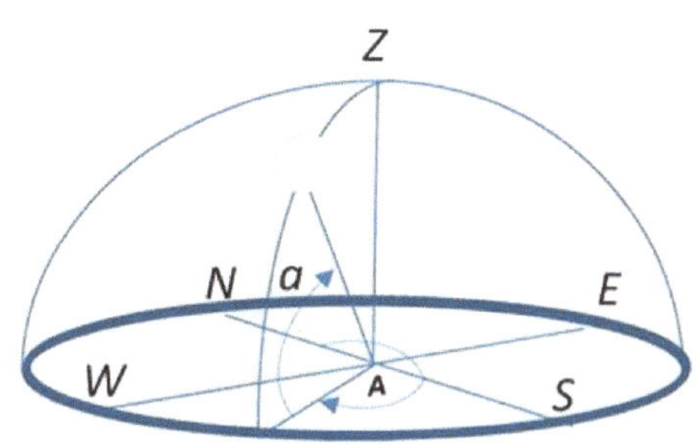

Figure 8.3-3 The Horizon Coordinate System

Many telescope mounts, called ***Alt-Az*** mounts, use the Horizon coordinate system to orient their optical system. They possess two degrees of freedom than run independently from one another. One degree of freedom runs in the sweeping Horizon plane rotating along the Azimuthal direction, and one degree of freedom runs in the elevation along the altitude direction. Some mounts are equipped with drive mechanisms that run automatically and independently in each degree of freedom to track the motion of celestial objects. They are equipped with special software that drives the attached optical system to the direction of the celestial object at a certain time and given the geographic location of the observer. To do this, the mount is first leveled by means of a bubble and oriented towards a couple of known stars. Some advanced models are equipped with GPS and a digital compass for automatic self-alignment, which eliminates the tedious requirement for manual alignment.

8.3.4 Equatorial coordinates

The Horizon coordinate system of the previous Section is defined with reference to the Horizon Plane *NESW*. The Equatorial coordinate system takes reference in the plane of earth's equator instead, as shown in Figure 8.3-4. The observer is again at point *O*. The equatorial plane *EϒRW* is used as a reference and is oriented 90°-ϕ with respect to the Horizon Plane *NESW,* where ϕ is the observer's geographic latitude. The line *OP* is the earth's rotation axis. Figure 8.3-4 shows the two coordinates used in the Equatorial coordinate system: the ***declination δ***, measured in degrees and is positive north of the equator, and the ***right ascension α***, is positive east of the fixed reference direction of the ***vernal equinox ϒ.***

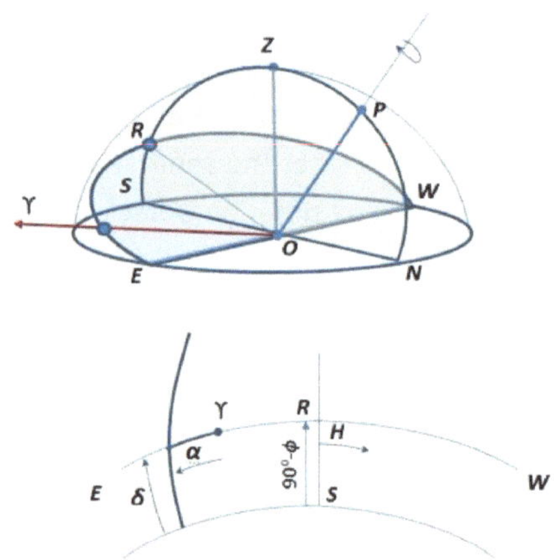

Figure 8.3-4. The equatorial coordinate system with ERW representing earth's equator

Another measure of how far an object in the sky has traveled in the equatorial plane is the Hour Angle, ***H***. This measure takes reference from the great circle NOSRZP (the meridian) where objects with declinations less than the geographic latitude are said to culminate, i.e., right ascension is maximum, and its azimuth is 180°. Both ***α***, and ***H*** are measured in degrees or in Hours with 24hr corresponds to 360°.

The hour angle *H* and *declination δ* form a variant of the equatorial coordinate system and is often referred to as the ***Hour angle coordinate system***. The hour angle and the right ascension coordinates are related through the ***sidereal time:*** *the hour angle of the equinox*. It should be noted that this is a time measure that refers to the stars instead of the sun as is the case of the civil timekeeping. To relate one observer to another, we differentiate between ***Local***

sidereal time (LST) for an observer located at some longitude ***L*** and the ***Greenwich sidereal time (GST)*** for a reference observer located at the ***Greenwich meridian.*** It immediately follows that

$$LST = GST + L = H + \alpha \qquad \textbf{8.3-1}$$

for an arbitrary celestial object. From the above definitions, the sidereal time has a period called a sidereal day, which is 24hr: 00 m every time earth makes a full turn. This differs from a solar day which lasts about 23hr: 56 m of solar time. This perfect time is due to the reference being fixed against the background sky versus the solar time, which refers to the sun which position varies with time.

Special mounts in use in astronomy track the rotating night sky using the German Equatorial mounts. These mounts are aligned to rotate about an axis parallel to earth's axis of rotation, in a plane parallel to the Equator. In this fashion, they can fix a celestial object in the field of view and track it over time to accumulating its dim light under long exposure.

8.3.5 Ecliptic coordinates

Planets of the solar system move in a near coplanar fashion around the sun. The plane in which earth, as well as these other planets, orbits the sun is called the Ecliptic plane. This coordinate system uses also the Vernal Equinox which intersects the traces of the ecliptic and equatorial planes which are inclined with respect to one another by the obliquity angle ε , which is equal to about 23°.26'.

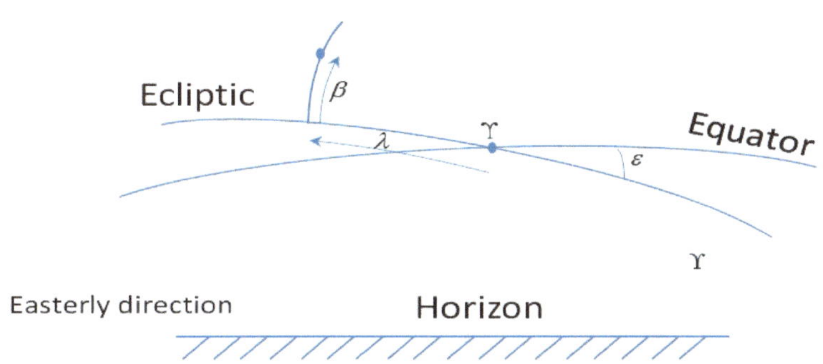

Figure 8.3-5 Ecliptic coordinate system

The coordinates λ and β are called the ecliptic longitude and the ecliptic latitude, respectively. In this coordinate system, shown in Figure 8.3-5, the motion of the sun is due East and happens to be at the Vernal Equinox on March 21, which corresponds to a zero longitude. The ecliptic latitude of the sun is always zero in this coordinate system. This coordinate system is important for a couple of reasons. The first one pertains to the astronomical New moon being defined to occur when the moon and the sun have the same ecliptic latitude. We can state that the astronomical New moon is defined to occur when $\lambda_m = \lambda_s$ and $\beta_s = 0$, where the indices *s* and *m* refer to the sun and to the moon respectively. The second reason involves some fundamental aspects of lunar and solar eclipses that we shall address in some later developments.

8.3.6 Coordinate system conversion

Having defined the different coordinate systems that are commonly used to position and track celestial objects in the sky, we present the transformations used to convert the coordinates from one system to another next. These transformations will prove extremely useful in addressing the relative moon/sun setting near the time of conjunction.

Given a celestial object having an altitude *a*, and an azimuth *A* in the Horizon coordinate system, its declination δ and hour angle *H* are given by the following transformations:

$$\sin\delta = \sin a \sin\varphi + \cos a \cos\varphi \cos A, \qquad 8.3\text{-}2$$
$$\cos H = \frac{\sin a - \sin\varphi \sin\delta}{\cos\varphi \cos\delta},$$

where φ is the latitude of the observer. The inverse transformations used to convert from Equatorial coordinates to Horizon coordinates are given by

$$\sin a = \sin\delta \sin\varphi + \cos\delta \cos\varphi \cos H, \qquad 8.3\text{-}3$$
$$\cos A = \frac{\sin\delta - \sin\varphi \sin a}{\cos\varphi \cos a}.$$

Note that to convert from hour angle to right ascension α, we use one of the two Equations in 8.3-1, e.g.

$$\alpha = LST - H, \qquad 8.3\text{-}4$$

where the parameter *LST* is the local sidereal time. To convert from Ecliptic coordinates to Equatorial coordinates, the following expressions are used

$$\lambda = \arctan\left(\frac{\sin\alpha \cos\varepsilon + \tan\delta \sin\varepsilon}{\cos\alpha}\right), \qquad 8.3\text{-}5$$
$$\beta = \arcsin\left(\sin\delta \cos\varepsilon - \cos\delta \sin\varepsilon \sin\alpha\right).$$

The above transformations need to be used with care, as they are not defined everywhere because the denominator cannot be equal to zero (e.g. $\cos\varphi = 0$, or $\cos a = 0$, etc.). Further, to convert from Horizon coordinates to Ecliptic coordinates, one can first convert from Horizon to Equatorial, then convert from Equatorial to Ecliptic, and vice-versa. For a derivation of these equations, the reader is referred to a standard textbook on spherical astronomy, e.g. (GREEN, 1985).

8.4 Time-reckoning

There are many time scales with names that are hard to keep track of. We shall summarize these scales in this Section. The perfection of the sidereal time, which we defined when we introduced the Equatorial coordinate system in Section 8.3.4 did not guarantee its dominance of time measurement. Humans use the cycle of the sun for day-to-day activities. The hour angle of the sun is used as a basis for civil time rather than the hour angle of the equinox as defined for the sidereal time.

The ***local apparent solar day***, defined as the time elapsed between two successive passages of the sun at the meridian of the observer, called ***apparent solar noon*** can be expressed simply as the hour angle of the sun H_s, or if we take origin at midnight for convenience, the ***local apparent solar time*** is defined as $12\,hr + H_s$, which is also $12\,hr + LST - \alpha_s$, after using Equation 8.3.1. earth's own rotation in the course of one year is quite regular, but the position of the sun from an observer on earth is less than uniform due to several reasons including earth's orbit being slightly elliptic, making earth speed non-uniform over the course of the year, and more importantly due to the obliquity of the ecliptic.

Using the second Equation of 8.3.5 for the sun, we have (after writing $\beta = 0$), a non-proportional relationship between the right ascension and the longitude of the sun of the form $\tan\alpha_s = \cos\varepsilon \tan\lambda_s$. This means the position of the sun may not be used as a reliable timekeeper for an earth observer. To tackle this issue, astronomers came up with the concept of ***fictitious sun.*** An imaginary object referred to as the ***dynamical mean sun***, which position coincides with that of the true sun at perihelion and aphelion and moves on the celestial sphere with a constant angular velocity equals the mean angular velocity of the earth around its orbit. The dynamical mean sun brings regularity to the longitudinal perturbations of the position of the true sun. In addition, to correct for the irregularities associated with the obliquity of the ecliptic plane, a second imaginary celestial body called the ***fictitious mean sun*** which moves along the equator, also at a constant speed and coinciding with the position of the dynamical mean sun at the equinoxes is adopted. As a result, we have a highly regular celestial body whose right ascension approximates that of the true sun and can be used to establish a consistent time scale which is free of irregularities: the ***local mean solar time***. The ***local mean solar time*** is therefore given by the hour angle of the fictitious mean sun in a way similar to the ***local apparent solar time*** described above for the actual sun. The difference between the two is called the ***equation of time*** and is equal to the difference between the right ascensions of the true sun and the fictitious mean sun.

8.4.1 Universal Time

Based on the previous information, we can finally define the Universal Time that most people are familiar with. The local mean solar time at the Greenwich longitude is defined as the

Universal Time or ***UT***. Astronomers rely mostly on the Universal time and the Ephemeris time which we shall address separately in Section 8.4.2. *UT* depends on the rotation of the earth. It is formally defined as H_G, the Greenwich hour angle of the fictitious ***Universal Mean sun***, which rotates around the equator in a uniform fashion. Starting the day at midnight, we have

$$UT = 12\ hr + H_G \qquad 8.4\text{-}1$$

The universal time is a solar time that can be used as a uniform scale around the globe. The rotation of the earth has certain variations and cannot be predicted ahead of time. A combination of multiple sources of observations is used to deduce *UT*, including the VLBI of extragalactic radio sources (Quasars), Lunar Satellite Laser Ranging, and GPS.

8.4.2 Ephemeris Time

Astronomers needed another solar time scale that did not depend on the perturbations of the earth rotation to deliver accurate almanacs. In 1960, they came up with the ***Ephemeris Time*** **(ET)**, a dynamical time measure that depends on some mean orbital parameters of the earth. Dynamical time scales use Newtonian celestial mechanics where time has an independent and absolute nature that follows conventional classical physics principles. *ET* was expressed as the hour angle of another fictitious sun: The ***Ephemeris Mean sun*** which rotates around the equator in a uniform fashion. The hour angle is measured at a fictitious meridian, a version of the Greenwich meridian that is free of the earth's perturbations. Starting the day at midnight, we have:

$$ET = 12\ hr + H_E \qquad \mathbf{8.4\text{-}2}$$

ET and *UT* are related through a quantity called ΔT that cannot be predicted in advance,

$$ET = UT + \Delta T \qquad \mathbf{8.4\text{-}3}$$

Estimates of future irregularities can be based on prior measurements. Such irregularities are not well understood; they are grouped in the quantity ΔT which can be obtained from the *Rapid/Prediction Center of the International Earth Rotation and Reference Systems Service (IERS)*, based at the *Observatoire de Paris*. In addition, the service provides Earth orientation parameters sooner than the publication of IERS final series (Bulletin B). The *U. S. Naval Observatory (USNO)* concurrently provides updates in Bulletin C about six months in advance through its Internet website (http://maia.usno.navy.mil/). The mission of the USNO includes the determination of the positions and motions of celestial bodies, measurement of the earth's rotation and orientation, the maintenance of the master clock for the U.S., and the dissemination of precise time (atomic and astronomical). The key parameters that are disseminated are universal time, precession, nutation, and polar motions. They are commonly used for high-accuracy navigation and positioning in the US. According to the service, ΔT is provided in two parts plus a bias in the form

$$\Delta T = 32.184\,s + (TAI - UTC) - (UT - UTC), \qquad 8.4\text{-}4$$

where *TAI* is the standard *Temps Atomique Internationale* (International Atomic Time), and *UTC* is the standard *Coordinated Universal Time. TAI* and *UTC* are defined in Sections 8.4.3 and 8.4.4, respectively *(TAI - UTC)* is the cumulative number of leap seconds Its evolution over several decades is shown in Figure 8.4-1. For example, at the end of 2015, the published difference *TAI-UTC* was equal to 36 seconds. The difference (*UT - UTC*) is a standard product referred to as ***DUT1***, published in *IERS Bulletin D* and is transmitted with the time signals to a precision of 0.1 seconds.

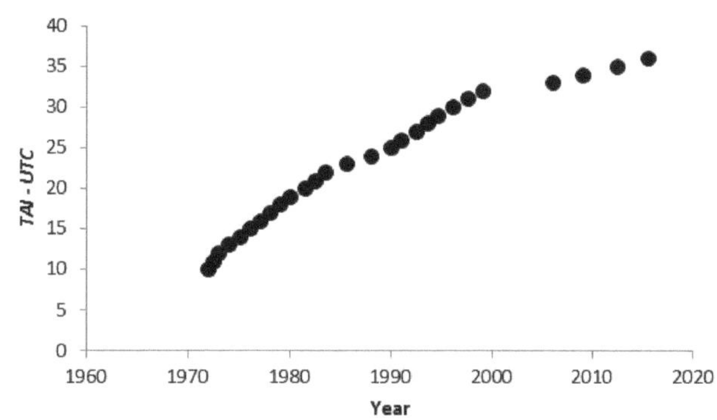

Figure 8.4-1 Historic Cumulative number of leap seconds.

8.4.3 International Atomic Time

The International Atomic Time (*TAI*) is an atomic time scale determined by the *Bureau International des Poids et Mesures* (BIPM). Its unit is exactly ***one*** Système Internationale (SI) second at mean sea level. The standard, adopted in 1976, relies on the oscillation of a transition of the cesium atom. The SI time unit is the second and is correlated to be identical to the Ephemeris unit of time. Atomic clocks have provided highly accurate measurements of celestial manifestations in recent years and have permitted the verification of many laws of non-Newtonian physics.

8.4.4 Coordinated Universal Time

The Coordinated Universal Time (*UTC*) designates the atomic time scale, which approximates the rotational time of the earth. *UTC* is today's worldwide standard for civil timekeeping. From inception, its rate and/or epoch has been adjusted to bring it close to *UT* and is continuously broadcast through radio signals. NASA uses *UTC* for flight and tracking operations. *UTC*, as defined by the International Radio Consultative Committee (CCIR) Recommendation 460-4, differs from *TAI* (Temps Atomique International, i.e., International Atomic Time) by an integral number of seconds. In 1958 *UTC* was referenced to *TAI* using Leap seconds which are introduced periodically so that the *UTC* remains within 0.9 s of *TAI*.

8.5 Rising and setting of celestial objects

The rising and setting times of celestial objects can be obtained from the following relations:

$$LST_{set} = \alpha + \cos^{-1}(-\tan\delta\tan\varphi), \text{ and} \tag{8.5-1}$$

$$LST_{rise} = \alpha - \cos^{-1}(-\tan\delta\tan\varphi). \tag{8.5-2}$$

Proper unit conversion is required when calculating the above expressions as the right ascension is usually given in hours. Because there are 24 hours in a 360-degree angle, a division by 360/24=15 is used to perform conversions. Also, adding or subtracting 24 is necessary whenever the computed time is outside the 24-hour range.

Over the 24-hour period, objects appear to move in circles about earth's axis. Depending on their polar distance, or how close they are to the axis of rotation, some objects never set or rise. ***Circumpolar stars*** never set or rise; they are in the sky, even if they are not visible during the day. There are also objects that never rise or set from an observer's perspective because they are always hidden by the earth's curvature.

To understand this concept, we use the ***celestial sphere*** shown in Figure 8.5-1 and consider an observer located in the Northern hemisphere. The conclusions to be drawn are equally valid for a Southern hemisphere observer. Shown are the points defining the horizon *NSW* of an observer *O*, the observer's Zenith *Z*, and the Northern and Southern celestial poles *P* and *Q*. Let us consider a celestial star *X* which is located on its declination δ and hour angle *H*, where $\delta = 90^\circ - PX$, and $H = ZPX$. The star *X* travels along *its parallel of declination XLRTX* over a 24-hour period, as the celestial sphere seems to rotate about the axis *PQ*, *rising* at point *R* and setting as it crosses the horizon plane again in its diurnal journey.

Also shown is a star *Y* with a higher declination angle which does not cross the horizon as it travels along its parallel of declination *YVU*, and therefore never rises or sets. The star *Y* is, therefore, a circumpolar star. By symmetry, in the opposite side of the celestial sphere, there are objects with lower parallel declinations that never come to cross the Horizon and therefore would not rise or set, making them invisible from the observer's vantage point.

As a minimum, a circumpolar star needs to have its lower transit *V* to occur north of the horizon *N*. At that point, we have the *polar distance PY* equals to the latitude of the observer

φ, that is $90^o - \delta = \varphi$. Therefore, it may be stated that for an object to be circumpolar, its declination must be such that $90^o - \delta \geq \varphi$. Similarly, using symmetry, when we have $-\delta \geq 90^o - \varphi$, an object will be in the invisible part of the celestial sphere. We shall revisit this topic to determine the bounds for crescent moonlight visibility in Chapter 9.

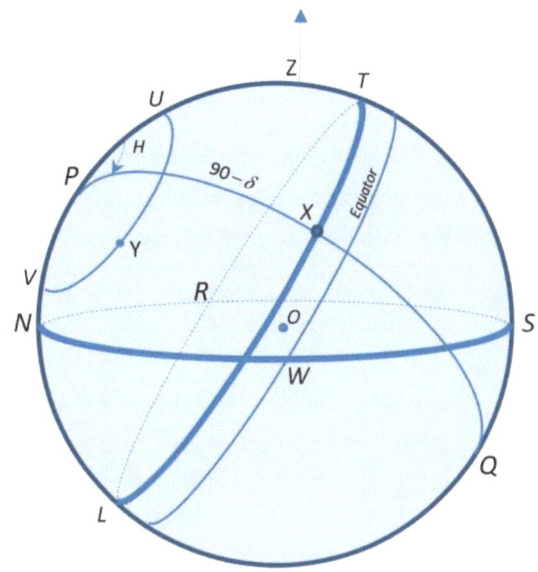

Figure 8.5-1 Celestial Sphere centered at point *O* with Horizon *NWS* and *E* (not shown)

8.6 The angular separation between celestial objects

One of the key parameters of the photometric law for first-crescent visibility is the apparent separation angle between the center of the sun and the center of the moon, as seen by an earth observer. Such distance is called ***elongation*** of the moon (or elongation for short in this text). This frame-invariant parameter should amount to a small value that increases continuously from not-necessarily zero at conjunction.

The following expressions will be used for the separation angle between the sun and the moon in subsequent developments. Using spherical geometry, it can be shown that two objects of equatorial coordinates (α_1, δ_1) and (α_2, δ_2) or ecliptic coordinates (λ_1, β_1) and (λ_2, β_2) are separated by an angle, *d* the cosine of which is given by

$$\begin{aligned}\cos d &= \sin\delta_1 \sin\delta_2 + \cos\delta_1 \cos\delta_2 \cos(\alpha_1 - \alpha_2) \\ &= \sin\beta_1 \sin\beta_2 + \cos\beta_1 \cos\beta_2 \cos(\lambda_1 - \lambda_2)\end{aligned} \quad \textbf{8.6-1}$$

It is important to note that the trigonometric functions need to be evaluated with high accuracy even if the exact equations given in Equations 8.6.1 are used. For objects with small separation angles, a third-order approximation of the angular distance can be derived, assuming $(\lambda_1 - \lambda_2)$ and $(\beta_1 - \beta_2)$ are small, (or $(\alpha_1 - \alpha_2)$ and $(\delta_1 - \delta_2)$ depending on the coordinate system in use). Under such conditions, we have:

$$d \simeq \sqrt{\cos^2 \beta \, (\lambda_1 - \lambda_2)^2 + (\beta_1 - \beta_2)^2}; \quad d \simeq \sqrt{\cos^2 \delta \, (\alpha_1 - \alpha_2)^2 + (\delta_1 - \delta_2)^2}. \qquad 8.6\text{-}2$$

where β and δ are the ecliptic latitude and the declination of either body, respectively. It can also be shown that for small separations, the distance between two celestial objects with coordinates (A_1, a_1), (A_2, a_2) in the Horizon coordinate system can be expressed as

$$d \simeq \sqrt{(a_1 - a_2)^2 + (A_1 - A_2)^2}. \qquad 8.6\text{-}3$$

Because $(a_1 - a_2)$ and $(A_1 - A_2)$ are small angles, Equations 8.6.2 and Equation 8.6.3 are equivalent and can be derived from one another using the coordinate system transformations of Section 8.3.6. They are also used instead of those of Equations 8.6-1, not only for simplicity but also for floating-point accuracy when the two objects are close to one another. Near conjunction, the moon is near the sun, and their separation angle is small enough to be properly approximated by Equations 8.6.2, which happens to be a third-order approximation.

8.7 New moon conjunction

The new moon conjunction presents a reference for the photometric lunar month. This is a well-established limiting condition that we shall refer to in establishing the parameters involved in first and limit new moonlight visibility conditions. To be clear about the nomenclature, we first list several definitions, and subsequently establish the lower bound parameters that will be used in the subsequent developments. We shall determine the elongation at conjunction and the moon-sun relative position. Astronomers have defined the beginning of the lunar month as the time of ***conjunction.*** This is the event which occurs when the sun and the moon have the same geocentric ***ecliptic longitude.*** This event is commonly referred to as the ***Astronomical New Moon*** or simply the ***New Moon*** In today's Almanacs. This event does not necessarily occur when the sun and the moon are closest together in the sky or when the right ascensions are equal. In fact, the moon and the sun can be up to several degrees apart when the Astronomical lunar month begins. To fully understand what this means from a photometric perspective, we shall consider the position of the moon to be near that of the sun and expand its coordinates about those of the sun in the ecliptic coordinate system. But before doing so, we shall define the frame invariant angular distance between the moon and the sun; we refer to such distance as the ***moon elongation,*** or simply ***elongation***.

8.7.1 Elongation at Conjunction

From Section 8.6, the angular distance between two close celestial bodies is given by

$$d \simeq \sqrt{\cos^2 \beta \, \Delta\lambda^2 + \Delta\beta^2}.$$

Here $\Delta\lambda$ and $\Delta\beta$ are small separations between the coordinates of the two bodies in the ecliptic coordinate system, and β is the ecliptic latitude of one of the two bodies. For the case of the sun-moon pair, we have $\Delta\lambda = \lambda_m - \lambda_s$ and $\Delta\beta = \beta_m - \beta_s$, where the indices *m* and *s* refer to the moon and the sun, respectively. For the ecliptic latitude of one of the two bodies, we conveniently choose that of the sun. The sun is located on the ecliptic plane and therefore its ecliptic latitude is conveniently chosen to be zero. In reality, β_s fluctuates by less than 1.2 arcseconds, but for the precision required herein this value is so small that it can be neglected and therefore we can use $\cos\beta_s \simeq 1$, and write that the elongation of the moon at conjunction,

$$d_c \simeq \sqrt{\Delta\lambda^2 + \beta_m^2}.$$

Finally, using the definition of conjunction, we have $\Delta\lambda = 0$, and we write that the ecliptic latitude of the moon β_m is equal to some initial value β_0 and we may state that ***at conjunction, the elongation is essentially given by the absolute value of the ecliptic latitude of the moon.*** That is

$$d_c \simeq |\beta_0|. \qquad 8.7\text{-}1$$

As we start looking into relating the photometric law to the astronomical elements of the sun and the moon, we keep in mind that there is a lower bound for the elongation, given by Equation 8.7.1, which is the ecliptic latitude of the moon. The ecliptic latitude of the moon varies each lunar month, not to exceed the inclination of the orbital plane of the moon which is about 5.145°. The largest value for d_c happens to be about 5.02° over 1250 lunations in the century 2015-2115.

8.7.2 The moon position at conjunction

In this section, we determine the relative equatorial position of the moon with respect to that of the sun at the time of conjunction.

The transformations from equatorial to ecliptic coordinates and vice versa, given earlier in Section 8.3.6, are:

$$\begin{bmatrix} \lambda = \arctan\left(\dfrac{\sin\alpha\cos\varepsilon + \tan\delta\sin\varepsilon}{\cos\alpha}\right) \\ \beta = \arcsin(\sin\delta\cos\varepsilon - \cos\delta\sin\varepsilon\sin\alpha) \end{bmatrix}, \quad \begin{bmatrix} \alpha = \arctan\left(\dfrac{\sin\lambda\cos\varepsilon - \tan\beta\sin\varepsilon}{\cos\lambda}\right) \\ \delta = \arcsin(\sin\beta\cos\varepsilon + \cos\beta\sin\varepsilon\sin\lambda) \end{bmatrix},$$

where λ and β are the ecliptic longitude and latitude, respectively, given in terms of the equatorial declination δ, right ascension α, and the obliquity angle ε. The declination is positive north of the equator.

For the case of the sun, the ecliptic latitude β_s is zero. From the second equation of the left transformation, we have the relationship between the declination of the sun, δ_s, and its right ascension α_s,

$$\tan\delta_s = \tan\varepsilon\sin\alpha_s. \qquad 8.7\text{-}2$$

Substituting back into the first equation of the right transformation, we obtain:

$$\tan\alpha_s = \cos\varepsilon\tan\lambda_s. \qquad 8.7\text{-}3$$

From Equations 8.7.2 and 8.7.3, the transformation of the coordinates of the sun between the ecliptic and equatorial coordinate systems and vice-versa are given by

$$\begin{bmatrix} \beta_s = 0 \\ \tan\lambda_s = \dfrac{\tan\alpha_s}{\cos\varepsilon} \end{bmatrix}, \quad \begin{bmatrix} \tan\alpha_s = \cos\varepsilon\tan\lambda_s \\ \sin\delta_s = \sin\varepsilon\sin\lambda_s \end{bmatrix}. \qquad 8.7\text{-}4$$

Next, we consider the transformations for the moon at the time of conjunction. The transformations between the ecliptic and equatorial coordinates of the are given by

$$\begin{aligned} \tan\alpha_m^s &= \frac{\sin\lambda_m^s\cos\varepsilon - \tan\beta_m^s\sin\varepsilon}{\cos\lambda_m^s}, \\ \sin\delta_m^s &= \sin\beta_m^s\cos\varepsilon + \cos\beta_m^s\sin\varepsilon\sin\lambda_m^s, \end{aligned} \qquad \mathbf{8.7\text{-}5}$$

where, the position of the moon in the equatorial coordinate system is given by (α_m, δ_m), while its coordinates in the ecliptic coordinate system are given by (λ_m, β_m). Expanding the position of the moon about the position of the sun, simultaneously in both coordinate systems up to the second-order in both Equations 8.7-5, leads to:

$$\tan\alpha_s + \frac{\tan\beta_s}{\cos\lambda_s}\sin\varepsilon - \tan\lambda_s\cos\varepsilon + (1+\tan^2\alpha_s)(\alpha_m-\alpha_s) +$$

$$(\lambda_m-\lambda_s)\left(\frac{\tan\beta_s\sin\varepsilon\sin\lambda_s}{\cos^2\lambda_s} - (1+\tan^2\lambda_s)\cos\varepsilon\right) + \frac{(1+\tan^2\beta_s)(\beta_m-\beta_s)\sin\varepsilon}{\cos\lambda_s} = 0 \text{ and}$$

$$\sin\delta_s - \cos\beta_s\sin\varepsilon\sin\lambda_s - \sin\beta_s\cos\varepsilon + (\sin\beta_s\sin\varepsilon\sin\lambda_s - \cos\beta_s\cos\varepsilon)(\beta_m-\beta_s)$$
$$+\cos\delta_s(\delta_m-\delta_s) = 0.$$

Setting the condition for new moon $\lambda_m = \lambda_s = \lambda_0$ and writing that the sun is in the ecliptic plane, i.e. $\beta_s = 0$, leads to:

$$\tan\alpha_s - \tan\lambda_s\cos\varepsilon + (1+\tan^2\alpha_s)(\alpha_m-\alpha_s) + \frac{\beta_m\sin\varepsilon}{\cos\lambda_s} = 0 \quad \text{and}$$
$$\sin\delta_s - \sin\varepsilon\sin\lambda_s - \beta_m\cos\varepsilon + \cos\delta_s(\delta_m-\delta_s) = 0.$$

Next, we use the two simple relations $\tan\alpha_s = \cos\varepsilon\tan\lambda_s$ and $\sin\delta_s = \sin\varepsilon\sin\lambda_s$ in Equations 8.7-4 to eliminate the two first terms in each one of the above two equations. The system of equations then reduces to

$$\alpha_m - \alpha_s = -\beta_0\frac{\cos^2\alpha_s}{\cos\lambda_0}\sin\varepsilon \quad , \qquad \delta_m - \delta_s = \beta_0\frac{\cos\varepsilon}{\cos\delta_s} \qquad \text{8.7-6}$$

where we use the notation $\beta_m = \beta_0$ and $\lambda_m = \lambda_s = \lambda_0$ at conjunction. If we use Equation 8.7-3, to eliminate the ecliptic longitude λ_0 from the first equation, this result means that at the time of conjunction, the equatorial coordinates of the moon are accurately determined by the separation angle and the position of the sun, providing a lower bound for the visible new moon which will be addressed next in the following Chapter. Numerical calculations show that the approximations given by Equations 8.6-2 provide accurate enough accuracies for our studies of the new moon. In general, the relative right ascension is less than 1-2 degrees in magnitude, while the relative declination is less than about 5 degrees in magnitude. Furthermore, from Equations 8.6-2, when the elongation *d=0*, we have to have both $\alpha_m = \alpha_s$ and $\delta_m = \delta_s$ which correspond to a condition when the moon and the sun occupy the same position in the sky, i.e., we have a solar eclipse. Such condition is a sufficient but not a necessary one for a new moon. That is every solar eclipse corresponds to a new moon, and not each new moon corresponds to a zero elongation.

8.8 Astronomical full moon

The astronomical full moon is defined as the time when the longitude of the moon in the ecliptic coordinate system is 180 degrees away from that of the sun. It is possible to derive a closed-form solution of the astronomical elements that lead to the condition of photometric Using the conversion from Ecliptic to Equatorial coordinates for the sun, we obtain

$$\tan\alpha_s = \frac{\sin\lambda_s\cos\varepsilon - \tan\beta_s\sin\varepsilon}{\cos\lambda_s} = \tan\lambda_s\cos\varepsilon$$
$$\sin\delta_s = \sin\varepsilon\sin\lambda_s$$

To convert from Ecliptic to Equatorial coordinates, the following expressions are used:

$$\tan\alpha_m = \tan\lambda_m\cos\varepsilon - \frac{\tan\beta_m}{\cos\lambda_m}\sin\varepsilon$$
$$\sin\delta_m = \sin\beta_m\cos\varepsilon + \cos\beta_m\sin\varepsilon\sin\lambda_m$$

Therefore, we have $\tan\lambda_s = \tan\lambda_m$, or using the first equation of the above system we have

$$\tan\alpha_m = \tan\alpha_s - \frac{\tan\beta_m}{\cos\lambda_s}\sin\varepsilon\ . \qquad \textbf{8.8-1}$$

Chapter 9 Astronomy of the new month

9.1 Introduction

The photometric law for crescent moon visibility was derived in Chapter 7 using the reflected sunlight during twilight, independently from the relative sun-moon position and irrespective of the position of the observer. In this chapter, we relate the photometric law to astronomical elements and formulate the problem to validate field sightings and ultimately provide predictions of where and when the new moonlight can be observed.

Setting celestial objects are first characterized in Section 9.2 using information from multiple coordinate systems. The reader is expected to understand commonly used coordinate systems of spherical astronomy, which were introduced in Section 8.3. We shall draw from the transformation between multiple coordinate systems to access the controlling elements of a new moon crescent visibility which are not immediately available from a single coordinate system. For example, no information on the onset of setting can be obtained directly from the ecliptic coordinate system where the astronomical new moon is defined. Therefore, we use the horizon coordinate system to express the setting event and use the transformation back from the ecliptic coordinate system to express positional coordinates right after conjunction. We shall differentiate between an observer located in the Northern hemisphere and an observer located in the Southern hemisphere. In the northern hemispheres, celestial bodies including the moon and the sunrise from the east and set in the west following a southerly direction that varies with the seasons from low to high in the sky. In the Southern hemisphere, celestial bodies also rise from the east and set in the west but follow a northerly direction. This difference will be relevant when we address the linear path of the setting bodies near the horizon.

Because of the proximity of the moon to the sun right after conjunction, as it sets during twilight, we use the multivariate linearization concept discussed in Section 8.2 to first relate the position of the moon to that of the sun, and subsequently connect with the photometry of the new moon. Sections 9.2 and 9.3 are devoted to the linear model which approximates the trajectories of setting celestial objects to straight lines. We also propose a higher order quadratic approximation of these trajectories at the end of the Chapter, in Section 9.8, where it is shown that the trajectories of setting objects are along arc of circles with large radii to which the straight lines of the linear model happen to be the tangents at setting points. For our purpose of identifying first visibility of the new crescent, the linear model has enough accuracy, in addition to its simple form.

In Section 9.4, we establish the parameters that define the location where first visibility occurs. In addition, we identify the region over the globe where optimal and limit visibility occur as time elapses past conjunction. Sections 9.5 and 9.6 are intended to help observers locate the crescent moon in the sky by deriving its orientation function of the moon coordinates. We also briefly discuss some important aspects of field sighting. Section 9.7 discusses records from open literature of new crescent moon sightings, which are considered reliable. We compare their results to the perspective of first visibility documented herein. Finally, in Section 9.8, we discuss sighting records which occur at large sun-moon relative setting and scatter angles. Such sightings do not qualify as first sightings over the globe in a new month, but are still considered first sightings for a particular geographic location, under special conditions of equatorial coordinates.

9.2 The setting of celestial bodies

In this section, we study certain geometric aspects of setting celestial bodies. We will show that the linearization of the transformation between the different coordinate systems is sufficient to fully characterize their position in the horizon sky.

Consider an object such as a planet, a star, the sun, or the moon setting on the western horizon. When the object is nearing the horizon, its altitude diminishes gradually, and its path in the sky gets to be close to a straight line. This can be immediately seen from a plot of the object's altitude as a function of its azimuth over about 30-45 minutes before it sets below the Horizon. An arbitrary celestial object may be represented by any one of the three coordinate systems discussed in Section 8.3. We recall that the coordinates of such an object can be transformed from one coordinate system to another. For example, the transformation between the horizon coordinates (azimuth A and altitude a) and the Equatorial coordinates (right ascension α, and declination δ) are given by:

$$\sin a = \sin\delta \sin\varphi + \cos\delta \cos\varphi \cos H, \tag{9.2-1}$$

$$\cos A = \frac{\sin\delta - \sin\varphi \sin a}{\cos\varphi \cos a}, \tag{9.2-2}$$

$$H = LST - \alpha, \tag{9.2-3}$$

where φ is the geographic latitude of the observer and LST is the local sidereal time, as discussed in Section 8.3.6. Denoting by $\left(A_{\swarrow}, a_{\swarrow} = 0\right)$, the coordinates of an object at the onset of setting, the second equation gives the azimuth of the setting point as a function of the declination and the geographic latitude,

$$\cos A_{\swarrow} = \frac{\sin\delta}{\cos\varphi}. \qquad 9.2\text{-}4$$

Typical declinations of the moon and the sun vary as illustrated in Figure 9.3-4. It was assumed that the geocentric altitude of the object is nil at the onset of setting. The definition of setting in classical astronomy is slightly different. The setting is defined as the onset of when the upper rim (of the sun or moon disk) is tangent to the horizon while rising is defined as the onset of lower tangency of the disk with the Horizon. In this text, we use the geocentric definition for rising and setting, i.e., when the geocentric coordinate is nil.

It is worth noting that the above equation is important for planning and designing observatories that are intended for optical studies of the horizon sky as the moon and the sunset at some geographic location. The minimum azimuthal field of view must consider the extreme declinations of the sun and of that of the moon (and perhaps of any other objects of interest) with an addition of a margin.

Taking the multivariate Taylor series expansion of Equation 9.2-1 near the setting point of the object, assumed to have the coordinates $\left(A_{\swarrow},\, a_{\swarrow} = 0\right)$, we obtain:

$$\cos\left(A_{\swarrow}\right)\cos(\phi) - \sin(\delta) - \sin\left(A_{\swarrow}\right) r \cos(\phi) + \sin(\phi) s = 0 \qquad 9.2\text{-}5$$

Where $r = A - A_{\swarrow}, s = a - a_{\swarrow} = a$. Alternatively, we can express the altitude in terms of the azimuth using:

$$a = \frac{\sin A_{\swarrow}}{\tan\varphi} A - \frac{A_{\swarrow} \sin A_{\swarrow}}{\tan\varphi} - \frac{\cos A_{\swarrow}}{\tan\varphi} + \frac{\sin\delta}{\sin\varphi} \qquad 9.2\text{-}6$$

Making use of Equation 9.2-4, the last two terms cancel out, and we have

$$a = \frac{\sin A_{\swarrow}}{\tan\varphi} A - \frac{A_{\swarrow} \sin A_{\swarrow}}{\tan\varphi} \qquad 9.2\text{-}7$$

And the linear expression relating the altitude to the azimuth becomes:

$$a = \frac{\sin A_{\swarrow}}{\tan\varphi}(A - A_{\swarrow}), \qquad 9.2\text{-}8$$

or we have the reciprocal form expressing the Azimuth as a function of altitude

$$A = \frac{\tan\varphi}{\sin A_{\swarrow}} a + A_{\swarrow} = \tan\mu\, a + A_{\swarrow}\,, \qquad 9.2\text{-}9$$

where $\tan\mu = \tan\varphi / \sin A_{\swarrow}$. The tangent of the angle that a setting celestial object makes with the horizon is thus given by $\sin A_{\swarrow} / \tan\varphi$. And because the setting point is in the west, around 270 degrees, $\sin A_{\swarrow}$ is near the value -1, which further simplifies the above equation to become

$$A = -\tan\varphi\, a + A_{\swarrow} \qquad 9.2\text{-}10$$

Because $\tan\varphi$ is positive in the northern hemisphere and negative in the southern hemisphere, we have the conditions illustrated in Figure 9.2-1 showing an example of a setting moon and a setting sun in the northern and southern hemispheres. The slope of the setting direction of a celestial body is negative in the northern hemisphere and positive in the southern hemisphere, and a good approximation of the set direction is given by the complementary of the latitude.

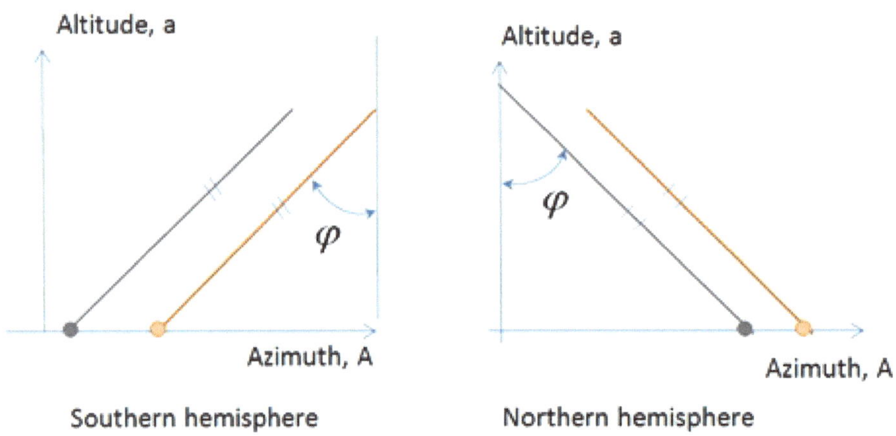

Figure 9.2-1 Sunset and moonset directions in the Northern (right) and Southern (left) hemispheres (the angle between the vertical and setting direction is about equal to the latitude of the observer)

As mentioned earlier, we have a departure from the true astronomical definition of a setting object which requires that the entire object goes under the horizon. The current assumption for a setting object corresponds to an elevation equals to zero instead. Also, we are using geocentric coordinates with no consideration is given to atmospheric refraction. Furthermore, the assumption that $\sin A_{\swarrow}$ is near -1 leads to tolerable errors for what we are trying to achieve. Table 9.2-1 shows the departure from the latitude angle assuming a setting location varying +/- 20 degrees from the west direction (i.e., 270 degrees), considering different geographic latitudes of the observer. The errors associated with the above approximations are considered small if one aims for a field of view of roughly 10 to 20 degrees.

For the sidereal time, we make the altitude nil in Equation 9.2-1 and dividing by the product $\cos\delta\cos\varphi$, we have

$$\cos H = -\tan\delta\tan\varphi \qquad 9.2\text{-}11$$

At the onset of setting, the hour angle H and the right ascension α are also fully determined for a given geographic latitude and a declination angle of a celestial object. After substituting Equation 9.2-11 into Equation 9.2-3, the local sidereal time for the setting is thus given by

$$LST_{set} = \alpha + \cos^{-1}(\tan\delta\tan\varphi)\,. \qquad 9.2\text{-}12$$

Before proceeding with the topic of relative moon/sun setting, we shall pause and discuss the practical verification of the result of Equation 9.2-10 through simple field measurements.

Table 9.2-1 Setting angle a celestial object within +/- 20 degrees of the western direction

		Observer's Lattitude (deg)					
		0.5	**10**	**20**	**30**	**40**	**50**
Setting Azimuth (deg)	**250**	0.53	10.63	21.17	31.57	41.76	51.74
	255	0.52	10.35	20.65	30.87	40.98	50.97
	260	0.51	10.15	20.28	30.38	40.43	50.43
	265	0.50	10.04	20.07	30.09	40.11	50.11
	270	**0.50**	**10.00**	**20.00**	**30.00**	**40.00**	**50.00**
	275	0.50	10.04	20.07	30.09	40.11	50.11
	280	0.51	10.15	20.28	30.38	40.43	50.43
	285	0.52	10.35	20.65	30.87	40.98	50.97
	290	0.53	10.63	21.17	31.57	41.76	51.74

9.2.1 Verification of setting parameters

Verification of the derived approximate setting parameters of celestial bodies can be achieved by a simple drive mount available from astrophotography. This is a field that has been developed by professional and amateur astronomers as well as enthusiast photographers. Driven by demand for better images of celestial objects, portable versions of the precision mounts used to accurately track moving celestial objects in observatories were produced and marketed to meet personal demands in the west. These mounts which are also referred to as drive mounts keep a celestial object from moving in the field of view preventing streaks of light against the background. Because many dim celestial objects have a low photon count, long exposures sometimes hours long are required to accumulate enough light so that discernable and blur-free images are formed.

Unlike it is the case in other imaging tasks in astronomy where tracking of celestial objects needs to be as precise as possible, we may tolerate some drift in new moon tracking, as we do not seek sharp images. Instead, all that is needed is a confirmation of appearance of a thin silver crescent light. To this end, we may borrow from methodologies used tracking objects that are not constant in the night sky such as comets. Other imaging methods used by star trackers, which are devices used to orient satellites, can also be adopted in our application. Star trackers are designed to deliberately set optical elements out of focus to pick key stars in their field of view. Without diving into the details, all we need to know for now is that there are two methods which can be used to track the new moon. One is to have a camera move with the motion of the sky, and the other is to translate and stack the images as time goes by. In classical nighttime astronomy, the dominant method is the former one.

There are several types of mounts that were engineered over the years to address a variety of needs. We shall not address the technical details of the precision mounting of optical devices. These mounts can track objects using two independently controlled. Because we do not have a need to track for a long time, the trajectory of the moving new moon is nearly a straight-line during observation. All that we need is a drive mechanism that provides rotation about one single axis, which when oriented parallel to earth's axis and driven at the rate of earth's rotation should keep the moon stationary in the field of view with small tolerable perturbations.

A decent quality drive system is used atop of a sturdy tripod and appended with a pointing laser devise to generate the moving field of view of a telescope system would have during the setting of an object.

After quick leveling of the tripod by a mean of a water bubble, the drive mount was tilted to the exact latitude of the test location, which happens to be at exactly 40-degrees north. A spherical joint, commonly used for mounting cameras on tripods for photography, was attached to the drive mount and provided a level plane over which the laser is placed and roughly oriented to the west. Using the spherical joint, the laser is tilted a number of degrees to simulate the tracking of a setting celestial object. The position of the laser spot is marked every 5 minutes on a graph paper, placed in front of the setup. A straight line going through the marks was drawn, and the slope was found to be 40.2 deg. This value compares well with the above approximation, which gives an angle $\beta = 40.7$ deg.

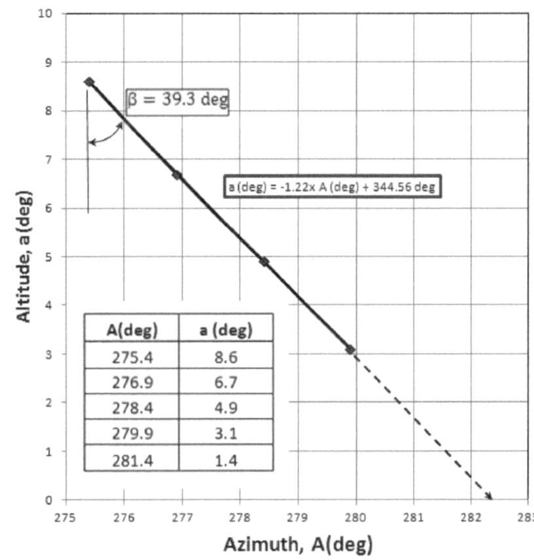

Figure 9.2-2 Example of moon setting at 40 deg latitude

Figure 9.2-2 shows the setting azimuth from the Naval Observatory ephemeris to be 282.4 deg and the angle $\beta = 39.3$ deg. Note that these values account for atmospheric refraction. Finally, the rate of descent was also measured and was found to be 15.08 arcsecond/s. This value is comparable to earth's angular rotation, which is about 15 arcsecond/s.

Given the fact that the moon crescent light has a large angular size (moon itself is 0.5 deg), this simple test shows that there is no need for high accuracy in tracking the new moonlight if the field of view is about or more than a couple of degrees. Nevertheless, we shall characterize the errors due to the above assumptions and truncation.

9.2.2 Approximation error

It can be shown that the error due to truncation that led to Equation 9.2-8 is of second order, i.e., a quadratic function of a and $(A - A_{\swarrow})$. In addition, we assumed that

$$\sin A_{\swarrow} \simeq -1$$

in the expression of the slope of the setting celestial bodies, which has led to Equation 9.2-10. This, in turn, has introduced the third-order position error, which is a function of the product

$$\sim (A - A_{\swarrow})(A_{\swarrow} - 3/2\,\pi)^2 .$$

This explains the satisfactory results of Table 9.2-1, showing that the errors from the approximations are less than a degree if the celestial object sets over a 30-degree span centered about the western direction.

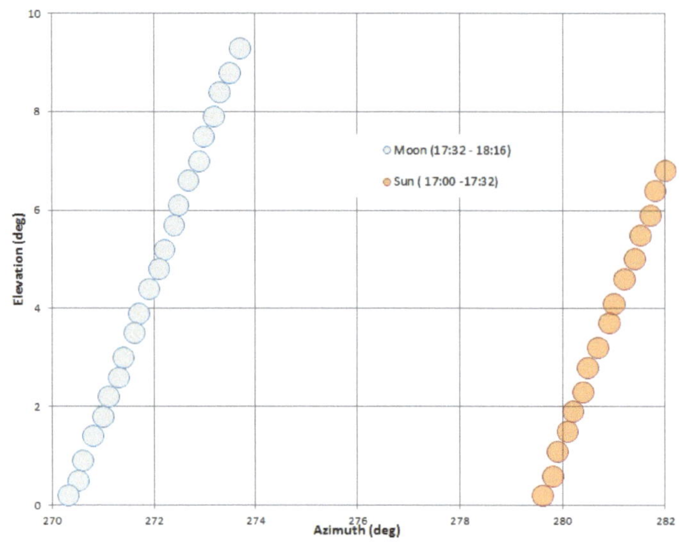

Figure 9.2-3 Sun and moon setting trajectories, Tacna, Peru (S° 18 W° 70, Aug 29, 2011)

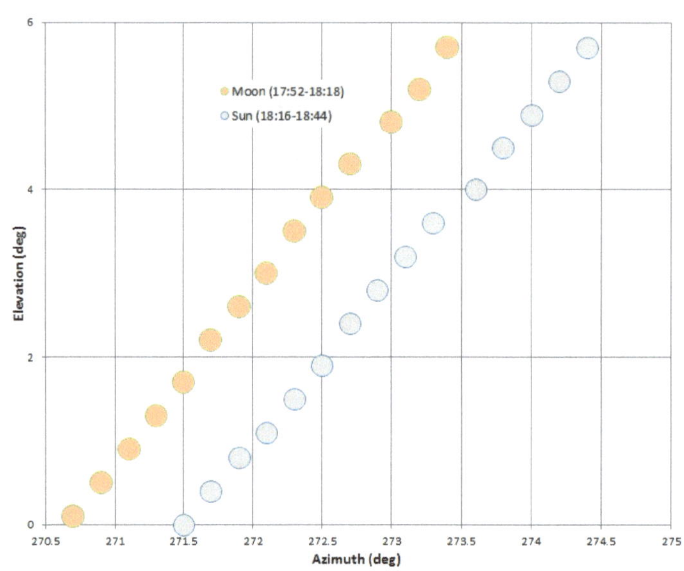

Figure 9.2-4 Sun/moon setting (S24° 0' W6° 6', October 20, 2017 –local time)

The approximation infers that all setting celestial objects have parallel trajectories near the horizon, a manifestation that can be verified by looking at the ephemeris of any setting object in the western direction. The examples in Figure 9.2-3 and Figure 9.2-4 illustrate this result, which can be interpreted in our case as the motions of the setting moon and setting sun over the 30-45 minutes of observation are mainly due to earth's rotation. The question then is how much error in the setting angle we have under the assumption of parallelism? Let us recall the angle that the trajectory of a setting object makes with the elevation axis. For the setting sun and a setting moon, we can write the respective slopes of the trajectories as

$$\mu_s = -\tan^{-1}\left(\frac{\tan\varphi}{\sin A_s}\right),\ \mu_m = -\tan^{-1}\left(\frac{\tan\varphi}{\sin A_m}\right), \qquad 9.2\text{-}13$$

where the azimuths of the sun and the moon at their setting points are A_s and A_m respectively. Figure 9.2-5 shows the possible configurations for the moon-sun pair, based on the sign of the setting slope. Expanding the moonset azimuth position around that of the sun in the first Equation in 9.2.13, we obtain

$$\frac{\tan\varphi}{\sin A_m} = \frac{\tan\varphi}{\sin A_s} - \frac{\tan\varphi}{\sin A_s}\cot A_s(A_m - A_s) + O(A_m - A_s)^2. \qquad 9.2\text{-}14$$

The term $\cot A_s(A_m - A_s)$ is of second order because both terms of the product are small due to the setting occurring in a westerly direction. We can then conclude that the descending trajectories are parallel to second order. From Table 9.2-1, it is also evident that linear approximations are the least accurate when the azimuth of the setting object is further away from the west. Also, the errors are larger at higher latitudes. Errors associated with linearization will be addressed again when we establish a formal linear model for the setting problem in the following section.

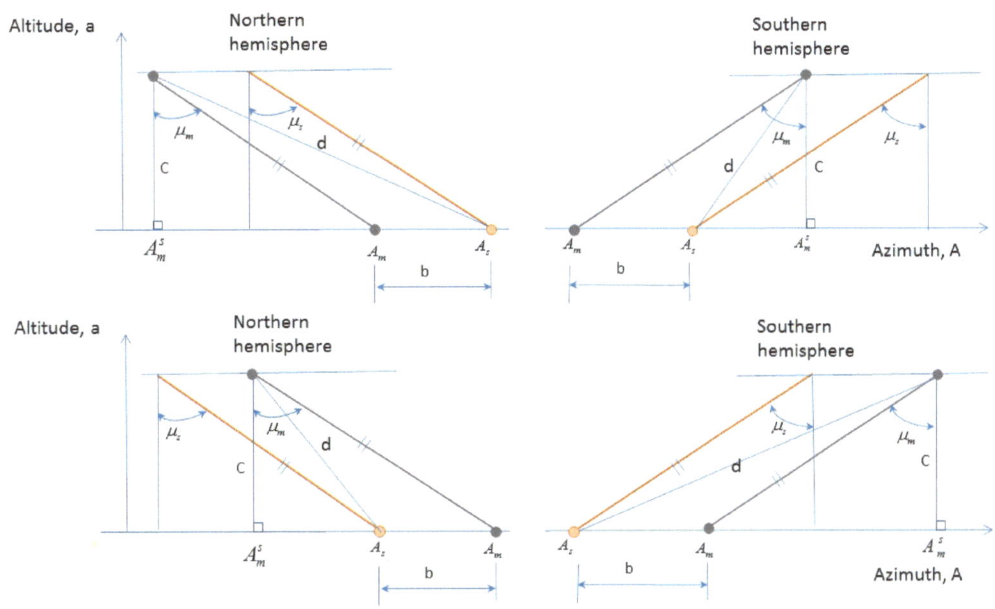

Figure 9.2-5 The setting configurations of the moon and the sun in Northern/Southern hemispheres

9.3 The solution of the setting moon problem

The relationship between the coordinates of the sun and the coordinates of the moon near conjunction as they set at some arbitrary latitude is the subject of this section. We shall extend the results of Section 9.2 to produce additional linear expansions of the position of the moon at the time of sunset about its moonset position in terms of two independent parameters: the elongation of the moon from the sun as seen from earth and the elevation of the moon at the time of sunset. We will present a solution in Section 9.3.1 based on geometry and a rigorous second solution derived from linearizing the transformation equations from horizon coordinates to equatorial coordinates in Section 9.3.3. Finally, we will assess the approximation introduced by the linearization at the end of the section and conclude this topic by discussing the trace of the terminator on a cylindrical projection of the earth's surface.

It can be shown that transformations between different coordinate systems conserve linearization. And in our case, the linear approximation during a setting event is frame-invariant.[17] Frame-invariance will be used together with the photometric sighting law to ultimately establish when and where the new moon can be observed.

9.3.1 Geometric approximate solution

Let us focus on a southern hemisphere observer (case in the left-hand side in Figure 9.2-1). The case of an observer in the northern hemisphere yields the same equations and conclusions. We further consider the case of a moonset trailing the trajectory of the sun in the Azimuth-Elevation plane, which means the Azimuth of the moonset point is smaller than that of sunset point. The alternative situation, shown in the example of Figure 9.2-4 and illustrated in the lower right of Figure 9.2-5, yields the same equations after adjusting for proper signs. Furthermore, we consider two situations depending on whether the azimuth of the moon at the time of sunset falls between or outside the azimuths of moonset and sunset. The geometric configurations are shown in Figure 9.3-1 and Figure 9.3-2. Finally, the limit configuration in which the Azimuths of the moon and the sun are equal at sunset, shown in Figure 9.3-3, will prove to be of relevance to the development herein. In Figures 9.3-1-9.3-3, the trajectories of the setting sun and setting moon are depicted by two parallel oblique lines with the positions of the moon and the sun at sunset highlighted by two large dots. The elevation of the moon at sunset is denoted by *C*, the distance between the azimuths at sunset and moonset *b*, and the angular distance between the sun and the moon which is the elongation of the moon, *d*. Also shown are the angle $\mu \approx |\varphi|$, the angle between the segment connecting the setting sun and the moon at sunset, and the trajectory of the moon α_1 , and the angle between the horizon and the segment connecting the sun and the moon at sunset,

---

[17] This can be established using multivariate Taylor series expansions of the coordinate transformations.

α_2. From geometry, it can be easily shown that both configurations, shown in Figures 9.3-1 and 9.3-2 yield the following three independent equations to four unknowns α_1, α_2, b and C.

$$\begin{cases} \sin\alpha_1 = \dfrac{b\cos\mu}{d}, \\ \sin\alpha_2 = \dfrac{C}{d}, \\ \alpha_1 + \alpha_2 = \dfrac{\pi}{2} + \mu. \end{cases} \qquad 9.3\text{-}1$$

The geometry of the problem has an indeterminacy of order one. Solving the system of equations for C as a function of b, d and μ, we obtain the relationship

$$C = \sin\left(\mu - \arccos\frac{b\cos\mu}{d}\right) d \qquad 9.3\text{-}2$$

Note that we kept the above expression in terms of the higher precision angle μ instead of the more relevant angle φ for now. An alternative to Equation 9.3-2, in which the non-dimensional parameters C/d and b/d, can be related from geometry can be established, but we keep the expression in the following simple form:

$$b = C\ \tan\mu \pm \sqrt{d^2 - C^2}. \qquad 9.3\text{-}3$$

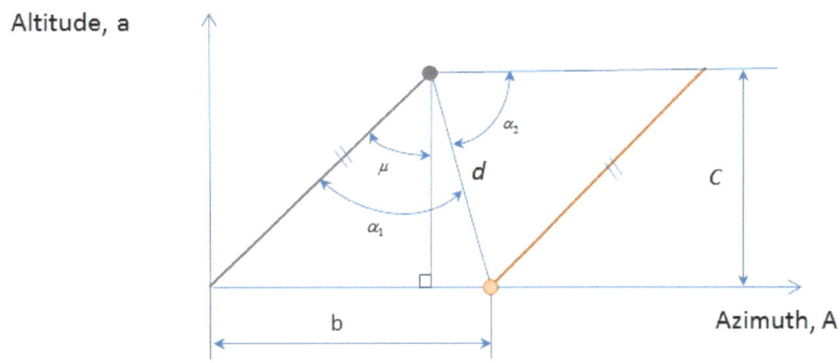

Figure 9.3-1 Scenario1: for sun & moon setting conditions in the Horizon coordinate system; sun Azimuth is west of that of the moon at sunset.

Furthermore, the angle traveled by the moon as it sets, starting from its position at sunset, is related to the time elapsed between the settings of the moon and the sun. This distance is referred to as the *moonset lagtime* or simply the *moon lag* and is given using geometry by

$$T = \frac{C}{\cos\mu} \simeq \frac{C}{\cos\varphi} \quad . \qquad 9.3\text{-}4$$

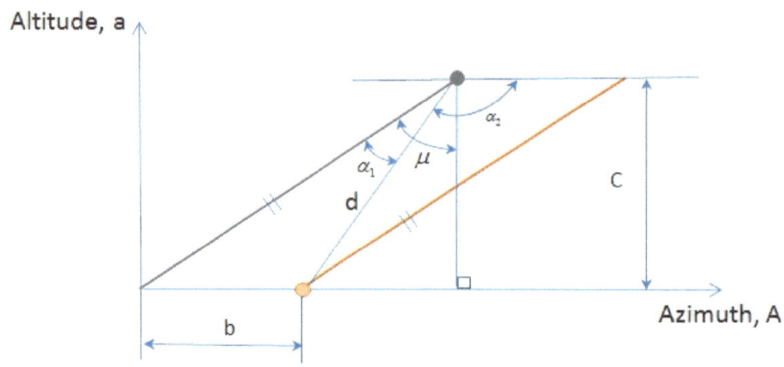

Figure 9.3-2 Scenario2: for sun & moon setting conditions in the horizon coordinate system; sun Azimuth is east of that of the moon at sunset in the Horizon coordinate system.

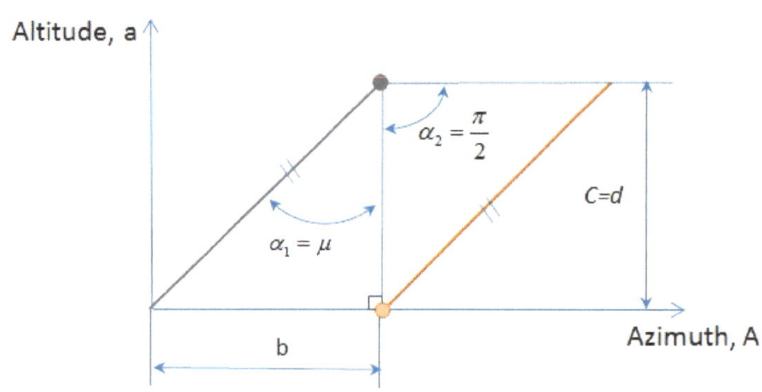

Figure 9.3-3 Limit condition of sun and moon settings which correspond to $d = C_{cr}$

Before proceeding, it is important to note that by examining the above two equations, we can state that given a location characterized by a latitude φ and at a particular time in which the elongation is *d*, there are only two independent parameters either (*C, b*) or (*T, b*). Equation 9.3-3 is that of a half circle in the (*b, C*) plane, centered at the point which coordinates are ($C\tan\mu$, 0) and of radius *d*. The maximum value for *C* is the elongation *d*, which corresponds to $b = C\tan\mu = d\tan\mu$. This is the case of Figure 9.3-3 in which the moon and the sun have the same azimuth at sunset. In addition, the two-parameter problem is reduced one parameter because *C* is defined and is equal to the elongation *d*. This is the condition of absolute first visibility of the new crescent. We will look at some of the observation data and assess the validity of the linear geometric model in Section 10.1.

9.3.2 Setting azimuths of the sun and the moon

Having shown that the moon and the sun have nearly parallel trajectories, we want to know next where do they set with respect to one another. Let us recall the transformation between equatorial and horizon coordinate systems given by Equations 8.3-3, and let us denote by δ_s and δ_m the declinations of the sun and the moon at their respective points of setting. We also denote by A_m^s, δ_m^s the azimuth and declination of the moon at the time of sunset. We denote by *C*, the elevation of the moon at sunset, and we denote by *b* the absolute value of the difference between the azimuth of the moon at moonset A_m and the azimuth of the sun at sunset A_s. Finally, *d* is the elongation of the moon as seen from an observer located at latitude φ. From Equation 9.2-4, we have the Azimuths of the sun and the moon settings given by:

$$\cos A_s = \frac{\sin\delta_s}{\cos\varphi} \quad \text{and} \quad \cos A_m = \frac{\sin\delta_m}{\cos\varphi}, \qquad 9.3\text{-}5$$

Figure 9.3-4 shows the sun declination angle undergoes a seasonal change between +/-23.4°, positive when the sun is north of the Equator in the year 2015. The extreme points correspond to the summer and Winter Solstices, which are characterized by the longest and shortest daylight time, respectively. The declination of the sun is zero twice a year, occurring around September 22/23 and March 19/20. These two dates are called the Vernal Equinox and the Autumnal Equinox, respectively.

The moon declination presents a regular cyclic variation. New moons, shown artificially larger occur near the sun direction with declination differences between the sun and the moon $\delta_s - \delta_m$ less than 10 degrees. Such differences are small, and therefore second-order expansions carry sufficient accuracies. The same can be shown true for right ascension differences.

When the declination angle of the moon is less than that of the sun at a time following a new moon, we have $\sin\delta_m \le \sin\delta_s$ and therefore $\cos A_m \le \cos A_s$ because $\cos\varphi > 0$, if we disregard the limiting case of the poles. This leads to $A_m \le A_s$ near an azimuth of 270 degrees. Conversely, when the declination angle of the moon is greater than that of the sun at a time following a New moon, we have $\sin\delta_m \ge \sin\delta_s$ and therefore $\cos A_m \ge \cos A_s$ because of the same arguments. This leads to $A_m \ge A_s$. Therefore, we conclude that following an astronomical new moon, the moonset azimuth may or may not trail the sunset azimuth from the southwest as it can be deduced from Figure 9.2-4. This conclusion is valid for both hemispheres. Figure 9.2-5 shows the schematics of the two cases. Figure 9.2-3 shows the sun/moon settings following the Astronomical New Moon of August 2011 as it would be seen in Tacna, Peru, illustrating the former case where $A_m \le A_s$, while Figure 9.2-4 shows the situation near the New moon of October 20, 2017 (Molad of the Month of Tishrei, year 5778 of the Jewish calendar), which illustrates the latter case where $A_m \geq A_s$. Note that both examples are from the southern hemisphere. There is a slight departure from perfect linearity in the Figures due to the modeling of the variation of the different layers of atmospheric refractions.

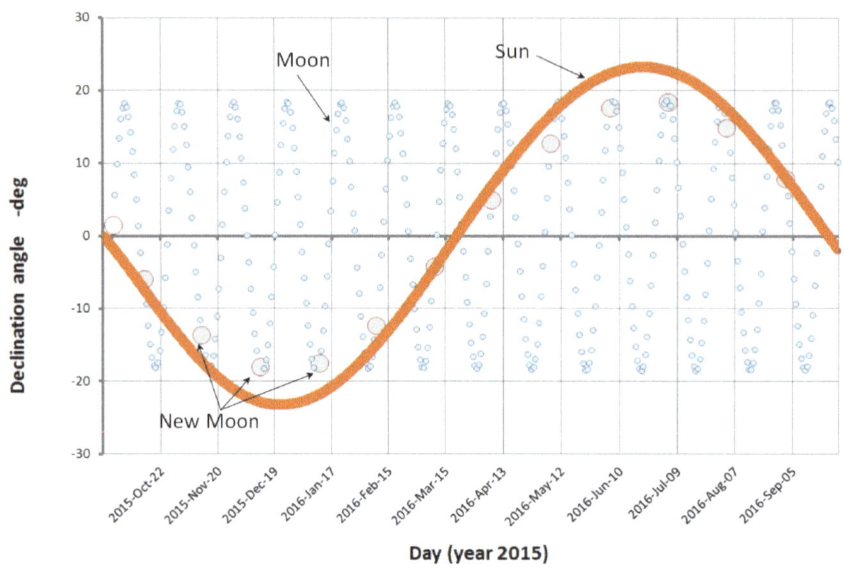

Figure 9.3-4 Declination of the moon and sun (δ_m, δ_s) -the year 2015

9.3.3 Analytical solution of the setting moon problem

In Section 9.3.1 we relied on geometry to relate the elevation of the moon at sunset to the elongation and established Equation 9.3-2, or its equivalent form given by Equation 9.3-3. We could rely on geometry because the approximations are good up to a second order, validating the assumption of linear trajectories near the setting points of the moon and the sun. In this section, we seek a rigorous solution in which we establish similar results directly from Taylor series expansions of the horizon to equatorial coordinates without reliance on geometry. In the process, we keep track of the needed assumptions that must be made, so that we have a good understanding of the applicability of the results. At the time of sunset, these transformations can be written as follows

$$\begin{aligned} &\cos A_m^s \cos\varphi \cos C + \sin C \sin\varphi - \sin\delta_m^s = 0, \\ &\sin\delta_m^s \sin\varphi + \cos\delta_m^s \cos\varphi \cos H_m^s - \sin C = 0. \end{aligned} \qquad 9.3\text{-}6$$

In the above system of equations, we use the same notation of the previous development and denote by (δ_s, δ_m) and (H_s, H_m) the declinations and the Hour angles of the sun and the moon at their respective points of setting. We also denote by A_m^s, δ_m^s and H_m^s the azimuth, declination and the Hour angle of the moon at the time of sunset and denote by C, the elevation of the moon at the observer's geographic latitude. Expanding the moon coordinates at sunset about those of the moon at the time of moonset up to second order and stating that C is small, the above system of equation is written as:

$$\begin{aligned}&\cos A_m \cos\varphi - \sin\delta_m - (A_m^s - A_m)\sin A_m \cos\varphi + C\sin\varphi - (\delta_m^s - \delta_m)\cos\delta_m = 0,\\&\sin\varphi\sin\delta_m + \cos\delta_m \cos\varphi\cos H_m - C + (\delta_m^s - \delta_m)(-\cos\varphi\sin\delta_m \cos H_m + \cos\delta_m \sin\varphi) -\\&(H_m^s - H_m)\cos\delta_m \cos\varphi\sin H_m = 0.\end{aligned} \quad 9.3\text{-}7$$

And using the moon's horizon to equatorial coordinate transformation, at the time of moonset, i.e., where the elevation of the moon is set to zero, we have the following two relations:

$$\begin{aligned}&\cos A_m \cos\phi - \sin\delta_m = 0,\\&\sin\delta_m \sin\varphi + \cos\delta_m \cos\varphi\cos H_m = 0.\end{aligned} \quad 9.3\text{-}8$$

which upon substitution in the system of equations 9.3-7, results in

$$\begin{aligned}&-(A_m^s - A_m)\sin A_m \cos\varphi + C\sin\varphi - (\delta_m^s - \delta_m)\cos\delta_m = 0,\\&(\delta_m^s - \delta_m)\frac{\sin\varphi}{\cos\delta_m} - (H_m^s - H_m)\cos\delta_m \cos\varphi\sin H_m = C.\end{aligned}$$

Adding and subtracting A_s in the first term and dividing by $\cos\varphi\sin A_s$ all terms of the first equation, we have

$$\begin{aligned}&C\frac{\tan\varphi}{\sin A_s} = (A_m^s - A_s) + (A_s - A_m) + (\delta_m^s - \delta_m)\frac{\cos\delta_m}{\sin A_m \cos\varphi},\\&(\delta_m^s - \delta_m)\frac{\sin\varphi}{\cos\delta_m} - (H_m^s - H_m)\cos\delta_m \cos\varphi\sin H_m = C.\end{aligned} \quad 9.3\text{-}9$$

Applying the formula for small separation angles in the horizon coordinate system given by Equation 8.6-3 to express the angle between the moon at sunset and the sun, we have

$$d \simeq \sqrt{C^2 + (A_m^s - A_s)^2}.$$

Also, recalling the notation from equation 9.2-13 that

$$\mu = -\tan^{-1}\left(\frac{\tan\varphi}{\sin A_s}\right),$$

and the positive measure $b = \pm(A_s - A_m)$ used in the geometry of the setting problem in Section 9.3.1, the first equation of system 9.3.9 is updated, and we have

$$C\tan\mu \pm \sqrt{d^2 - C^2} = b + (\delta_m^s - \delta_m)\frac{\cos\delta_m}{\sin A_m \cos\varphi},$$
$$(\delta_m^s - \delta_m)\frac{\sin\varphi}{\cos\delta_m} - (H_m^s - H_m)\cos\delta_m \cos\varphi \sin H_m = C.$$

Over the three-quarter of an hour during which sighting takes place, the assumption that $\delta_m^s \approx \delta_m$ generates a tolerable small error and the above system of equations reduces to:

$$C\tan\mu \pm \sqrt{d^2 - C^2} = b,$$
$$(H_m - H_m^s)\cos\delta_m \cos\varphi \sin H_m = C. \qquad 9.3\text{-}10$$

We finally reproduced the result that we have established from simple geometry. The first equation of the above system is a duplicate of Equation 9.3.3, while the second equation needs an interpretation which we will address later.

9.3.4 Applicability of the approximate solution

The linear model is second-order accurate. To have compatibility with the geometric representation, we assumed that $\delta_m^s \approx \delta_m$ and adopted a small separation angle.

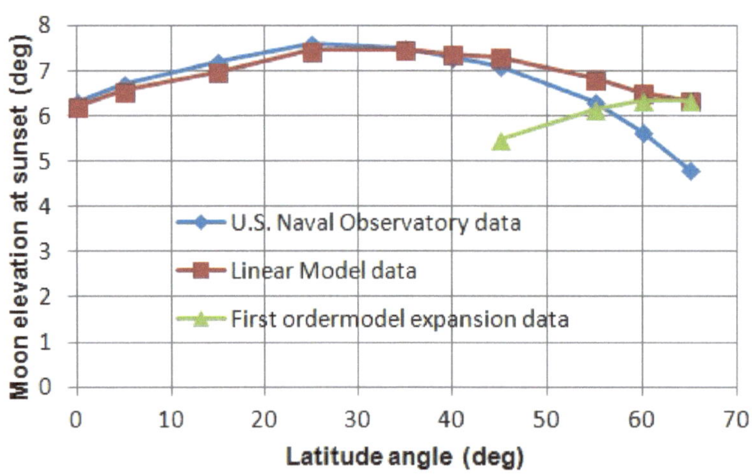

Figure 9.3-5 Linear model prediction of moon elevation versus Naval Observatory data

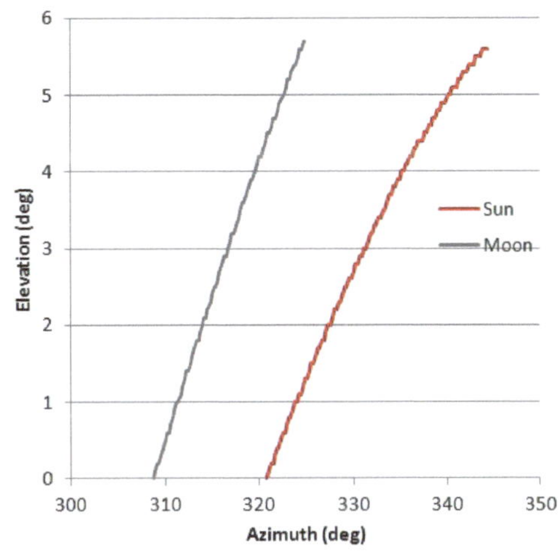

Figure 9.3-6 Settings at higher latitude along the meridian W105° 16', S60° 01'

For illustration, we consider the moonset condition of June 27, 2014. And consider an observer placed on the meridian W105 16'. Conjunction occurred on this day at 8:08 UTC. Figure 9.3-5 shows a good fit between the naval observatory data and the expression for *C* derived based on the linear assumption up to the latitude of about S 55°. The correlation starts breaking down as we go further north. Departure from linearity is clear as illustrated in Figure 9.3-6 below, showing the settings of the sun and the moon for the higher latitude of S60° 01' at the same meridian W105° 16'. Note that the different points in the plot of Figure 9.3-5 are not taking place at the same Universal time and sunset and moonset Azimuths are far from the western direction. Most of the world population lives below 60° latitude, and therefore we have no serious limitation to what we are trying to achieve.

9.3.5 Sun and moon terminators

Recall the coordinate transformations between the horizon and ecliptic reference frames from Section 8.3.6. For a celestial body with altitude *a* and an azimuth *A* in the horizon coordinate system, its declination δ and hour angle *H* are given by Equations 8.3-2, we have:

$$\sin\delta = \sin a \sin\varphi + \cos a \cos\varphi \cos A,$$
$$\cos H = \frac{\sin a - \sin\varphi \sin\delta}{\cos\varphi \cos\delta},$$

When the object is at the onset of rising or setting, its altitude *a=0*, and we have

$$\varphi = \arctan\left(-\frac{\cos H}{\tan\delta}\right). \qquad 9.3\text{-}11$$

This Equation relates the hour angle H *to* the latitude angle φ of the points where such an event gives a trace on the surface of the globe for some declination δ .

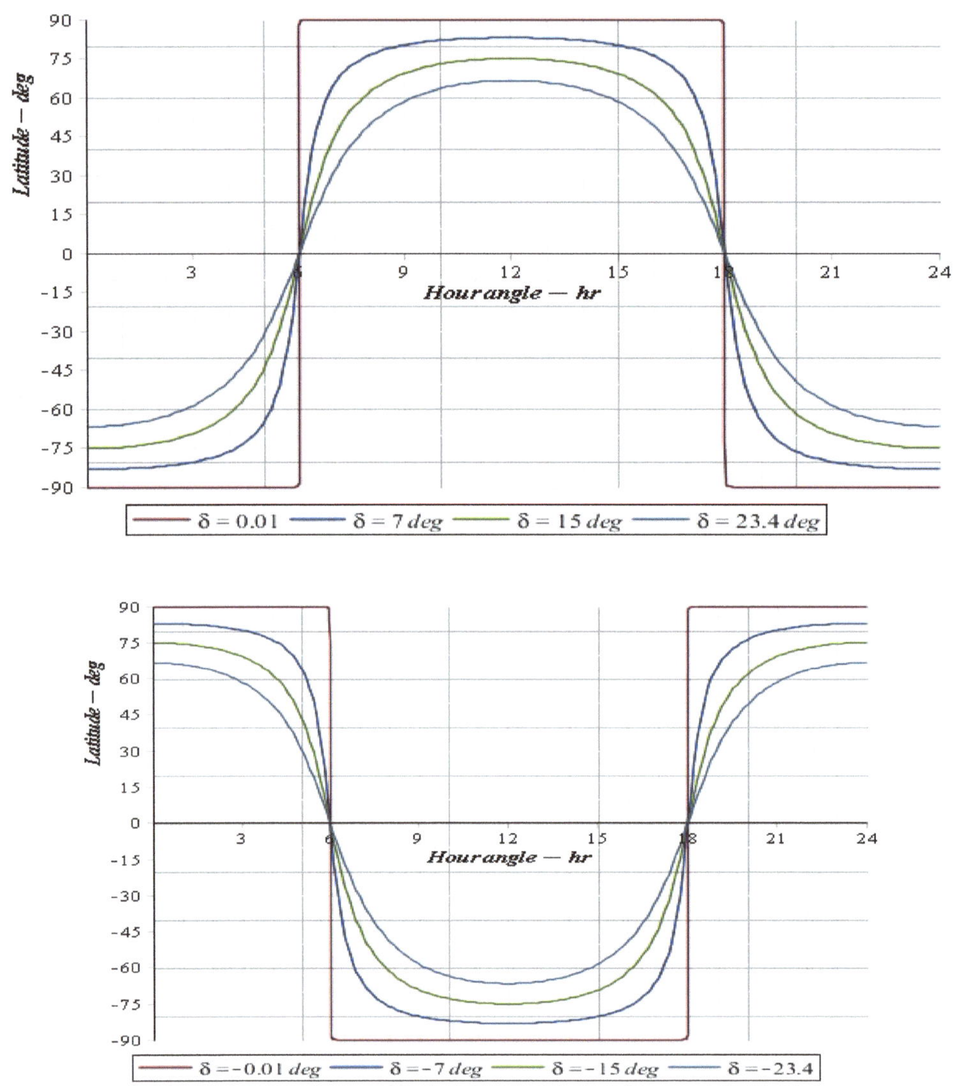

Figure 9.3-7 Terminator traces for several positive declinations (Top) and for several negative declinations (Bottom)

Figure 9.3-7 shows the sun terminator traces for positive and negative declinations between 0.01 and 23.4°. There are several characteristics these curves have. First, all terminators cross at two equatorial nodes (H=6^h, φ =0) and (H=18^h, φ =0). Second, the terminators behave quasi-linearly from these two nodes going either north or south and have the following slopes: $\mathrm{sgn}(\delta)\ \pi/2-\delta$ at (H=6^h, φ=0), and $-\mathrm{sgn}(\delta)\ \pi/2+\delta$ at (H=18^h, φ=0^h). Also, the bell-shaped curves happen to have their extrema at H=0^h and H=12^h. The extrema are given by the same values as the slopes: $\mathrm{sgn}(\delta)\ \pi/2-\delta$ and $-\mathrm{sgn}(\delta)\ \pi/2+\delta$. These are the visibility latitude limits of the celestial sphere discussed in Section 8.5, and to be addressed again in Section 9.5.2. The Hour angle is positive westward and is given in terms of the right ascension using

$$H=\tau-\alpha=\tau_0-L-\alpha \qquad 9.3\text{-}12$$

where τ and τ_o are the local sidereal time and the local sidereal time at Greenwich respectively, and L is the observer's longitude, measured positive west and negative east of Greenwich. The *right ascension α* is positive east of the fixed reference direction of the *vernal equinox ϒ.* If the celestial body happens to be the setting sun ($a=0$ and $H=H_s$), we have from Equation 9.3-11 the latitude of the observer φ at which sunset occurs, given a particular sidereal time τ_0-L . Applying Equation 9.3-11, for the case of the sun, we have

$$\varphi=\arctan\left(-\frac{\cos H_s}{\tan\delta_s}\right) \qquad 9.3\text{-}13$$

It is important to note that the moon and the sun Hour angles are not measured at the same meridian and therefore cannot be plotted in the same reference frame so that a global view is generated. To have the moon terminator in the same diagram as that of the sun, we may seek its relative position from that of the sun as follows:

$$\varphi=\arctan\left(-\frac{\cos H_m}{\tan\delta_m}\right)=\arctan\left(-\frac{\cos(\mathrm{H}_s+\Delta H)}{\tan(\delta_s+\Delta\delta)}\right)=\arctan\left(-\frac{\cos(\mathrm{H}_s-\Delta\alpha)}{\tan(\delta_s+\Delta\delta)}\right) \qquad 9.3\text{-}14$$

where $\Delta H=H_m-H_s=-(\alpha_m-\alpha_s)=-\Delta\alpha$ and $\Delta\delta=\delta_m-\delta_s$. Given the differences in Hour angles and declinations are small right after conjunction, expanding Equation 9.3.14 for the variable pair $(\Delta H,\Delta\delta)$ near (0, 0), we obtain after eliminating the latitude φ between Equations 9.3-13 and 9.3-14

$$(1+\tan\delta_s{}^2)\Delta\delta+\tan H_s\tan\delta_s\Delta H=0,\text{ or} \qquad 9.3\text{-}15$$

$$2\,\Delta\delta+\tan H_s\sin 2\delta_s\ \Delta H=0. \qquad 9.3\text{-}16$$

The above represents a linear relationship between the hour angle separation ΔH and the declination separation $\Delta\delta$ between sunset and moonset in terms of the position of the sun. Recall now Equation 8.6-2, we can write

$$d^2\simeq\cos^2\delta_s\ \Delta\alpha^2+\Delta\delta^2.$$

For a known position for a setting sun and a known angular separation of the moon, the neighboring position of the setting moon is uniquely determined using Equation 8.6-2 and Equation 9.3.16. Furthermore, the points where the lines of moonset and sunset intersect can be obtained using

$$\tan H_s^* = -2\frac{\Delta\delta}{\Delta H \sin(2\delta_s)}, \text{ and } \tan\varphi^* = -\frac{\cos H_s^*}{\tan\delta_s} \qquad 9.3\text{-}17$$

The above equation indicates that there are two roots one for the rising and one for setting sides. The sunset-moonset intersection point is highlighted in Figure 9.3-8 for the particular astronomical elements of $\delta_s = 23^o$, $\delta_m = 17^o$, $\Delta\delta = -6^o$, $d = Ccr = 6.35^o$ and happens to have the coordinates 5.48 Hr and -17.5 deg. This point is important as it provides the necessary but not sufficient limit condition for visibility, due to the sun having to set before the moon. Also shown is the point of first visibility plotted using the coordinates given in Equation 9.5.7 to be covered in Section 9.5.1. Finally, the *Conjunction Visibility Limits* given by the latitude bounds $-\pi/2+\varepsilon<\varphi<\pi/2-\varepsilon$ are also highlighted in the graph.

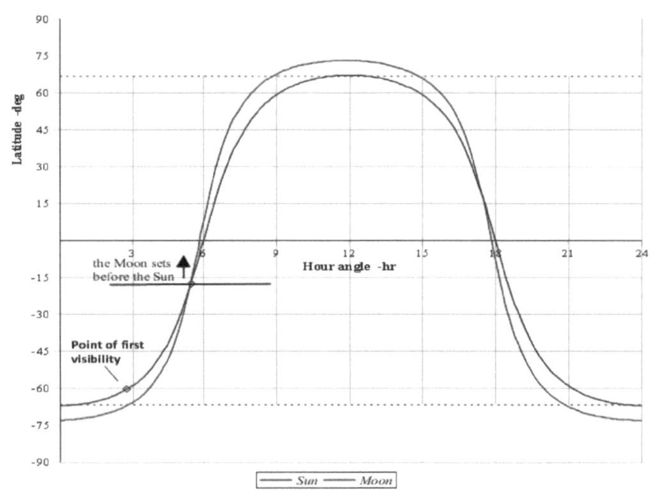

Figure 9.3-8 Analytical terminators of the sun and the moon with setting cross-over point and circumpolar limit visibility lines

The Terminators of the sun and the moon can be reproduced on a cylindrical projection earth map by using the variable change

$$R = H - \tau_0 + \alpha = -L \qquad 9.3\text{-}18$$

with R, the negative longitude direction as abscissae as established in Equation 9.3.12. In Figure 9.3-9, the solid lines are plotted using the ephemeris, and the dots are plotted using Equations 9.3-13 for the sun and 9.3-14 for the moon terminators. Good correlation is obvious from this example.

We have shown that the moon and sun terminators can be produced analytically through linearization together with other pertinent characteristics of nodal intersection points of limit and first visibility.

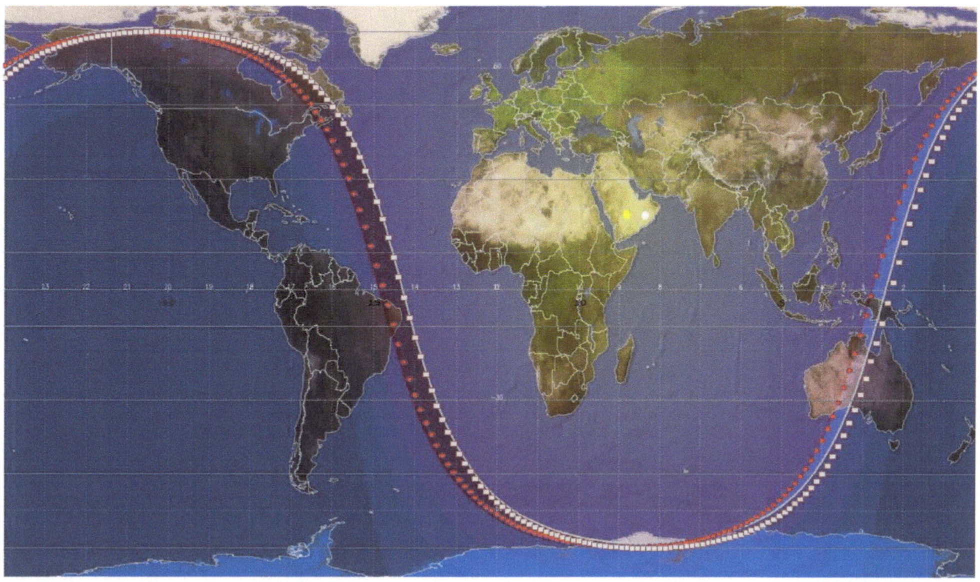

Figure 9.3-9 The variable change given by Equation 9.3.18 is used to plot the terminators of the sun and the moon over a cylindrical map (dots) while continuous lines are generated using ephemeris data on May 21 20:50 UTC 8:46.

9.4 New moon visibility

The physical sighting model of Chapter 7 resulted in a simple expression controlling new moonlight visibility which relates the sum of the angular position of the setting moon and the depression angle of the sun below the horizon to a critical threshold value, *Ccr,* beyond which the contrast increases and therefore visibility starts to be possible. The sum of these two angles is nearly constant and can be identified as the moon elevation at sunset.

Separately, In Section 9.3 we derived a relationship between the elevation of the moon at sunset, *C*, the azimuthal separation between the setting points of the sun and the moon, the latitude of the observer observed φ and how far the moon has separated from the sun via the elongation d. Let us relate the sighting model of Chapter 7 to the fundamental astronomical variables. Recall Equation 9.3.2, which we reproduce here for clarity:

$$C = \sin\left(\mu - \arccos\frac{b\cos\mu}{d}\right)d \ .$$

The altitude of the moon at sunset C is always less or equal to the elongation of the moon d because the sine function is less or equal to unity. This can be verified geometrically by examining the right triangle in the plots shown in Figure 9.3-1 and Figure 9.3-2. In addition, from the second Equation 8.6-1, the angular separation between the moon and the sun will always be greater than the absolute value of the ecliptic latitude of the moon $|\beta_m|$ because the ecliptic longitudes of the sun and the moon are equal and the ecliptic latitude of the sun $\beta_s = 0$ at conjunction. At the start of a lunation, the angle d monotonically increases from $\beta_0 = |\beta_m|$ as $|\lambda_1 - \lambda_2|$ increases from zero at Conjunction. Therefore, the following inequalities hold during any sunset in the hours that follow conjunction:

$$|\beta_m| \le C \le d.$$

As time passes, the moon elongation d increases gradually from its minimum value β_0 at conjunction and reaches the critical value C_{cr} that we established in the derivation of the photometric law. In either one of the scenarios of sun-moon settings, shown in Figures 9.3-1 and 9.3-2, the elongation increases in time while the terminator travels on the earth's surface. When the elongation becomes equal to the critical elevation of the moon altitude at sunset C_{cr} , the relative sun-moon configuration culminates to the limit condition of Figure 9.3-3. The sine term in Equation 9.3-2 is then equal to unity, which also means that there exists latitude μ at which C reaches C_{cr}. This latitude is given by:

$$\mu - \arccos\left(\frac{b\cos(\mu)}{d}\right) = \pi/2 \ .$$

That is

$$\mu = \arctan\left(\frac{b}{d}\right) = \arctan\left(-\frac{\tan\varphi}{\sin A_s}\right), \text{ or } \frac{b}{d} \simeq \tan\varphi \qquad 9.4\text{-}1$$

This is the condition for new moon visibility which is illustrated in Figure 9.3-3, and in which

$$A_m^s = A_s \text{ and } d = C = C_{cr} .$$

It can be easily checked from the geometry of Figure 9.3-2 that the only configuration for which $d = C_{cr}$ is that of Figure 9.3-3. The new crescent moon visibility occurs at the latitude at which sunset occurs when the moon is at elevation $C = C_{cr} = d$. And the determination of the latitude where first visibility occurs together with the time of sunset yields a unique longitude and latitude for the point of first visibility on the surface of the globe. The preceding development assumes that the elongation is smaller than C_{cr} at conjunction. When the elongation is already greater than C_{cr} when at the first sunset $C = C_{cr}$ following conjunction (case occurs when C_{cr} is very small with the new moon near Equinoxes -not of relevance but addressed for completeness), there will not be *necessarily* latitude μ for which *d is* exactly equal to C_{cr} . Instead, the elongation is slightly larger than C_{cr} by a fraction of a degree and visibility is considered to occur immediately in such limiting case.

We have finally reached the stage to tackle the most important topic of when and where the new moonlight visibility occurs on the earth's surface. We shall start by establishing the time and location of ***first new moonlight visibility*** as time elapses past conjunction. Because the new crescent moon ***visibility is limited to a geographic region***, from the time of first visibility, we shall identify the limit points of this region on the path of the sun terminator as it travels westward over time. And finally, we shall determine the ***most favorable*** geographic location of visibility as time progresses following *first visibility*. Even though the moon elongation has exceeded the critical minimum value for visibility, there is a point on the sun terminator, between the two points defining the visibility limits where the elevation of the moon at sunset is maximum.

9.4.1 First visibility of the crescent moon

We have established that the elongation is approximated by the absolute value of the ecliptic latitude of the moon at the time of conjunction. As time elapses, and the terminator moves westward, the elongation keeps on increasing from the absolute value of the angle β_0 . If we consider either one of the configurations shown in Figure 9.3-1 and Figure 9.3-2, maintaining *d* constant and rotating the segment connecting the setting sun to the moon about the point where the sun sets, it can be immediately seen that the configuration of Figure 9.3-3, which we reproduce here as Figure 9.4-1 presents the limit case for which the elongation ***d is***

minimum and, at the same time, the elevation of the moon at sunset C over all sun terminator points is maximum.

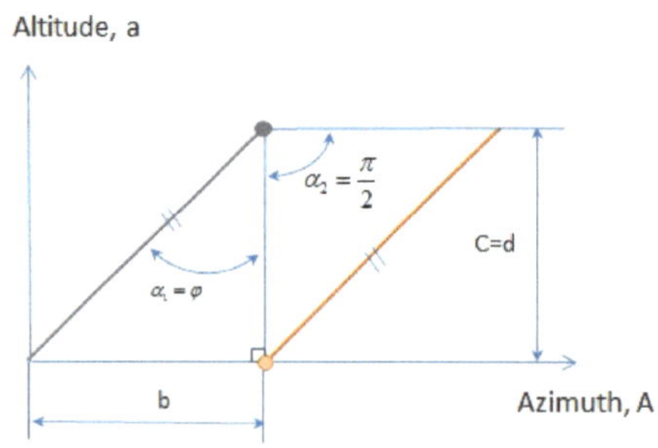

Figure 9.4-1 Limit condition between scenarios (1) and (2) for which C is maximum

That is, because the possible latitudes vary in absolute value between 0 and 90°, at a certain time, there exists a point on the sunset terminator of some latitude $\varphi = \breve{\varphi}$ and therefore a single point on the globe at which *C* is first to reach a maximum value equals C_{cr} for the first time. Therefore, we may write:

$$C = C_{cr} = d; \qquad b = b_{\max} = C_{cr} \tan\breve{\varphi} = d \tan\breve{\varphi}$$

Furthermore, at sunset, we have the elevation is zero and the azimuths of the sun and the moon are given by

$$\cos A_s = \frac{\sin\delta_s}{\cos\breve{\varphi}}, \quad \cos A_m^s = \cos A_s = \frac{\sin\delta_m^s - \sin\breve{\varphi}\sin d}{\cos\breve{\varphi}\cos d}, \qquad 9.4\text{-}2$$

where δ_m^s is the declination of the moon at sunset. Because for this limit case we have $A_s = A_m$, we can write the cosines of the two angles are also equal, and therefore we have:

$$\frac{\sin\delta_s}{\cos\breve{\varphi}} = \frac{\sin\delta_m^s - \sin\breve{\varphi}\sin d}{\cos\breve{\varphi}\cos d}, \qquad 9.4\text{-}3$$

Or after multiplying by $\cos\varphi$ right and left and rearranging, we obtain the relation

$$\sin\breve{\varphi} = \frac{\sin\delta_m^s}{\sin d} - \frac{\sin\delta_s}{\tan d}, \qquad 9.4\text{-}4$$

which gives the latitude at which *C* is at its maximum value *d*. Because *d* is a small angle, we have $\sin d \simeq \tan d \simeq d$, which reduces the above relation to

$$\sin\breve{\varphi} \simeq \frac{1}{d}\left(\sin\delta_m^s - \sin\delta_s\right) \qquad 9.4\text{-}5$$

On the other hand, because both d and $\delta_m^s - \delta_m$ are small, we can take a bivariate Taylor series expansion of Equation 9.4.4 around the point of moonset, i.e., around (d, $\delta_m^s = \delta_m$), we obtain

$$\sin\breve{\varphi} \simeq \frac{\cos\delta_s}{d}\left(\delta_m^s - \delta_s\right). \qquad 9.4\text{-}6$$

Therefore, the latitude $\breve{\varphi}$ at which visibility should occur first is given by

$$\sin\breve{\varphi} = \frac{\sin\delta_m^s}{\sin d} - \frac{\sin\delta_s}{\tan d} \simeq \frac{\cos\delta_s}{d}\left(\delta_m^s - \delta_s\right) = \frac{\cos\delta_s}{C_{cr}}\left(\delta_m^s - \delta_s\right). \qquad 9.4\text{-}7$$

Any one of the three expressions can be used. Note that because the left equality is derived straight from the relation between the Horizon and Equatorial coordinate system, the latitude is exact, as it was established without approximation.

Having determined the latitude of the point where visibility occurs for the first time, we can determine the longitude where such event is bound to happen by expressing the sunset condition, $a=0$ in the expression of the hour angle $\breve{H}_s$ given by second Equation of 8.3.2. The time of sunset is given by Equation 8.5.1. In summary, we have:

$$\begin{aligned} &\sin\breve{\varphi} \simeq \frac{\cos\delta_s}{C_{cr}}\left(\delta_m^s - \delta_s\right), \\ &\cos\breve{H}_s = -\tan\breve{\varphi}\tan\delta_s,\ LST_{\breve{s}} = \alpha_s + \cos^{-1}(-\tan\delta_s\tan\breve{\varphi}) \end{aligned} \qquad 9.4\text{-}8$$

The above result can be established separately by solving simultaneously the quadratic equation giving the moon elongation d when small angles are considered in the Equatorial coordinate system, and the equation giving the elevation of the moon at the time of sunset.

Assuming small angles, we can rewrite Equation 8.6.2 in a normalized form for the particular time when the elongation d increases from its conjunction value β_0 to reach a critical value C_{cr}, and obtain

$$\cos^2\delta_s\left(\frac{\Delta H}{C_{cr}}\right)^2 + \left(\frac{\Delta\delta}{C_{cr}}\right)^2 = 1 \qquad 9.4\text{-}9$$

In the above expression, we have $\Delta H = H_m^s - H_s = -(\alpha_m^s - \alpha_s)$ and $\Delta\delta = \delta_m^s - \delta_s$. Next, for an observer placed on the sun terminator at some latitude φ on the earth surface, we apply Equation 8.3.3 to convert the Equatorial coordinates of the moon to the Horizon ones at the time of sunset and while the moon is slightly above the Horizon. We can write

$$\sin\delta_m^s \sin\varphi + \cos\delta_m^s \cos\varphi \cos H_m^s = \sin a_m^s .$$

By taking the multivariate Taylor series expansion of the position of the moon $(\delta_m^s, H_m^s, a_m^s)$ about that of the setting sun $(\delta_s, H_s, a_s = 0)$ up to second order, we have

$$\begin{aligned} &\sin\delta_s \sin\varphi + \cos\delta_s \cos\varphi \cos H_s - a_m^s - \cos\delta_s \cos\varphi \sin H_s (H_m^s - H_s) + \\ &(-\cos\varphi \cos H_s \sin\delta_s + \sin\varphi \cos\delta_s)(\delta_m^s - \delta_s) = 0 \end{aligned}$$

The first two terms cancel because at sunset, using Equation 8.3.3 for the sun, we have

$$\sin\delta_s \sin\varphi + \cos\delta_s \cos\varphi \cos H_s = 0$$

In addition, multiplying the last equation by $\tan\delta_s$, we obtain (even if $\delta_s = 0$, we have the same result below)

$$\frac{\sin^2\delta_s}{\cos\delta_s}\sin\varphi = -\cos\varphi \cos H_s \sin\delta_s \text{, or}$$

$$\frac{\sin\varphi}{\cos\delta_s} - \cos\delta_s \sin\varphi = -\cos\varphi \cos H_s \sin\delta_s \qquad 9.4\text{-}10$$

Making use of this Equation in the last term of the expansion and simplifying yields

$$-\cos\delta_s \cos\varphi \sin H_s (H_m^s - H_s) + \frac{\sin\varphi}{\cos\delta_s}(\delta_m^s - \delta_s) = -a_m^s \qquad 9.4\text{-}11$$

And finally, writing that $a_m^s = d = C_{cr}$, we have the normalized Equation

$$-\cos\delta_s \cos\varphi \sin H_s \frac{\Delta\alpha}{C_{cr}} + \frac{\sin\varphi}{\cos\delta_s}\frac{\Delta\delta}{C_{cr}} = 1 \qquad 9.4\text{-}12$$

The pair of Equations 9.4.9 and 9.4-12 are solved simultaneously for the normalized unknowns ΔH and $\Delta\delta$. Figure 9.4-2 illustrates the solution as the intersection of the ellipse given by Equation 9.4.9, with the straight line, given by Equation 9.4-12, approximating the relationship between the difference in sun-moon declinations and the separation of hour angles (or right ascensions) at sunset. The following expressions for the intersection point coordinates are obtained (note that the declination separation is established to be independent of the hour angle using the relation $\cos H_s = -\tan\delta_s \tan\varphi$ for a setting sun (see Equation 8.3.2).

$$\frac{\Delta H}{C_{cr}} = -\frac{\Delta\alpha}{C_{cr}} = -\frac{\cos^2\delta_s - \sin^2\varphi}{\cos^3\delta_s \cos\varphi \sin H_s} = -\frac{(1-\sin^2\varphi/\cos^2\delta_s)^{1/2}}{\cos\delta_s},$$
$$\frac{\Delta\delta}{C_{cr}} = \frac{\sin\varphi}{\cos\delta_s} \qquad 9.4\text{-}13$$

In the above solution, the second Equation is identical to the first Equation of System 9.4.8, giving the latitude at which the first new moon crescent visibility occurs. The first Equation gives the corresponding change in right ascension as constrained by the elongation is equal to a critical value, i.e. $d = C_{cr}$. Therefore, we can summarize the above solution as follows. ***The location of first visibility depends on three parameters: the sun elongation d which is also equal to the critical moon height at sunset C_{cr}, the declination of the sun, δ_s and the separation in declination between the moon and the sun, $\delta_m - \delta_s$.***

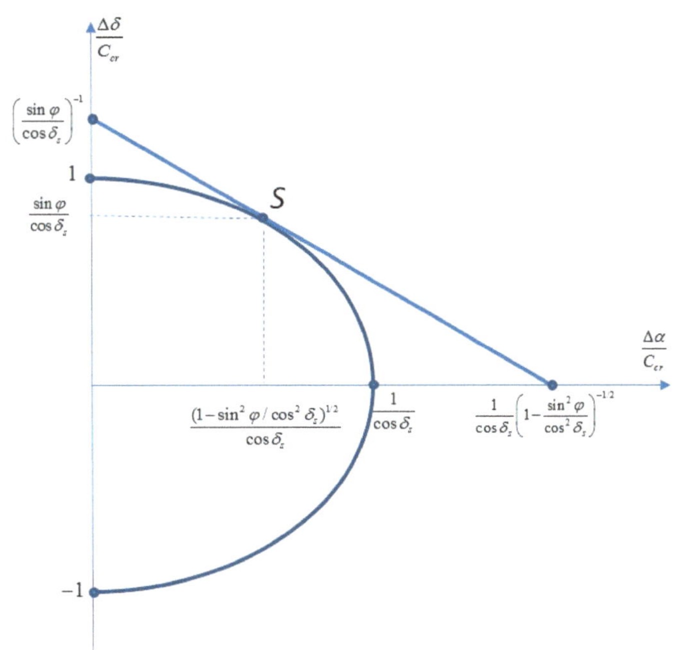

Figure 9.4-2 Geometric interpretation illustrating the point for which $C = C_{cr} = d$, as an intersection between a partial ellipse representing the moon elongation and a linear function representing the expansion of the moon relative position to the sun when its elevation at sunset is equal to its elongation. The results are limited to small separation angles and are second-order accurate.

Before closing, there are three remarks that need to be made. First, the solution illustrated in Figure 9.4-2, which depicts the intersection point *S* of a linear curve with an elliptic curve is also a point of tangency. Written independently of H_s, the slope of the line given by

$$Slope@S = -\frac{\cos^2\delta_s}{\sin\varphi}\left(1-\frac{\sin^2\varphi}{\cos^2\delta_s}\right)^{1/2}.$$

Second, the shape of the ellipse representing the elongation of the moon depends only on the declination of the sun. The ellipse becomes a perfect circle of a unit radius at the Equinoxes. Third, when the latitude is zero, the tangent is vertical, and the point S is at the apex of the ellipse. Finally, we have identified the geographic coordinates of the point where the moon elevation is equal to the elongation *d* and at the same time equals the critical value for visibility C_{cr}. The coordinates of the sun and the moon are known at any time, and the same is true for the elongation *d*. Prior to conjunction, the moon terminator is west of the sun terminator, and as time elapses and while both terminators are traveling westward, the moon terminator starts lagging the terminator of the sun. The elongation starts increasing from the conjunction minimum value β_0 until it reaches C_{cr}, at which time there exists a geographic location on the surface of the earth with latitude $\varphi = \breve{\varphi}$ over which the line of the setting sun of the terminator is passing and the moon elevation in the observer's Horizon coordinate system is also equal to C_{cr}. This latitude is given by $\sin\breve{\varphi} = \cos\delta_s\,\Delta\delta / C_{cr}$.

9.4.2 Geographic visibility limits

There are two geographic limits on new moon visibility. The first one is imposed by the limit visibility of the Celestial Sphere. This limit is called the *Celestial Sphere Visibility Limit*. The second limit is dictated by the fact that the new photometric month never starts prior to the conjunction. We call this limit the *Conjunction Visibility Limit*. We shall see that the astronomical new month characterized by the equal ecliptic latitude of the moon and the sun introduces a limit on the latitude beyond which visibility is not possible.

Celestial Sphere Visibility Limit

In Section 8.5, we discussed the stars that set below the horizon, called circumpolar stars, and by symmetry, there are stars that never rise above the horizon when the observation is carried at certain latitudes. The visibility of the new moon is not spared by such limits. Next, we shall establish the visibility limits for a northern and southern hemisphere observer.

Let us apply the expression for the separation angle in the Equatorial coordinate system, given by Equation 8.6.1, to the moon-sun pair. We have

$$\cos d = \sin\delta_m^s \sin\delta_s + \cos\delta_m^s \cos\delta_s \cos(\alpha_m^s - \alpha_s). \qquad 9.4\text{-}14$$

And for small angles, its tri-variate expansion leads to the following expression:

$$d^2 = \left(\delta_m^s - \delta_s\right)^2 + \cos^2\delta_s(\alpha_m^s - \alpha_s)^2 \ . \qquad 9.4\text{-}15$$

Taking the square of Equation 9.4.7, we have (after dropping the breve symbol from the latitude at which first visibility occurs for simplicity)

$$d^2 = \frac{\cos^2\delta_s}{\sin^2\varphi}\left(\delta_m^s - \delta_s\right)^2 .$$

Eliminating *d* from the last two equations and rearranging, we obtain

$$\left(\frac{\cos^2\delta_s}{\sin^2\varphi} - 1\right)\left(\delta_m^s - \delta_s\right)^2 = \cos^2\delta_s(\alpha_m^s - \alpha_s)^2 \qquad 9.4\text{-}16$$

Because both right-hand side terms are positive, we must have either

$$\frac{\cos\delta_s}{\sin\varphi} \geq 1, \ \text{ or } \ \frac{\cos\delta_s}{\sin\varphi} \leq -1 .$$

For the northern hemisphere, i.e. $\varphi \geq 0$, we have either $\cos\delta_s \geq \sin\varphi$, or $\cos\delta_s \leq -\sin\varphi$. Only the former case is possible because $\cos\delta_s \geq 0$ and $-\sin\varphi \leq 0$. Therefore, we can write that $\cos\delta_s = \sin(\pi/2 - \delta_s) \geq \sin\varphi$, which implies either $\pi/2 - \delta_s \geq \varphi$ when $\pi/2 - \delta_s \leq \pi/2$, i.e., when $\delta_s \geq 0$, or $\pi/2 - \delta_s \leq \pi - \varphi$ when $\pi/2 - \delta_s \geq \pi/2$, i.e., when $\delta_s \leq 0$.

Therefore, in the ***northern hemisphere***, first visibility is bound by:

$$0 \leq \varphi \leq \frac{\pi}{2} - \delta_s, \ \text{when } \delta_s \geq 0 \ \text{ and } \ 0 \leq \varphi \leq \frac{\pi}{2} + \delta_s, \ \text{when } \delta_s \leq 0 \qquad 9.4\text{-}17$$

For the southern hemisphere, i.e. $\varphi \leq 0$ and $\sin\varphi \leq 0$, we have either $\cos\delta_s \leq \sin\varphi$, or $\cos\delta_s \geq -\sin\varphi$. Only the latter case is possible this time because $\cos\delta_s \geq 0$ and $\sin\varphi \leq 0$. Therefore, we can write that $\cos\delta_s = \sin(\pi/2 - \delta_s) \geq -\sin\varphi = \sin(-\varphi)$, which implies either $\pi/2 - \delta_s \geq -\varphi$ when $\pi/2 - \delta_s \leq \pi/2$, i.e., when $\delta_s \geq 0$, or $\pi/2 - \delta_s \leq \pi + \varphi$ when $\pi/2 - \delta_s \geq \pi/2$, i.e., when $\delta_s \leq 0$.

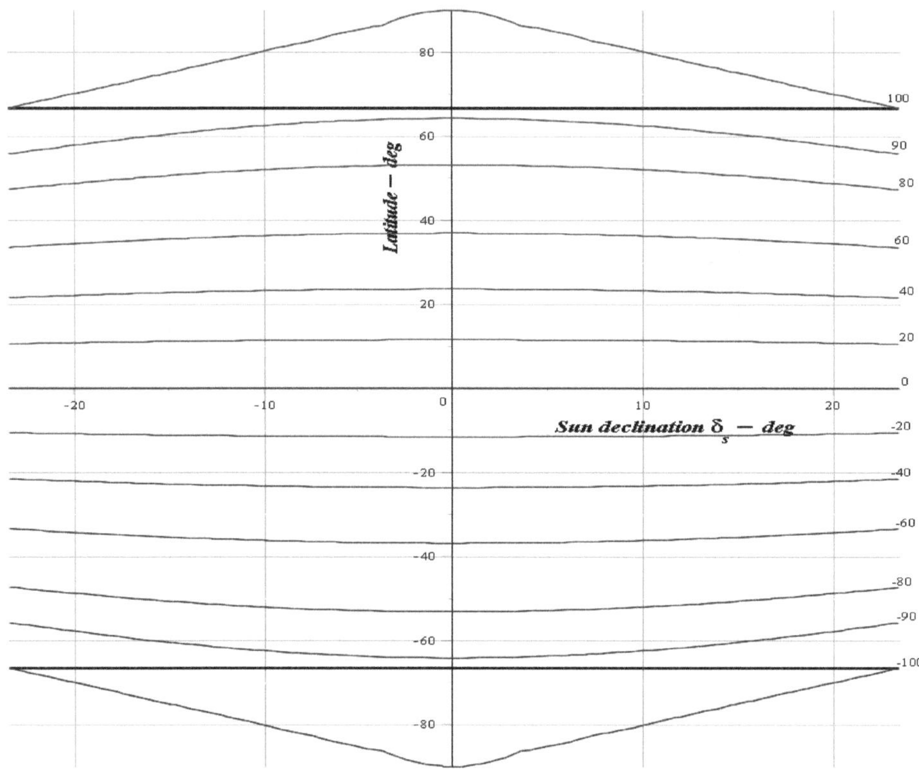

Figure 9.4-3 Contour lines of first New moon visibility for equal values of $(\delta_m^s - \delta_s)/d$, from 0 to 100 %. Also shown are the celestial sphere limit visibility curves $\varphi = \pi/2 \pm \delta_s$ and $\varphi = \pm\delta_s - \pi/2$. Finally, the conjunction 'visibility' limits given by $\varphi = \pi/2 - \varepsilon$, and $\varphi = -\pi/2 + \varepsilon$, i.e. $\varphi = \pm 66^o\ 34'$ are also shown in the plot.

Therefore, in ***the southern hemisphere***, first visibility is bound by:

$$\delta_s - \frac{\pi}{2} \le \varphi \le 0, \text{ when } \delta_s \ge 0 \quad \text{and} \quad 0 \le \varphi \le -\frac{\pi}{2} - \delta_s, \text{ when } \delta_s \le 0 \qquad 9.4\text{-}18$$

For the case of a northern hemisphere observer of an object having a declination $\delta \le 0$, the limit for visibility was established in Section 8.5. It is Understandable that the visibility limit of the celestial sphere comes into the picture, see Figure 9.4-3 because we are dealing with the setting of the sun and the moon at the time of sunset and for the two objects to set, they need to be visible in the first place for a period of time. The figure shows the contour lines for the latitude at which crescent visibility occurs for the first time. It can be immediately verified from the right Equation in 9.4.7 when $\delta_m^s - \delta_s = \pm d$, we have $\sin\varphi = \pm\cos\delta_s$ or $\varphi = \pi/2 \pm \delta_s$

when φ is positive and $\varphi=\pm\delta_s-\pi/2$ when φ is negative, which correspond to the inclined straight limit curves of Figure 9.4-3. The bumpy shape near the poles is due to error round off.

Conjunction 'visibility' Limits

In what follows, we show that the Celestial Sphere visibility limits are not the controlling bounds at higher latitudes. We already showed that, at conjunction, we have the elongation is equal to the ecliptic latitude of the moon, which is not necessarily zero. In addition, in Section 8.7.2, we established Equation 8.7-6 relating the difference between the declinations of the sun and the moon, the ecliptic latitude of the moon and the obliquity angle, $\delta_m^s-\delta_s=\beta_0\cos\varepsilon/\cos\delta_s$. At conjunction, we can write $(\delta_m^s-\delta_s)/d=\pm\cos\varepsilon/\cos\delta_s$ and therefore, we have

$$\frac{(\delta_m^s-\delta_s)}{d}=\pm\frac{\cos\varepsilon}{\cos\delta_s}=\frac{\sin\varphi}{\cos\delta_s}$$

That is, at conjunction we either have:

$$\sin\varphi=\cos\varepsilon\text{, i.e. }\varphi=\pi/2-\varepsilon\text{,}$$

Or for the southern hemisphere,

$$\sin\varphi=-\cos\varepsilon\text{, leading to }\varphi=-\pi/2+\varepsilon\text{ .}$$

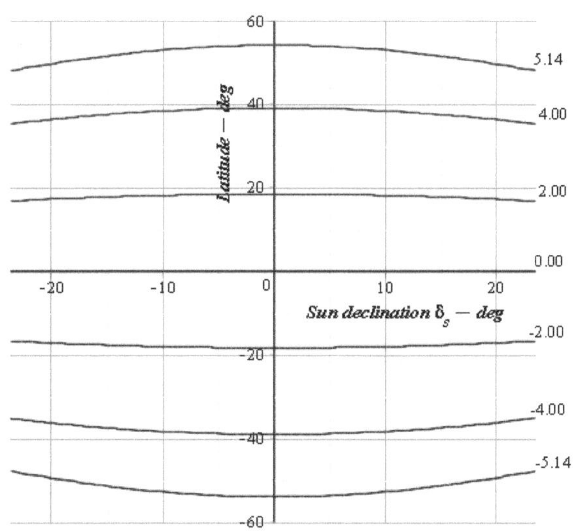

Figure 9.4-4 Contours of first visibility for equal values of d_m^s - d_s. Plot for $d = C_{cr}$ *6.35°*

Because $\varepsilon=23^o\ 26'$, these limits are $\varphi=\pm66^o\ 34'$, and are smaller in absolute value than the visibility limits given by inequalities 9.4.17 and 9.4.18, and therefore should control the extreme bound on visibility. In other words, should visibility be considered possible when the elongation is all the way down to conjunction where $d = |\beta| = \beta_0$ the latitude at which visibility can occur is $\pm 66^\circ\ 34'$. This latitude may be considered as the one where the astronomical new moon is born. And the longitude would be the one at which sunset occurs at the instant of conjunction.

As time passes, the elongation increases gradually and reaches the critical value, e.g., $C_{cr} = 6.35$ degrees as adopted in this text. Depending on the time of the solar year, the sun declination is at some value between - 23.5° and +23.5° and the moon declination is following the path of the sun, also varies due to the inclination of its orbit with respect to the ecliptic by a small angle up to about 5.14°. As a result, in Figure 9.4-4, we have the contour plots for equal values of $\delta_m^s - \delta_s$ between 0 ° and 5.14° and $d = C_{cr} = 6.35°$ giving a maximum latitude value of 54.07° when $\delta_m^s - \delta_s = 5.14°$ and $\delta_s = 0$.

It can be stated that the latitude at which first visibility occurs is mainly a function of the difference between the declinations of the sun and the moon. It can also be stated that:

1. First visibility occurs in the northern hemisphere when $\delta_m^s \geq \delta_s$ and in the southern hemisphere when $\delta_m^s \leq \delta_s$. In addition,
2. The larger the difference between the moon-sun declinations, the further away from the equator is the latitude of first visibility.

Next, from Equations 8.3-2 and 8.3-4, the Hour angle and local sidereal time of the setting sun are given by:

$$H_s = \cos^{-1}\left(-\tan\varphi\tan\delta_s\right) \text{ and}$$
$$\alpha_s = LST_s - H_s = LST_s - \cos^{-1}\left(-\tan\varphi\tan\delta_s\right) \qquad 9.4\text{-}19$$

And the Hour angle and local sidereal time of the moon at sunset are:

$$H_m^s = \cos^{-1}\left(\frac{C}{\cos\varphi\cos\delta_m^s} - \tan\varphi\tan\delta_m^s\right)$$
$$LST_m^s = H_m^s + \alpha_m^s = \cos^{-1}\left(\frac{C}{\cos\varphi\cos\delta_m^s} - \tan\varphi\tan\delta_m^s\right) + \alpha_m^s$$

Or when $C = C_{cr} = d$, at some geographic location where the sun is setting, we have

$$H_m^s = \cos^{-1}\left(\frac{d}{\cos\varphi\cos\delta_m^s} - \tan\varphi\tan\delta_m^s\right), \text{ and}$$

$$LST_m^s = H_m^s + \alpha_m^s = \cos^{-1}\left(\frac{d}{\cos\varphi\cos\delta_m^s} - \tan\varphi\tan\delta_m^s\right) + \alpha_m^s \qquad 9.4\text{-}20$$

From Equations 9.4.19 and 9.4.20, we evaluate the difference in right ascensions and obtain

$$\alpha_m^s - \alpha_s = H_s - H_m^s = \cos^{-1}(-\tan\varphi\tan\delta_s) - \cos^{-1}\left(\frac{d}{\cos\varphi\cos\delta_m^s} + \tan\varphi\tan\delta_m^s\right)$$

Which upon substitution back into the expression for the separation angle between two objects given by Equation 8.6.1 and rearranging, provides the approximation-free relation

$$\cos^{-1}(-\tan\varphi\tan\delta_s) - \cos^{-1}\left(\frac{d}{\cos\varphi\cos\delta_m^s} + \tan\varphi\tan\delta_m^s\right) - \cos^{-1}\left(\frac{\cos d - \sin\delta_m^s\sin\delta_s}{\cos\delta_m^s\cos\delta_s}\right) = 0$$

Taking the multivariate Taylor series expansion of the above expression in the variables (d, δ_m^s) about $(0, \delta_s)$ up to second order, we have

$$2\arcsin(\tan\varphi\tan\delta_s) + \frac{d + \dfrac{\sin\varphi}{\cos\delta_s}(\delta_m^s - \delta_s)}{\sqrt{\cos(\varphi-\delta_s)\cos(\varphi+\delta_s)}} = 0$$

Eliminating $(\delta_m^s - \delta_s)$ by using the approximation given by 9.4.7, we have the following equation,

$$d = \frac{-2\arcsin(\tan\varphi\tan\delta_s)\sqrt{\cos(\varphi-\delta_s)\cos(\varphi+\delta_s)}}{1 + \dfrac{\sin^2\varphi}{\cos^2\delta_s}} \qquad 9.4\text{-}21$$

when $d = C_{cr}$ and $\varphi = \breve{\varphi}$

And at the time of sunset, using the second equation of 8.3.2, we have

$$\sin\frac{d(\cos^2\delta_s + \sin^2\varphi)}{2\cos^2\delta_s\sqrt{\cos(\varphi-\delta_s)\cos(\varphi+\delta_s)}} = -\tan\varphi\tan\delta_s = \cos H_s,$$

when $d = C_{cr}$ and $\varphi = \breve{\varphi}$

or simply

$$H_s = \frac{\pi}{2} - \frac{d(\cos^2\delta_s + \sin^2\varphi)}{2\cos^2\delta_s\sqrt{\cos(\varphi-\delta_s)\cos(\varphi+\delta_s)}} \quad . \qquad 9.4\text{-}22$$

when $d = C_{cr}$ and $\varphi = \breve{\varphi}$

Consider a universal time past the time of conjunction. We focus on the moving solar setting terminator on the globe and assume that the sun has the equatorial coordinates (δ_s, α_s). These coordinates shall remain nearly constant while the elongation increases from the critical value for first visibility over the subsequent few hours. The latitude angle for which the altitude of the moon at sunset C culminates to the elongation is readily obtained from Equation 9.4-21. Also, the local sidereal time or the Hour angle is also readily available from Equation 9.4-22. Figure 9.4-5 shows the contour lines of first visibility for multiple values of critical elevation between 1 and 12 degrees, for positive declinations of the sun, while Figure 9.4-6 presents the same data but considering both positive and negative sun declinations. There are two limit points where visibility is possible which correspond to $C = C_{cr}(\lambda)$. Consider the right-angle triangle connecting the point of sunset, the position of the moon at elevation C and the point of intersection of the line dropped from the moon on the horizon line.

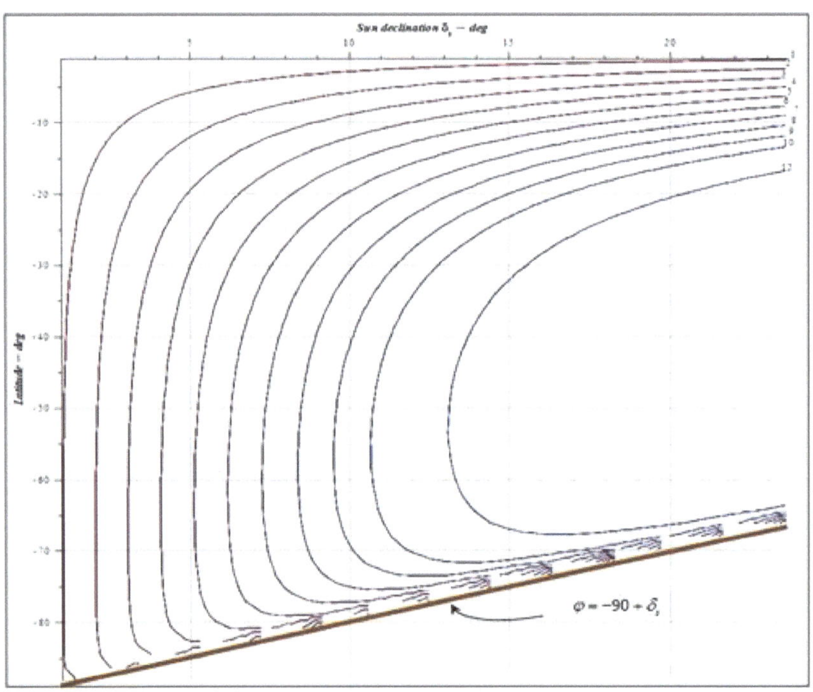

Figure 9.4-5 Latitude angle of point of first visibility, as a function of positive sun declination for different values of C=Ccr=d=1°-12°.

The distance between the setting sun and the third point of this triangle is given by $\sqrt{C^2 - d^2}$, and we have:

$$A_s - A_m^s = \pm\sqrt{d^2 - C_{cr}(\lambda)^2} \,,$$

where (+) is for scenario 1 and the (-) sign is for scenario 2. Rearranging the above expression, we obtain the pair of limit latitude points for visibility as solutions for the angle φ from the nonlinear equation

$$\cos^{-1}\left(\frac{\sin\delta_s}{\cos\varphi}\right) - \cos^{-1}\left(\frac{\sin\delta_m^s - \sin\varphi\sin C_{cr}}{\cos\varphi\cos C_{cr}}\right) = \pm\sqrt{d^2 - C_{cr}(\lambda)^2} \,.$$

An explicit expression for the angle φ may be obtained, it is however quite lengthy, and we use the above form to infer the actual solutions.

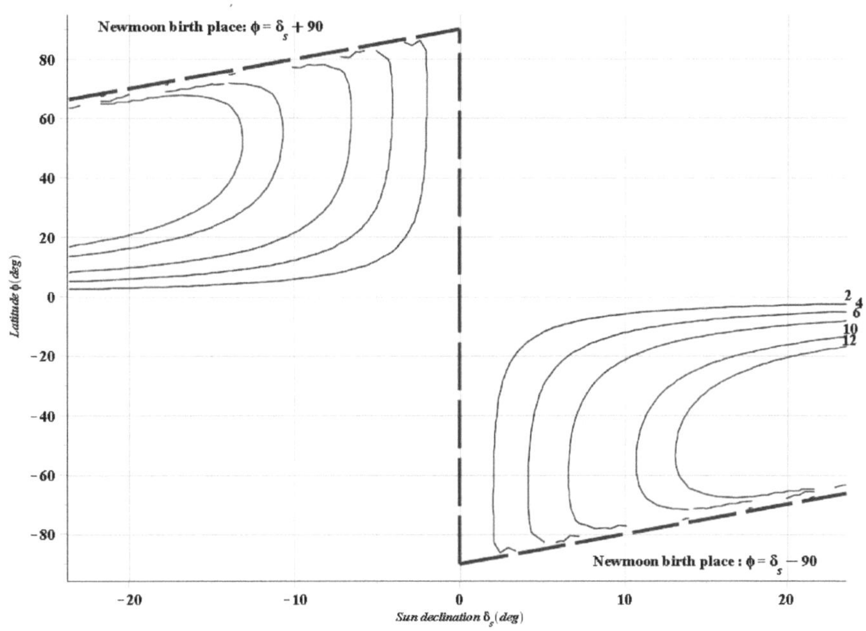

Figure 9.4-6 Latitude angle of crscent first birth point when C=Ccr=2°,4°,6.35°,10°, and 12°. Also shown is the New moon birth line (in dashed line)

9.4.3 Geographic location for optimal new moon visibility

Having established when and where a new moon can be first seen, the next important task is to locate the geographic path for optimal visibility and have the best chance to track the new moonlight. The logical answer is to position the observer on the path where the elevation of the moon at sunset is equal to the elongation over time, as the elongation increases from its minimal value β_o past conjunction. Such path should guarantee the optimal condition, as the elevation of the moon at sunset cannot be larger than the elongation, and visibility is better achieved for higher values of *C* from a photometric standpoint. Figure 9.4-7 shows the latitudes for such paths by plotting the latitudes obtained implicitly in terms of *d* when *d=Ccr* obtained for negative values of δ_s . The corresponding geographic longitudes are those of the points of sunset on the terminator that can be computed from Equation 9.4-22. Also, shown in the Figure is the limit latitude for visibility, dictated by conjunction. The upper region above 6.35° is the region where visibility is possible. The plot is limited 10-degree elongation as the approximation starts to lose accuracy. Finally, when δ_s is below a value little above 10 degrees, there is no longer a single track over which visibility is optimal. For higher declination angles, the progress of new moon visibility is illustrated in Figure 9.4-8 with an example from the lunation of July 27, 2014, in Figure 9.4-9.

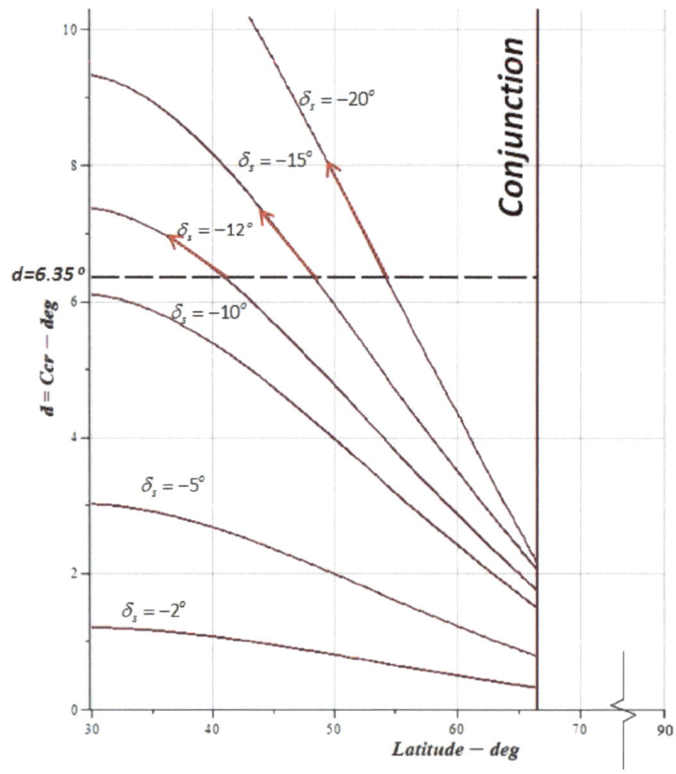

Figure 9.4-7. Optimal paths (arrows) for initial visibility when the sun declination is negative

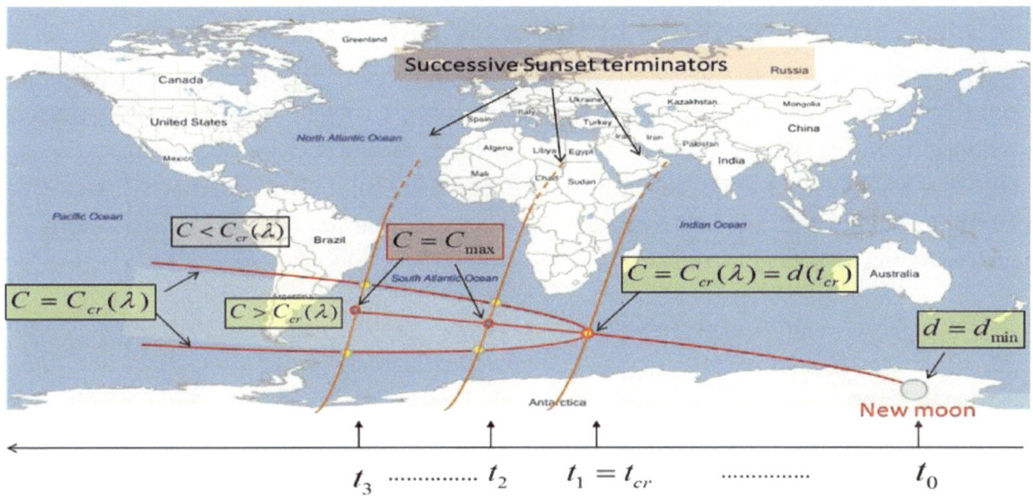

Figure 9.4-8 Point of conjunction, crescent birth track, and domain of first visibility

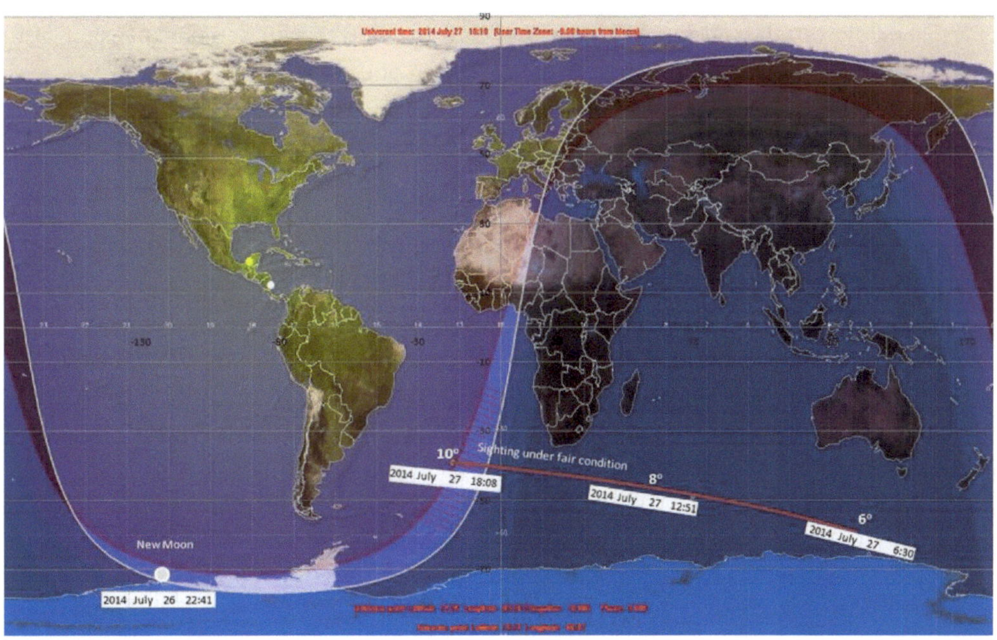

Figure 9.4-9 Point of Conjunction, new moon optimal path for best visibility, July 27, 2014

9.5 Position angle of the new moon crescent

All analyses relied on the geocentric position of the moon. However, knowledge of the position of the limb is important for advanced instrument-based precise tracking of the new moonlight. The position of the center point of the crescent is given by the angle *L* in the spherical triangle *NSM* of the celestial sphere shown in Figure 9.5-1. Point *N* is the celestial North Pole and points *M* and *S* represent the moon and the sun respectively. Therefore, we may write the following relations

$$\begin{aligned} \sin E \sin L &= \cos\delta_S \sin(\alpha_S - \alpha_M) \\ \sin E \cos L &= \sin\delta_S \cos\delta_M - \cos\delta_S \sin\delta_M \cos(\alpha_S - \alpha_M), \\ \cos E &= \sin\delta_S \sin\delta_M + \cos\delta_S \cos\delta_M \cos(\alpha_S - \alpha_M) \end{aligned} \qquad 9.5\text{-}1$$

Dividing the first Equation of 9.5-1 by the second one, we obtain

$$\tan L = \frac{\cos\delta_s \sin(\alpha_s - \alpha_M)}{\sin\delta_S \cos\delta_M - \cos\delta_S \sin\delta_M \cos(\alpha_s - \alpha_M)}. \qquad \mathbf{9.5\text{-}2}$$

The above expression is not restricted to the moon; it is valid for any celestial object, such as a partially illuminated planet. For the case of a new moon, Taylor series expansion of the position angle when the latitude of the sun is near that of the moon gives

$$\tan L = \frac{\cos\delta_S}{\sin(\delta_M - \delta_S)}(\alpha_M - \alpha_S) + O(\alpha_M - \alpha_S)^3. \qquad 9.5\text{-}3$$

That is for small sun-moon relative declination, we have the second-order approximation of the angle given by

$$\tan L \simeq \cos\delta_S \frac{\alpha_M - \alpha_S}{\delta_M - \delta_S}. \qquad 9.5\text{-}4$$

The direction of the bright limb measured from the North when the new crescent moon near the horizon following sunset is shown in Figure 9.5-2. Next, we recall Equation 9.4-7, giving the approximation of latitude $\breve{\varphi}$ at which first visibility occurs, together with the quadratic elliptic Equation 9.4-9. We have

$$\sin\breve{\varphi} \simeq \cos\delta_s \frac{\Delta\delta}{C_{cr}}, \text{ and } \cos^2\delta_s \left(\frac{\Delta H}{C_{cr}}\right)^2 + \left(\frac{\Delta\delta}{C_{cr}}\right)^2 = 1,$$

where $\Delta H = H_m^s - H_s = -\Delta\alpha = -(\alpha_m^s - \alpha_s)$, and $\Delta\delta = \delta_m^s - \delta_s$.

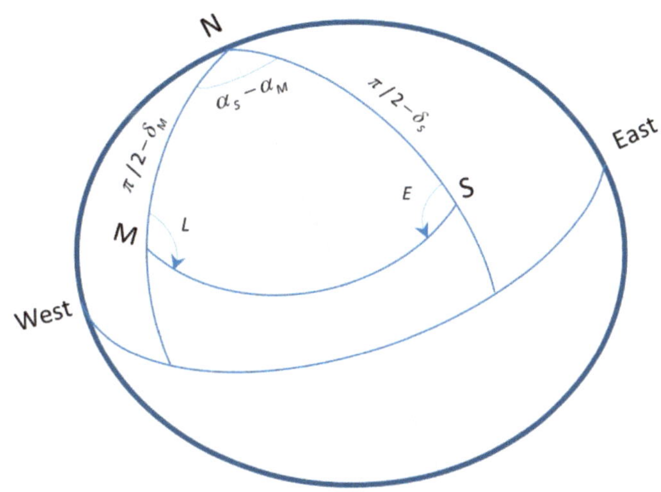

Figure 9.5-1 Celestial sphere, the north celestial pole and the apparent equatorial position of the moon and the sun

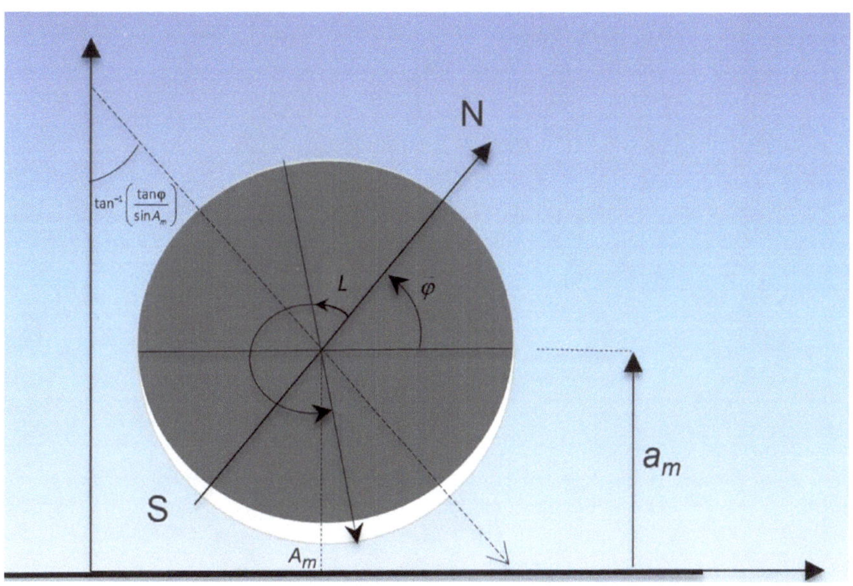

Figure 9.5-2 Position angle of a new crescent moon in a setting moon in the western horizon

Rearranging the above relations, we have

$$\Delta\delta \simeq \frac{C_{cr}\sin\breve{\varphi}}{\cos\delta_s},\ \Delta\alpha \simeq \frac{C_{cr}}{\cos\delta_s}\left(1-\frac{\Delta\delta^2}{C_{cr}^{\ 2}}\right)^{1/2} = \frac{C_{cr}}{\cos\delta_s}\left(1-\frac{\sin^2\breve{\varphi}}{\cos^2\delta_s}\right)^{1/2}.$$

Substituting the last two relations into Equation 9.5-4, we obtain

$$\tan L \simeq \cos\delta_S \frac{\alpha_M - \alpha_S}{\delta_M - \delta_S} = \frac{\left(\cos^2\delta_s - \sin^2\breve{\varphi}\right)^{1/2}}{\sin\breve{\varphi}} = \left(\frac{\cos^2\delta_s}{\sin^2\breve{\varphi}} - 1\right)^{1/2}. \quad 9.5\text{-}5$$

We can also show that

$$\sin L \simeq \cos\delta_s \frac{\Delta\alpha}{C_{cr}} \text{ and } \cos L \simeq \frac{\sin\breve{\varphi}}{\cos\delta_s} \simeq \frac{\Delta\delta}{C_{cr}}, \quad 9.5\text{-}6$$

which can be derived from the angular separation formula for small angles, i.e.

$$d^2 = C_{cr}^{\ 2} = \cos^2\delta_s \Delta\alpha^2 + \Delta\delta^2 .$$

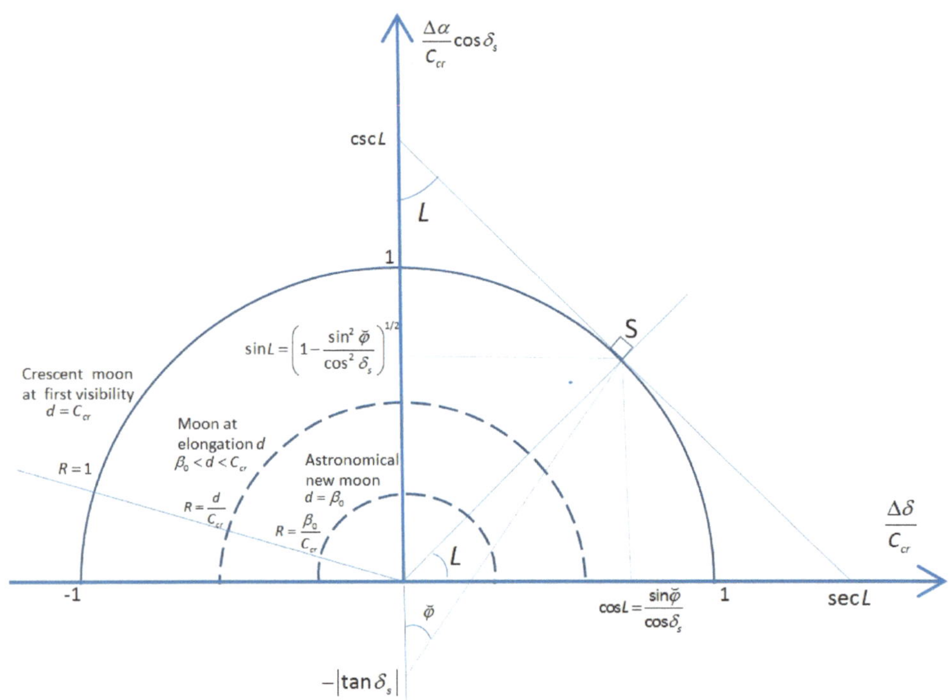

Figure 9.5-3 Direction of the new crescent at first visibility in a normalized coordinate system

It is worth noting that the angle L does not depend on C_{cr} as it can be concluded from the expressions given in 9.5-6; a result that is physically expected. Also, the orientation of the crescent is directly related to the latitude of first visibility. Figure 9.5-3 presents an updated version of Figure 9.4-2, showing the relationship between the parameters for which first visibility occurs together with the orientation of the new crescent. The three concentric circles represent the normalized angular separation expressions at the onset of new crescent visibility, at an astronomical new moon and at an intermediate elongation *d*. The orientation angle *L* of the new moon crescent happens to be the angular coordinate of the unit circle representing first visibility. Point *S* corresponds to the latitude of first visibility $\breve{\varphi}$ as discussed earlier while describing Figure 9.4-2. The points (0,1), (1,0) and (-1,0) correspond to conditions where $\breve{\varphi}$ =0, $\breve{\varphi}=\pi/2-\delta_s$ and $\breve{\varphi}=-\pi/2+\delta_s$ respectively, assuming $\delta_s > 0$. A similar illustration can be obtained by symmetry when $\delta_s < 0$.

9.6 Introduction to new moon field detection

Practical aspects of new moonlight detection are important topics outside the scope of this work. The subject spans multiple disciplines, including instrumentation for imaging, mounting, tracking, image processing, atmospheric optics, weather forecasting, site selection, and procedural logistics, to name a few. Special-purpose instrumentation and methods need to be developed for reliable twilight lunar photometry, detection and imaging of the new moon. From the optical perspective, the polarization and spectral content of new moonlight, in combination with the optical properties of a twilight background sky create a set of unique requirements for detectors and optical systems. Some technological solutions used in conventional nighttime astronomy can be used detection and verification of the start of the month. However, these solutions are often not optimized for the task or do not have the necessary reliability. To highlight some of the differences, the primary requirement for the main optical system in our case is not to gather as much light as possible, like in the case of observing dim celestial objects at night. Instead, the main task is to extract the new silver thin moonlight from diffuse scattered sunlight. This type of astronomy, which we have coined *Twilight Astronomy*, requires a set of optimized tools for detection and verification in the visible and near-infrared spectra of a moving dim moonlight crescent in a horizon sky characterized by a diffuse haze of scattered sunlight.

As new moon observers are set to track the first light in a remote area away from dusty cities, they immediately realize the multiple special challenges they must overcome. First, there is a need to travel to the general area where detection needs to take place which is in the path of first and most favorable visibility, as discussed in Section 9.4. For example, between 2007 and 2017, the first visibility of the new moon of Ramadan occurred in the southern hemisphere during winter. Each year, the task entailed traveling to either Australia, South Africa, South America or the Pacific Islands. Once arrived in the general area, an open site for tracking is identified depending on weather conditions. Flying, driving hundreds of miles and sailing between islands is commonly required to ensure cloud-free and open western skies over hundreds of miles west of the point of observation.

After identifying an open site with a clear view of the western horizon, there is a need for equipment alignment and calibration. Like in the case of nighttime astronomy, the optical system is pointed at the object being tracked over the observation time, so that photons from the new moon are gathered and accumulated. To this end, a drive mount system is required to move the optical assembly which is usually composed of an optical tube which collects light using optics, filters, and imaging hardware. In turn, the drive mount is usually made of two independent subsystems and comes in two flavors, either an Altitude-Azimuth configuration or an Equatorial configuration. Each includes a gear motor, position sensor, and a controller. The controller has built-in knowledge of the locations of the celestial objects as seen from any location and at any time. The mounting requires alignment for precise positioning, usually using well known celestial objects. Obviously, there is no light from the new moon or the stars to go by and therefore when a setup is erected just before sunset, and therefore one could not use techniques from nighttime astronomy. Techniques from solar astronomy and proper equipment capable of handling the pointing at the sun could be adapted. In such case, and with proper safety measures, to avoid permanent damage to equipment and more importantly potential loss of sight, the path of the sun prior to the setting can be used to orient a uniaxial drive mount system. Common biaxial mounts are not necessarily needed for the task. Because the motion of the moon over the several minutes of observation is essentially due to the earth's rotation and the motion of the moon itself can be neglected. After all, interest is not in image quality in this case, and the chief purpose is to photon detection. Finally, alignment using well-defined stars during the night prior to observation requires camping at the same site, which is usually not practical.

Capabilities with precise tracking are available. They are usually heavy and expensive. Lighter systems consist of accurate one axis drive systems, geared to rotate with earth's rotation. They can be used with an optical system with specialized polarization. Such drives maintain the new moonlight in the same position of the field of view in a way similar to nighttime astronomy. As an alternative, we could use a stationary platform for observation with an orientation that encompasses the path of the moon over the few degrees in its final descent to the horizon. The knowledge of the sunset Azimuth and the moon Azimuth at sunset together with the latitude of the observation are enough to properly position the observation system with the field of view centered at the moon trajectory during the critical sighting time.

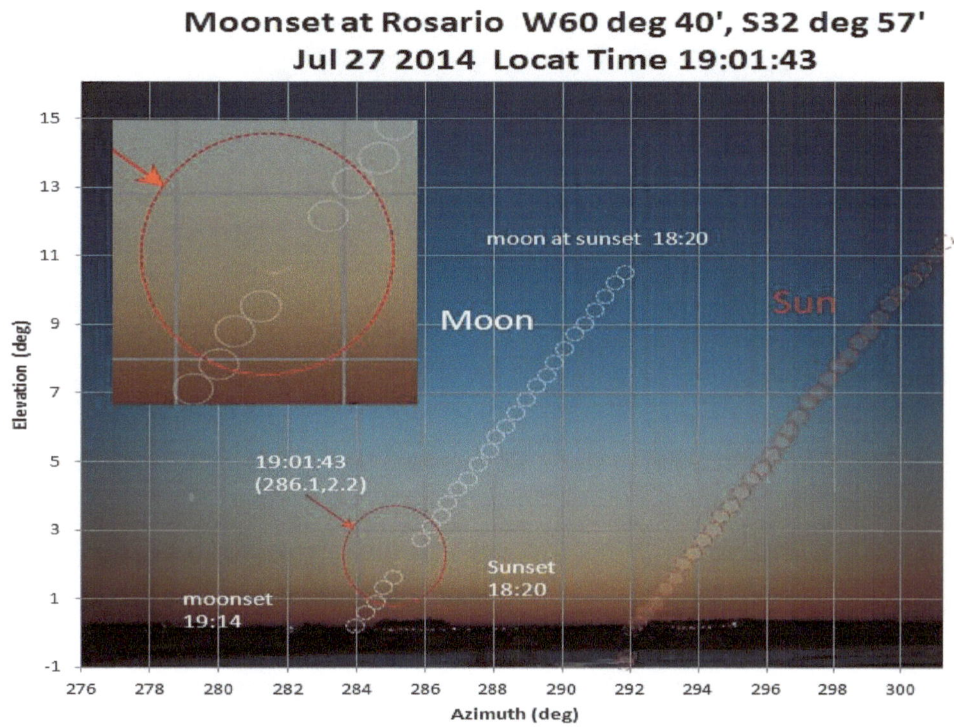

Figure 9.6-1 Rosario W 60deg 40', S32 deg 57' July 27, 2014 19:01:43 (local time)

Shown in Figure 9.6-1 is the case of a new moon sighting just outside the city of Rosario, Argentina on July 2014. The moon and the sun trajectories are shown against the actual background sky. The observation station is kept stationary throughout the observation time and beyond into the Astronomical Twilight, at which time the stars became visible. Images of the background sky were recorded every five seconds. Table 9.6-1 shows the astronomical objects that were observable in the field of view at the time. These objects are also shown in Figure 9.6-2 together with the plane of the ecliptic. The stars in constellation Leo lid up and of prominence are alpha Regulus, Algieba near the Ecliptic plane. These objects registered in multiple frames provided highly accurate post-resolution of the thin moonlight as a function of time, establishing the path of the moon track across the field of view. The field of view of the observing instrument is then mapped using a minimum of two stars with a certain degree of reliability to sub-arcsecond level, and all prior images can be evaluated at the pixel level to address the manifestation of the new moonlight. And as a byproduct, local atmospheric refraction can be determined with a high degree of accuracy from the multiple images.

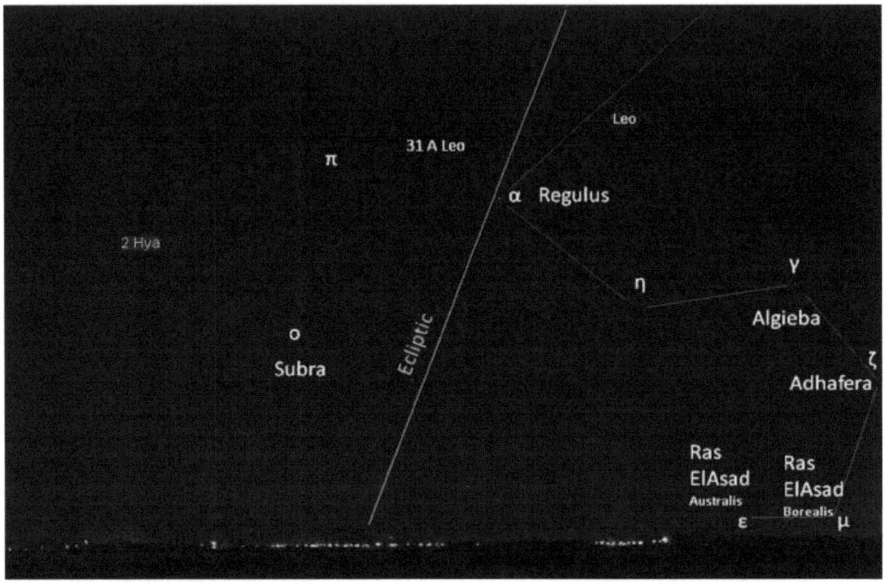

Figure 9.6-2 Objects in the field of view 27-Jul at 19:30:03

Table 9.6-1 Object list in the field of view on 27-Jul 2014 at 19:30:03

Catalogue	RA	DEC	Name	Mag		Common name
Star	09h46m39.58s	+23°42'23.4"	Eps Leo	mV: 2.97	HD:84441	Ras Elased Australis
Star	09h41m54.84s	+09°49'29.6"	Omi Leo	mV: 3.52	HD:83808	Subra
Star	09h53m34.38s	+25°56'16.3"	Mu. Leo	mV: 3.88	HD:85503	Ras Elased Borealis
Star	09h39m12.10s	+04°34'55.8"	2 Hya	mV: 4.68	HD:83425	Fl:2
Star	10h09m08.04s	+11°53'43.3"	Alp Leo	mV: 1.36	HD:87901	Regulus
Star	10h20m45.59s	+19°46'03.5"	Gam01 Leo	mV: 2.01	HD:89484	Algieba
Star	10h17m29.00s	+23°20'40.7"	Zet Leo	mV: 3.43	HD:89025	Adhafera
Star	10h08m06.69s	+16°41'28.1"	Eta Leo	mV: 3.48	HD:87737	Fl:30
Star	10h00m58.22s	+07°58'24.3"	Pi. Leo	mV: 4.68	HD:86663	Fl:29

9.7 Modern new moon sighting data

In this section we review published new moon sighting data and investigate how well the development which we presented in the previous sections of this chapter fits the reported sighting data. Two databases are retrieved for comparison. The first one is from Caldwell and coworkers from the South African Astronomical Observatory (SAAO); they maintained a database of credible and critical observations which was obtained from the literature, the Internet and members of the SAAO's own efforts (Caldwell, 2001). The second database has been maintained by the Islamic Crescent Observation Project (ICOP), a group of observers interested in astronomical topics related to Islamic applications.

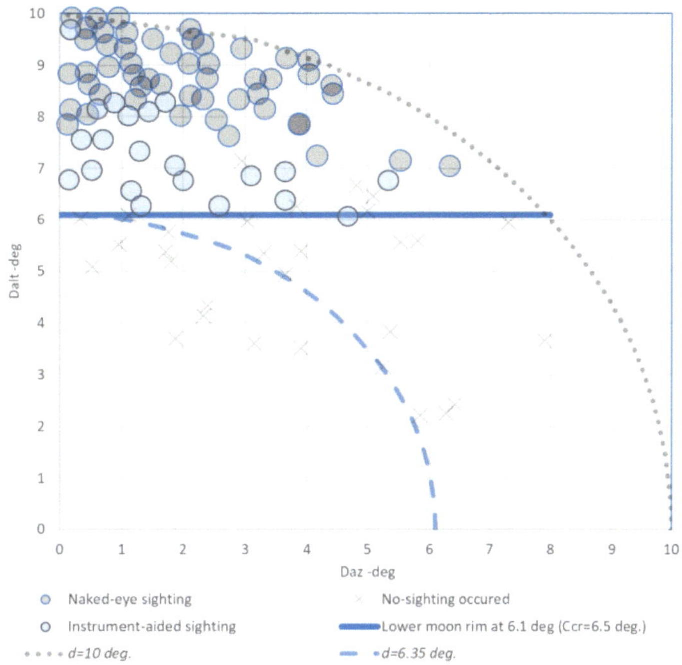

Figure 9.7-1 Sighting records from the SAAO database **(Caldwell, 2001)**, restricted to moon elevations and elongations less than 10 degrees. All successful sightings correspond to moon elevations at sunset above 6.35 deg (lower moon limb above 6.1 deg).

The SAAO database is limited to crescents within a restricted altitude range relative to the setting sun, which excludes all relatively trivial sighting events. Reports from multiple observers at the same event and at nearly the same location are condensed to one entry. The basic sources for the "historical" sightings are the compilations by Schaefer (Schaefer B. , 1988), Schaefer et al. (Schaefer B. A., 1993), Doggett and Schaefer (Doggett L.E. and Shaefer, 1994), Ilyas (Ilyas, 1994), and Schaefer (1996). The positions of the moon and the sun in the SAAO database were obtained using the Interactive Computer Ephemeris (ICE) program from the US Naval Observatory Almanac Office.

The SAAO sighting data subset with less than 10-degrees elevation of the moon at sunset, less than 10-degree relative sun-moon azimuth at sunset, and less than 10 deg elongation is retained for analysis. While this restriction is necessary to retain second order accuracy of the linear model of Section 9.3 there is no limitation on the prediction of first-visibility events which occur at values below the 10-degrees limits. The dataset is shown in Figure 9.7-1 and it includes 73 reliable sightings of the crescent moon (52 with the naked-eye and 21 instrument-aided), and 30 negative sightings, for a total of 103 observation events. The figure represents entries of the elevation of the new moon lower rim at sunset, Dalt, which is equal to the parameter *C* minus the radius of the moon, as a function of the relative moon-sun azimuth at sunset, Daz, which is the square root of $(d^2 - C^2)$, as it can be deduced from Figures 9.3-1- 9.3-3.

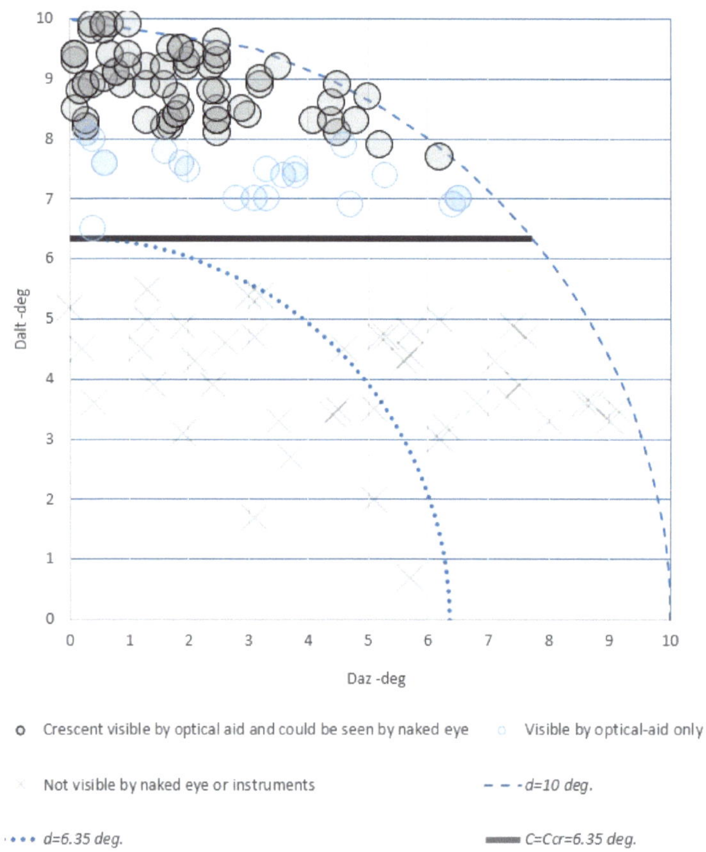

Figure 9.7-2 Sighting records from the ICOP database **(Odeh, 2004)** restricted to moon elevations and elongations less than 10 degrees.

All confirmed sightings correspond to moon elevations above the 6.35° limit which we established in Section 7.9 . Note that the figure shows a limit of 6.1° which is equal to 6.35° minus half the radius of the moon, which is about 0.25° , because the observations of the database refer to the lower rim of the moon and not to its center. The figure also shows that the negative attempts for sighting are mainly below the 6.35° limit. The few cases above the limit could be due to any one or combination of the factors preventing sighting that we discussed earlier. It can also be noted that most positive naked-eye sightings took place for lower values of Daz (closer to the vertical axis), i.e. when *C* is closer to *d*, which is in-line with the current finding that states first-sighting occurs when *C=d=Ccr*. Furthermore, the higher the value of C, the easier the sighting is possible, while the lower the value of C is the more helpful instrumentation is in confirming positive sighting. Note the instrument-aided sighting are closer the 6.35° limit, and most of the naked-eye are at higher values of C. The ICOP database expands the SAAO database to include additional sightings from undocumented private communication sources and ICOP project sightings (Odeh, 2004). The sightings are illustrated in Figure 9.7-2 showing similar trend and conclusions to those of the SAAO. It should be noted that both the SAAO and ICOP databases include few positive sightings where C is less than the 6.35° limit, but at values of d greater than 10°. Such sightings -if credible- correspond to either larger azimuthal separations from the sun, and therefore do not correspond to first visibility events, or could have been obtained using infrared equipment operating outside the visible spectrum. Infrared observations are excluded from what is deemed sighting as discussed earlier. It is unfortunate that both databases are missing information on the nature of equipment used in sightings.

9.8 Higher order crescent moon sighting model

In this section, we shall introduce a second order accurate sighting model of a new crescent moon. This model is useful when the relative setting angles of the sun and the moon become large; a condition which can arise at some geographic location past the time of first visibility. In addition, we provide an assessment of the scattering angle and phase function when such angle becomes large enough and violates the limits of the linear approximation model which was introduced earlier in this chapter.

The linear model of first crescent moon visibility is valid when the combined geometry of sunset and moonset involve small angles. These angles include the moon elevation at sunset, the moon elongation, and sun-moon azimuth difference. From basic trigonometry, it can be shown that accuracy is maintained when such angles are limited to about 10 degrees, which is luckily the case when considering first sighting over the globe. From numerical data, the angle that is most affected by the limitation of linearity is the sun-moon setting azimuth difference. In Section 9.2.2, we saw that the setting azimuth angle which happens to be in a westerly direction is second order accurate. But the difference between the sun and the moon azimuth angles can introduce loss of accuracy; a situation which may occur in a sighting of the crescent moon at some geographic location at a time past that of the first global sighting.

At times, we also assumed that the setting direction is within ten degrees from the west, so that we have a formulation which is independent of the azimuthal location of the setting point as discussed in Section 9.2.2. All these assumptions were shown to be second order accurate. The elongation which we discussed in Section 8.6 can be represented by third order accurate expressions given by equations 8.6-2 and 8.6-3, depending which coordinate system is used.

The published data combining Babylonian records with modern observation, which will be discussed in Section 10.1 has two examples of such observations where the moon elevation at sunset is below the **6.35°** limit and the relative moon-sun azimuth at sunset is beyond the 10 degree limit as highlighted in Figure 9.8-1. Obviously, these two observations are not close to the vertical axis which correspond to first visibility in a global sense (*d=C*. Note the large number of confirmed sightings are concentrated near the vertical axis).

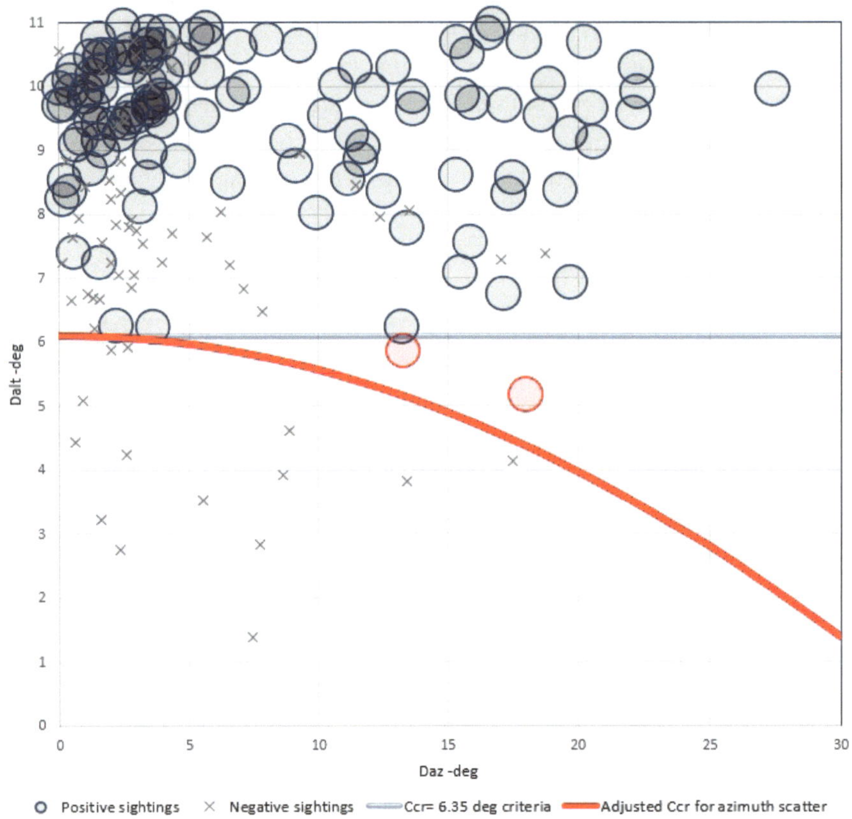

Figure 9.8-1 Sighting records combining Babylonian records with modern records published in Fatoohi (Fatoohi, 1999) showing two positive observations with elevation at sunset below the 6.35° limit and large relative moon-sun azimuth at sunset, Daz, which is $(d^2 - C^2)^{.5}$.

The second equation of 8.3-3 is one of the two equations that relate the horizon coordinates to the equatorial coordinates. We rewrite this equation in the following form:

$$\cos A \cos\varphi \cos a - \sin\delta + \sin\varphi \sin a = 0$$

Expanding the left-hand side about the set point on the horizon, given by the coordinates

$$\left(A_{\swarrow} = \cos^{-1}\left(\frac{\sin\delta}{\cos\varphi}\right),\ a = 0 \right),$$

Has led to the linear expression $a = \dfrac{\sin A_{\swarrow}}{\tan\varphi}(A - A_{\swarrow})$, given in equation 9.2-8. A second order expansion of the second equation of 8.3-3 leads to the following equation of a circle

$$\left(A-(A_{\swarrow}-\tan A_{\swarrow})\right)^2+\left(a-\frac{\tan\varphi}{\cos A_{\swarrow}}\right)^2=\left(\frac{\tan A_{\swarrow}}{\cos\mu_{\swarrow}}\right)^2 \qquad 9.8\text{-}1$$

which center *Cs* and radius *S* are given by:

$$Cs=\left(A_{\swarrow}-\tan A_{\swarrow},\frac{\tan\varphi}{\cos A_{\swarrow}}=-\tan A_{\swarrow}\tan\mu_{\swarrow}\right)\text{, and}$$

$$S=\left(\frac{\tan^2\varphi}{\cos^2 A_{\swarrow}}+\tan^2 A_{\swarrow}\right)^{0.5}=\left|\frac{\tan A_{\swarrow}}{\cos\mu_{\swarrow}}\right|.$$

Differentiating the equation of the circle with respect to the azimuth and the elevation to obtain a tangent vector to the trajectory at the setting point, we obtain the following: $\left[\tan\varphi/\cos A_{\swarrow},\ \tan A_{\swarrow}\right]^T=\left[-\tan A_{\swarrow}\tan\mu_{\swarrow},\ \tan A_{\swarrow}\right]^T$, or $\left[-\tan\mu_{\swarrow},\ 1\right]^T$, after eliminating the factor $\tan A_{\swarrow}$. Figure 9.8-2 shows the arc circles of the trajectories of a setting moon (left highlighted arc above the horizon) and a setting sun (highlighted arc below the horizon on the right circle). The setting moon trajectory is shown to emanate from the position of coordinates (A_m^s, a_m^s) occupied by the moon at the time of sunset and ends at its own setting point $(A_m, 0)$. The highlighted part of the solar trajectory is shown to start from the point of sunset $(A_s, 0)$ and ends at some position that corresponds to the onset of moonset. Also shown in the figure are the approximate linear model trajectories with tangencies to the arc circles at the lunar and solar setting points, given by $A_{\swarrow} = A_m$ and $A_{\swarrow} = A_s$, respectively. The tangents make the angles μ_m and μ_s with the vertical (altitude axis), given by the same angles of the linear models. In other words, the tangents constitute the linear setting trajectories of the moon and the sun, given by

$$\tan\mu_m=-\frac{\tan\varphi}{\sin A_m},\ \text{and}\ \tan\mu_s=-\frac{\tan\varphi}{\sin A_s}.$$

We have so far provided the second order relationship between the elevation and the azimuth of a setting celestial body. The other two parameters are the elongation d at sunset, and the azimuthal difference between the points of sunset and moonset, b. Luckily the expressions given in equations 8.6-2 for the elongation happen to be third-order approximations and therefore at a minimum second order accurate. We have

$$d \simeq \sqrt{\cos^2 \beta \, (\lambda_1 - \lambda_2)^2 + (\beta_1 - \beta_2)^2} \simeq \sqrt{\cos^2 \delta \, (\alpha_1 - \alpha_2)^2 + (\delta_1 - \delta_2)^2} \simeq \sqrt{(a_1 - a_2)^2 + (A_1 - A_2)^2}.$$

The parameter b is given from Equation 9.3-5:

$$b = A_s - A_m = \cos^{-1} \frac{\sin\delta_s}{\cos\varphi} - \cos^{-1} \frac{\sin\delta_m}{\cos\varphi}$$

And its second order accurate approximation can be written as

$$b \simeq \frac{1}{\tan\delta_s \tan A_s}(\delta_s - \delta_m) - \frac{1}{2} \frac{\tan\varphi^2}{\tan A_s \sin A_s^{\,2}}(\delta_s - \delta_m)^2 + O(\delta_s - \delta_m)^3.$$

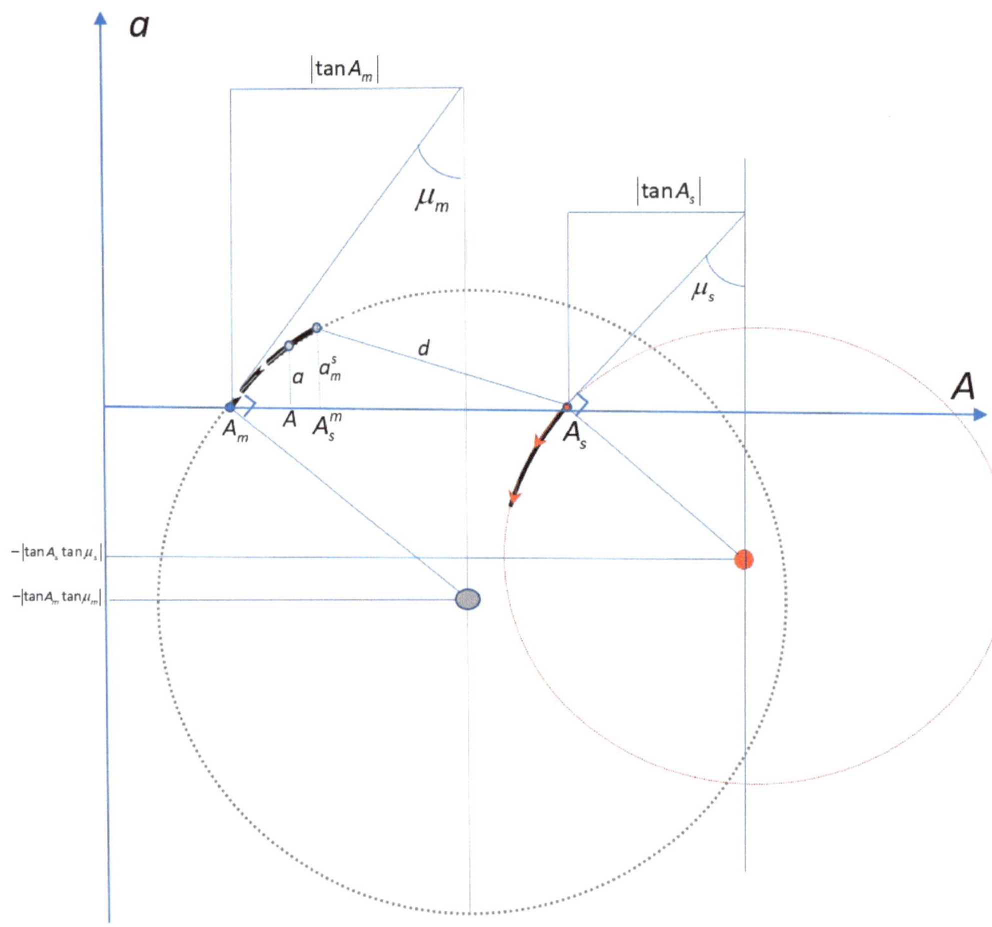

Figure 9.8-2 Second order approximation of a setting moon (left arc circle) and a sun below the horizon (right arc circle)

Geometrically, using the same argument as in the linear model, as the elongation d increases from its value at an astronomical new moon, there will necessary be a time when d reaches

C_{cr}, the critical value for the moon elevation at sunset. And at that time, there is a unique location on the globe at which the elevation of the moon C is first to reach C_{cr}. Back to the second equation of 8.3-3, if we write the expression at the lunar position during sunset, we have

$$\cos A_m^s \cos a_m^s = \frac{\sin\delta_m^s}{\cos\varphi} - \tan\varphi \sin a_m^s \text{, and}$$

using the same assumption of Section 9.3.3 that $\delta_m^s \approx \delta_m$ over the few minutes of sighting, generates a tolerable small error, leads to

$$\cos A_m^s \cos a_m^s - \cos A_m + \tan\varphi \sin a_m^s \simeq 0$$

because $\sin\delta_m / \cos\varphi = \cos A_m$. Next, expanding the above Equation about the lunar setpoint given by the coordinates $(A_m, a_m = 0)$ up to a second order, leads to

$$+a_m^s \tan\varphi - \frac{1}{2}\cos A_m a_m^{s\,2} - \sin A_m^s (A_m^s - A_m) - \frac{1}{2}\cos A_m (A_m^s - A_m)^2 \simeq 0\,,$$

which a second order polynomial in $(A_m^s - A_m)$ with the solution

$$(A_m^s - A_m) = -\tan A_m \pm \left(\tan A_m^2 + 2a_m^s \frac{\tan\varphi}{\cos A_m} - a_m^{s\,2} \right)^{0.5} \text{, and finally}$$

From the geometry of Figure 9.8-2, we have $b = A_m^s - A_m + (d^2 - a^2)^{.5}$, which leads to

$$\begin{aligned} b &= (A_m^s - A_m) = -\tan A_m \pm \left(\tan A_m^2 + 2a_m^s \frac{\tan\varphi}{\cos A_m} - a_m^{s\,2} \right)^{0.5} + \sqrt{d^2 - a_m^s}, \quad \text{if } A_m \neq \frac{3}{2}\pi \text{ and} \\ b &= -a_m^s \tan\varphi + \sqrt{d^2 - a_m^s}, \quad \text{if } A_m = \frac{3}{2}\pi. \end{aligned} \qquad 9.8\text{-}2$$

If we examine the above bivariate expansion, we can immediately notice that the quadratic terms are actually small (near 3rd order, which explains the success of the linear model) because $\cos A_m$ is near zero in a westerly direction. Dropping the quadratic azimuth term only would result in what we call a *serendipity quadratic model* with an immediate extension to the linear model form Section 9.3.3 :

$$b = a_m^s \frac{\tan\varphi}{\sin A_m} - \frac{1}{2}\frac{\cos A_m}{\sin A_m} a_m^{s\,2} + (d^2 - a_m^{s\,2})^{.5} \,. \qquad 9.8\text{-}3$$

Examination of the quadratic model terms does noes yield an interpretation for local first visibility observations when *d* and *b* are large. In other words, adjustment of the model $C = C_{cr} = d$ criteria through exploiting the nonlinearity of Equation 9.8-2 or Equation 9.8-3 would not explain situations where local sighting occurs for the first time. However, the light scattering angle from an already set sun can no longer be approximated by the elevation of the moon at sunset when b and d are large. And this, in turn, makes the valued scattering phase function different from the one adopted in Section 7.3.

To evaluate the effect of larger scattering angles, let us assume a setting moon subjected to a scattered sunlight in which the scattering angle C_1 is small, and vary the azimuthal sun-moon difference by an angle ω. It can be shown that the scattering varies from angle C_1 to an angle C_2, were

$$\cos C_2 = \cos\omega \, \cos C_1 \qquad 9.8\text{-}4$$

To illustrate this effect numerically, we assume an initial situation in which $C_1 = 6.35^o$ and $b = 0$. Then, we plot the variation of C_2 as a function of the sun-mon azimuth difference, *b*. The result is shown in Figure 9.8-3. The effect of the azimuthal distance on the scattering angle is obviously negligible when such angles are small. However, when such angles are large, the scattering angles increases from the elevation angle at sunset, *C* to reach the asymptotic value of the sun-mon azimuth difference, *b*.

The corresponding phase function, given by Equation **7.4-6** which is illustrated in Figure 9.8-4, does not vary much either at small angles. No appreciable changes can be observed when the azimuthal sun-moon difference is small (less than 10 deg). The effect is however important at larger azimuth difference angles.

Next, we recall the contrast illustrated in Figure 7.8-1. When the elevation of the moon becomes larger than a critical value, the contrast starts increasing at a near-constant slope over a range of sun declinations until the moon approaches the horizon, at which time the contrast declines until the moon finally sets.

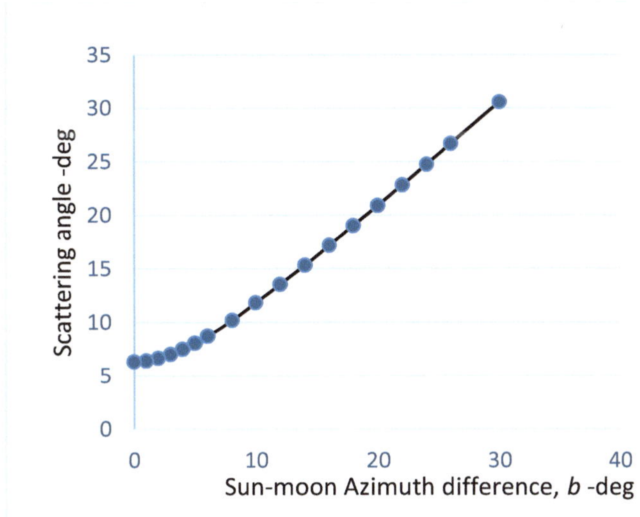

Figure 9.8-3 Scattering angle variation with sun-moon azimuthal difference, *b*

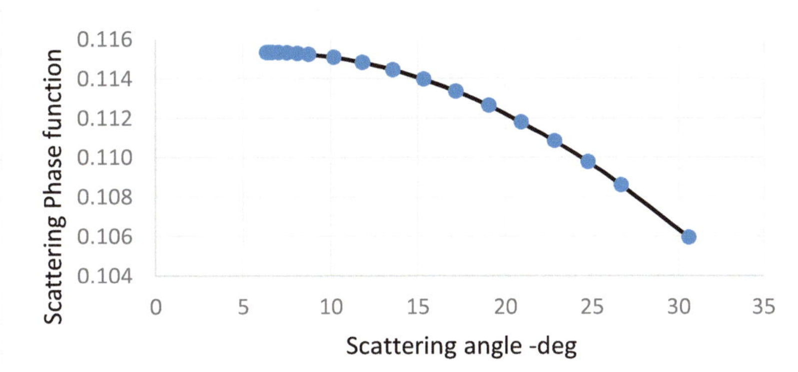

Figure 9.8-4 Variation of the scattering phase function with sun-moon azimuthal difference, *b*

The model is geared for small angles but for larger angles b, a corrective normalized function for the contrast variation with moon-sun azimuthal separation illustrated in Figure 9.8-5 can be updated to reflect the b-effect. Applying this correction to the $C = C_{cr}$ leads to the adjustment highlighted in Figure 9.8-1, which interpret the two observations at times past first global sightings, as the new crescent moon starts being visible elsewhere and for which the elevation of the moon at sunset could be below the limit of 6.35 deg.

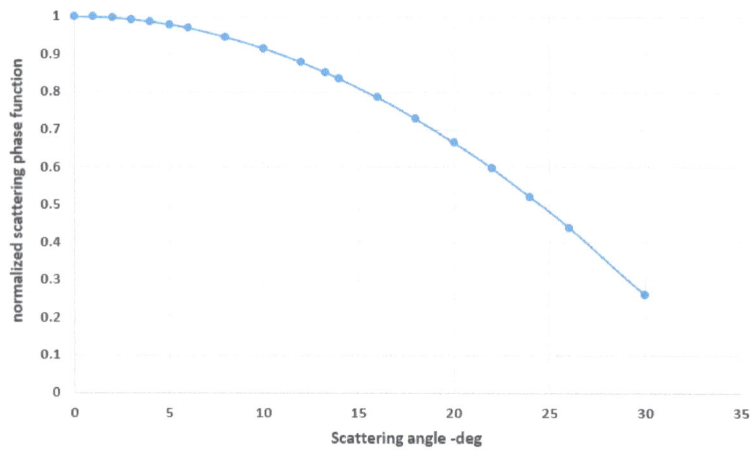

Figure 9.8-5 Normalized variation of scattering phase function with sun-moon azimuthal difference, *b*

There is no need to adjust our 6.35 deg limit for new moon visibility to account for larger *b* values and therefore larger scattering angles, as it can be shown that the angle *b* is always small (less than 10 deg) when *C*=*d*=*Ccr*=6.35°, as illustrated in Figure 9.8-6. And therefore, the the scattering angle is also small, resulting in less than 5 % adjustment in the value of scattering function. We carried the analysis here to interpret the documented sightings that occurred with larger values of *b*. If one is interested in higher accuracy, adjustment of the 6.35 deg limit angle using the curve in Figure 9.8-5 would result in a very close number for all practical cases.

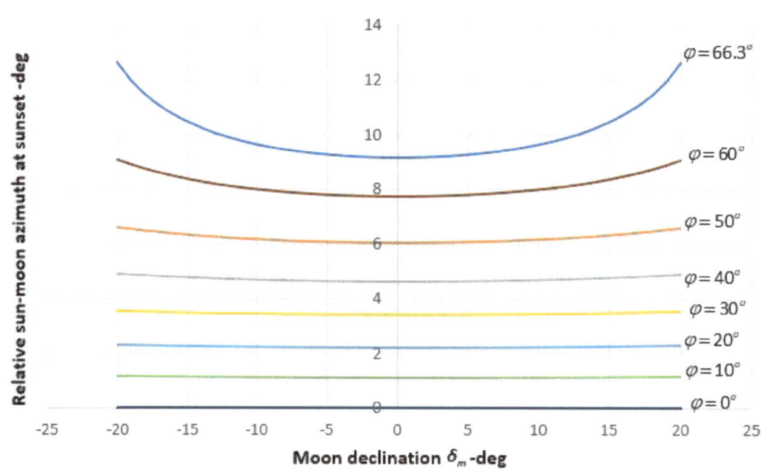

Figure 9.8-6 Relative moon-sun setting azimuth, *b*, as a function of the moon declination for different latitudes, when *C*=*Ccr*=*d*=6.35 deg

Chapter 10 Historical events and records

In this chapter, we address several applications of the results obtained so far in characterizing the start of the lunar month from a photometric perspective. We revisit several calendars that we have discussed in Chapter 2 .

The first two sections revisit documented sightings of the new moon carried by ancient astronomers. In Section 10.1, we address records of confirmed new crescent moon visibility which occurred during the last five hundred years BCE. The data was obtained from Babylonian tablets retrieved from Iraq in the 19th century (see Section 2.6.2). Several investigators provided reliable visibility information which we shall relate to our first order model which was presented in Chapter 9. Subsequently, we evaluate how successful the Mayans were at predicting the lunar month in Section 10.2. We also discuss the nature of their lunar calendar in relation to other calendars from the old world.

Next, we revisit the Jewish calendar and establish the date and time of the Jewish New Year as a non-local universal time that can be celebrated concurrently at the same instant around the world. Subsequently, the timing of the month of Ramadan is established so that it is observed and celebrated universally. Finally, the date, time and circumstances surrounding crucifixion are addressed to illustrate the photometric particularities of its timeline.

The Islamic lunar calendar, which we discussed in Section 2.7.4 is revisited to present an analysis of the fallout of the shift from the pre-Islam lunisolar Arabic calendar to a purely lunar calendar. Because the shift kept the same names of the months, there is a misconception that the historical events referred to during the life of the prophet pertain to the Islamic calendar, while they actually refer to the lunisolar Arabic calendar which has led to inaccuracies in the records which we shall discuss. As we correct the dates, some interesting new facts emerge. Even though these applications relate to religious faiths, there is no inherent claim for historical or religious opinions, nor there is an intent to influence certain belief systems.

We shall also address the puzzle of why the Arabic names of the months of the Islamic and Arabic calendars occur in the wrong seasons? We shall see that the only explanation is that the Calendar must have had a solar origin and was converted to a lunisolar one, probably at the time when Judaism was introduced to the south of the Arabian Peninsula.

Detailed work by specialists and consensus are needed to establish final opinions and draw conclusions about the complex historical, social and religious matters addressed. The intent here is to provide clarity into information we already know based on what we have learned so far from the effort presented in the previous chapters. Nevertheless, it is our intention to start a dialogue on the matters covered in the book.

10.1 First visibility in the Babylonian era

Two semi-empirical models for predicting first visibility of the lunar crescent from the Babylonian era data were proposed by P. V. Neugebauer **(Neugebauer P. V., 1929)** and Schoch **(Schoch, 1928)**, and later vetted by Fatoohi and coworkers **(Fatoohi, 1999)** against 209 separate observations. Both models consist of two parameters: the elevation of the moon at sunset and the azimuth difference between the sun and the moon. The Schoch's model covers values up to 19 degrees, while Neugebauer's model extends to 23°, as shown in Table 10.1-1 and in Figure 10.1-1. At any specified azimuthal difference from the sun, the crescent is expected to be visible when the moon is not lower than a critical true altitude at sunset. Any pair of azimuth difference and moon elevation values above the curves means visibility is probable, while no visibility can occur for pairs below these two curves.

Table 10.1-1 Schoch and Neugebauer criteria for Babylonian new moon sighting records against the linear model of Section 9.3

True Azimuth Difference (deg)	Minimum True Lunar Altitude *C* (deg)		*d* (deg)
	Schoch	Neugebauer	
0	10.7	10.4	10.5
1	10.7	10.4	10.5
2	10.6	10.3	10.5
3	10.5	10.2	10.4
4	10.4	10.1	10.4
5	10.3	10	10.3
6	10.1	9.8	10.2
7	10	9.7	10.1
8	9.8	9.5	10.0
9	9.6	9.4	9.9
10	9.4	9.3	9.7
11	9.1	9.1	
12	8.8	8.9	
13	8.4	8.6	
14	8	8.3	
15	7.6	8	
16	7.3	7.7	
17	7	7.4	
18	6.7	7	
19	6.3	6.6	
20		6.2	
21		5.7	
22		5.2	
23		4.8	

The Figure shows also a plot of the elongation *d* for the corresponding values of *b* and *C* given from the linear model of Section 9.3 is valid for small angles. This means that the model parameters must be less than about 10-12 degrees to ensure accuracy**.** The relationship between these angles is given by Equation 9.3-3:

$$b = C \tan\mu \pm \sqrt{d^2 - C^2}.$$

The angle μ is about equal to the latitude φ, which is about 23.6° for Babylon, *C* is the elevation of the moon at sunset, *d* is the elongation of the moon and *b* is the difference in azimuth. The above equation can be rewritten as

$$d = \sqrt{(b - C\tan\varphi)^2 + C^2} \qquad 10.1\text{-}1$$

The value of *d* is plotted in Figure 10.1-1 for given values of *b* and *C* less than 10° from either model (used Neugebauer's data for illustration). When angles are larger than 10°, the approximation is no longer valid and deviates from the Neugebauer and Schoch models. The figure shows a clear flattening of both the Schoch and Neugebauer curves as well as the curve representing *d* from the linear model, which means *C* is about equal to *d* for new moon sighting in both models for small angles. Which confirms the result obtained in the linear model of Section 9.3 .

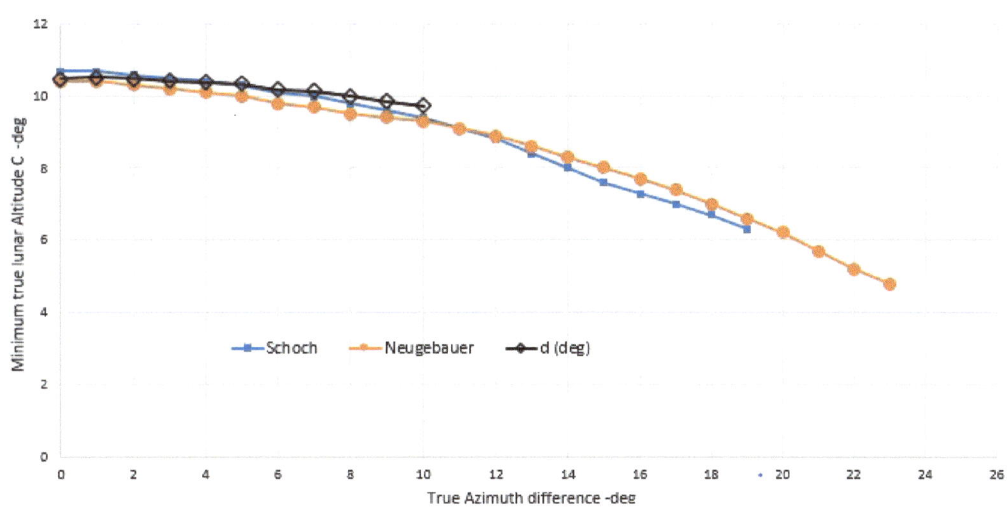

Figure 10.1-1 Schoch and Neugebauer criteria for Babylonian new moon sighting records against the elongation *d* of the linear model of Section 9.3

From the above assessment, Schoch and Neugebauer models infer that first visibility is when *C* and *d*, are about equal to 10.5 degrees.

Fatouhi (Fatoohi, 1999) combined modern crescent moon observational data, compiled by Schaefer and Doggett, over the period from 1859 to 1996 CE, with 294 Babylonian observations from the year 568 B.C.E. to 74 B.C.E. The modern data include 271 evening observations, 23 last visibility of the old moon, and 81 negative sightings in which the crescent could not be visible. Most of the positive sightings at low elevations angles, shown in Figure

10.1-2, are from modern sightings which took place at northern and southern latitudes. The figure represents entries of the elevation of the new crescent moon, The 6.35 deg criteria is shown as the parameter C minus the radius of the moon, displayed as a function of the relative moon-sun azimuth at sunset, Daz, which is the square root of $(d^2 - C^2)$.

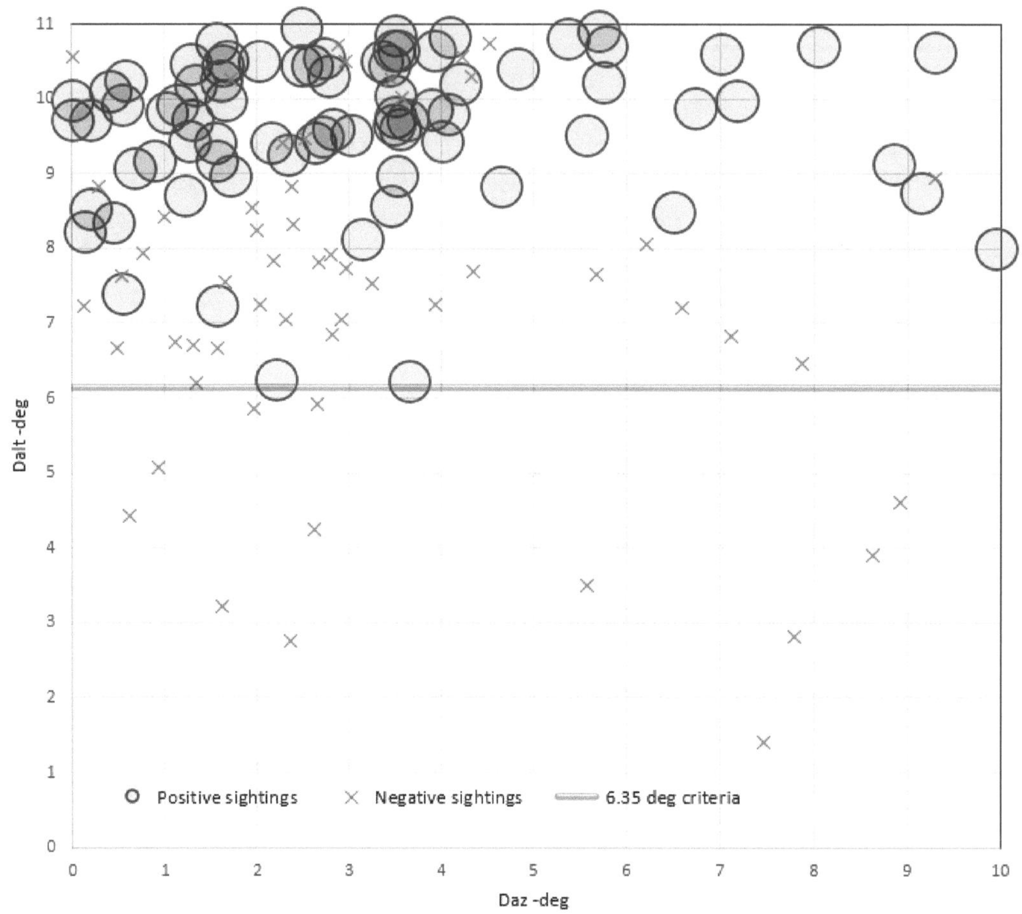

Figure 10.1-2 Crescent visibility over small angles, following conjunctions in combined Babylonian records with modern sighting records collected by Fatoohi (Fatoohi, 1999).

10.2 Mayan records

In Section 2.6.3, we discussed the independent historical achievements of the Mayan people in identifying the lunar cycle and other celestial bodies. The Maya kept an extensive record of the timing of the moon (see the moon in the Mayan hieroglyphic script in Figure 10.2-1). Their meticulous records over centuries ultimately led to the development of the most advanced calendars of the time. Teeple (Teeple, 1931) discussed the accuracy of the Mayan Calendar and referred to many of their recorded dates. We use fourteen of such dates, spanning over more than four centuries to get an idea about how astronomers of this unique civilization dealt with the matter of the new month. The randomly selected fourteen dates are listed in the first column of Table 10.2-1 in Mayan Long Count, which was discussed in Section 2.6.3, and are shown in the hieroglyphic form in Figure 10.2-2.

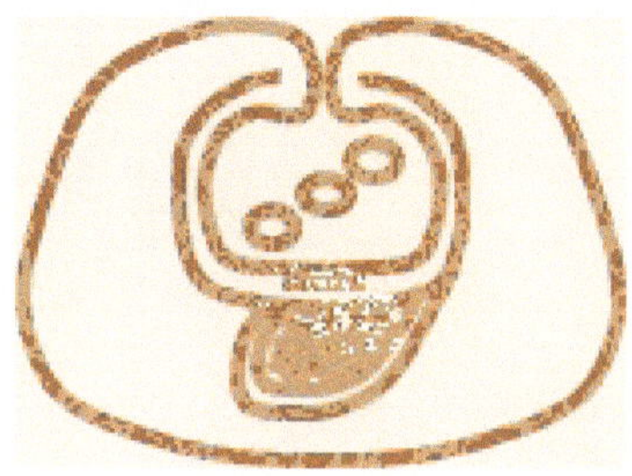

Figure 10.2-1 The moon in Mayan hieroglyphic form

These Mayan dates are converted to Gregorian equivalent and are investigated for the photometry of the moon as it would have been seen in Mesoamerica around what is today Guatemala. Mayans refer to the old moon in their calendar computation. Unlike the dating system of the old world, records show that Mayan dates are given in a number of days since the last moon. These dates are provided in the third column of Table 10.2-1 and start from day zero[18]. We also consider here both old and new moon to see which origin really fits each of the fourteen dates. Depending on the elevation of the moon at sunset for a new moon or its elevation at sunrise, in the case of an old moon, we provide classifications for potential sighting: Low, Fair, high and very high. This classification is shown in the last column for both old moon and new moon. The Mayan date is also highlighted in this classification.

The Mayan count is confirmed to be from old moon and counts are in elapsed time rather than the current time. Furthermore, the Maya choose the old moon because it provides a better chance for sighting than the new moon. The light from the moon does not go through a heated atmosphere and ground and therefore, the chance for sighting is much better than that of a new moon. It is also known that the Maya kept a record of whether the month that elapsed is 29 or 30. They probably regulated the count through the fact the observations that two consecutive months would nearly equal to a little more than 59 days, and therefore adjustments are made if no visibility is achieved[19].

[18] The Mayans independently discovered the concept of zero

[19] The 19-year lunar cycle was also independently established by the Mayan

Table 10.2-1 Fourteen Mayan days confirm the use of old moon for calendar reference

Mayan Calendar Date						Gregorian date (C.E.)			Mayan record	Visibility date and time								
Long Count date	Long Count details									Old moon					New moon			
	Baktun	Katun	tun	uinal	k'in	Year	Month	Day	Day since last Moon	Low	Fair	High	Very Highy	Time (UT)	Low	Fair	High	Time (UT)
8.16.0.0.0	8	16	0	0	0	357	Apr	1	25		7-Mar	6-Mar		12:00	9-Mar		10-Mar	0:00
9.3.0.0.0	9	3	0	0	0	495	Jan	28	17		12-Jan	11-Jan		12:00	14-Jan		15-Jan	0:00
9.5.0.0.0	9	5	0	0	0	534	Jul	3	5	29-Jun	28-Jun	27-Jun		12:00		30-Jun	1-Jul	0:00
9.7.0.0.0	9	7	0	0	0	573	dec	5	25	12-Nov	11-Nov	10-Nov		12:00		14-Nov	15-Nov	0:00
9.8.0.0.0	9	8	0	0	0	593	Aug	22	19	4-Aug		3-Aug		12:00	5-Aug		6-Aug	0:00
9.9.0.0.0	9	9	0	0	0	613	May	10	13	28-Apr		27-Apr		12:00			30-Apr	0:00
9.10.0.0.0	9	10	0	0	0	633	Jan	25	8	18-Jan		17-Jan		12:00			20-Jan	0:00
9.11.0.0.0	9	11	0	0	0	652	oct	12	4 and 5		10-Oct	9-Oct	8-Oct	12:00		13-Oct	14-Oct	0:00
9.12.0.0.0	9	12	0	0	0	672	Jun	29	28		3-Jun	2-Jun	1-Jun	12:00		5-Jun	6-Jun	0:00
9.14.0.0.0	9	14	0	0	0	711	Dec	3	17		18-Nov	17-Nov	16-Nov	12:00		20-Nov	21-Nov	0:00
9.15.0.0.0	9	15	0	0	0	731	Aug	20	10	11-Aug	10-Aug	9-Aug		12:00		13-Aug	14-Aug	0:00
9.16.0.0.0	9	16	0	0	0	751	May	7	5	3-May		2-May		12:00		5-May	6-May	0:00
9.17.0.0.0	9	17	0	0	0	771	Jan	22	0		23-Jan	22-Jan		12:00		26-Jan	27-Jan	0:00
9.18.0.0.0	9	18	0	0	0	790	Oct	9	24		16-Sep	15-Sep		12:00	18-Sep		19-Sep	0:00

Figure 10.2-2 Mayan date record of Table 10.1-1 illustrated in the hieroglyphic form (ordered top-down and left-to-right -Graphics generated courtesy of the Smithsonian National Museum of the American Indian)

It is however certain at least from the record studied here that the Maya did not have a mean to detect an early new moon where the crescent is between 6-9 degrees above the horizon. Mayan astronomers were certainly facing the same controversy that the old world has been dealing with over thousand years. They seem not better at detecting the beginning of the month than the people of the old world, as shown in Table 10.2-1. In fact, there is evidence of discrepancies in the date records from site to site, as discussed in (Teeple, 1931). It must be mentioned that the average length of the lunar month was very accurate in the Maya calendar and there are no cumulative errors carried from month to month, very much like it was the case in the Babylonians and the Egyptian lunar calendars.

10.3 Molad, Ramadan, and Good Friday

Using the photometric calendar, we shall revise certain celebration dates in the three major monotheistic religions: Judaism, Christianity, and Islam. The proposed new dates are tabulated separately in Chapter 11 for readability.

The Jewish calendar can have a unified date and time for its New Year across the globe. Unification is possible by making the time of first visibility of the new light from the moon the origin of the calendar, irrespective of the location where it occurs. The Molad of the month of Tishrei is identified in Table 11.2-1 for 32 years from 2014 to 2045 CE. There are three sections in the Table. The first section gives the Jewish lunisolar year number, the number of days in the year, whether the year is a leap and the corresponding Gregorian date. The time of conjunction, as computed by the NASA Goddard ephemeris, for the lunation of Tishrei is given in the second section of the Table while the date, time and location of new moon first visibility are made available in the third section. In addition, the time elapsed since the time of conjunction is displayed. There are several shortcomings that we can extract from the Table. It can be noted that first visibility of the month varies from about 7 hours to almost 24 hours with an average of about 12 hours. In addition, because the Tishrei lunation occurs near the Fall Equinox, the small inclination of the lunar orbit tends to statistically favor higher southern latitudes over higher Northern ones.

Unlike the later versions of the Hebrew lunisolar calendar, the Islamic lunar calendar is not rooted in a rule-based construct. It is fundamentally based on new moon sighting because of religious references. Nevertheless, beside the process of intercalation, the first of the month is treated the same way as in the Jewish calendar. As a result, the photometric calendar can synchronize observation of the month of Ramadan, Eid[20] celebration around the world and pinpoint the dates of Hajj (Pilgrimage to Mecca. For illustration, Table 11.3-1 gives the conjunction date and time (UTC) for the start of the lunation of Ramadan followed by that of the subsequent month of Shawwal using data from the NASA Goddard ephemeris between the years 2015 and 2113 CE. Also, shown is the data from our computation for comparison and the "location" of lunations based on the definition of Section 9.4. Table 11.4-1 gives the

---

[20] Marks the end of the month of Ramadan and of fasting

date, time (UTC) and location corresponding to the start and end of the photometric month of Ramadan for C=Ccr=6.35°, 8° and 10° also between the years 2015 and 2113 CE. While the start of the lunar month should be always based on a molecular atmosphere, free of moisture and aerosol. i.e., Ccr=6.35°, the values for C=Ccr= 8° and 10° are given in the Table just in case there is a need for conservatism[21]. Should that be the case, we have conservatism both at the start and at the end of the month. The start of fasting would take place the morning which follows the start time of the month which is shown in the Table (devoted people would probably start fasting exactly at the start time of the month even if it occurs in the middle of the day for them). The end of fasting should take place right at the time shown for the start of Shawwal, even if it is during daylight (there is a precedence to such practice before the time of telecommunication). Under such scenario, a numerical simulation in which the globe is partitioned in a 2° by 2° grid shows that people would fast either 29 or 30 days. Also, the photometric month is shown to start between about 6.5 hours and 14 hours following conjunction, as illustrated in Figure 10.3-1. There is an obvious cyclic behavior that can be identified using a fast-Fourier analysis of the data.

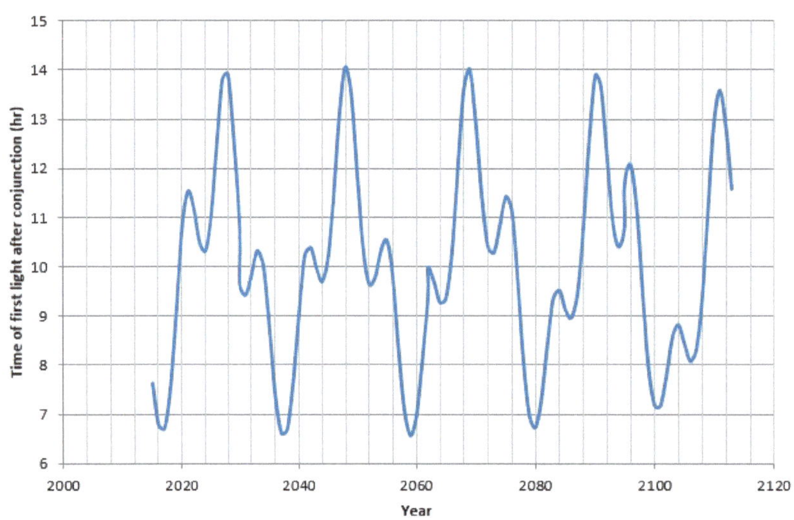

Figure 10.3-1 Elapsed time from conjunction for the start of the month Ramadan (6.35°)

The timeline of the crucifixion of Christ was investigated by many researchers using clues from Scriptures and other historical documents. To do this, we look at the relative sun-moon photometric situation during this event using the results of the preceding chapters We shall examine the timeline of crucifixion, given the historical references to the lunar dating. The

[21] It may be argued that there is a need for conservatism and fasting should start based on Ccr=6.35°, while the Eid comes on C=Ccr= 8° or 10°.

sighting of a lunar eclipse over Jerusalem the day of crucifixion is discussed considering the current knowledge of earth's rate of rotation. An eclipse was seen in Jerusalem right after sunset as the full moon was rising for several minutes. The timeline of the umbral eclipse is calculated and addressed in conjunction with the reported events in question. Finally, the lunisolar photometric condition which occurred during the crucifixion day was found to exhibit coincident solar and lunar terminators, a condition occurring only 160 times during the first three millennia of the Common Era.

To identify the date and year of the crucifixion, the main arguments cited by researcher in the literature refer to Matthew 26:3-4 and John 11:49-53. Both indicate that hostility towards Jesus was instigated by the high priest Caiaphas who was in charge from 18 to 36 CE, on the orders of the Governor Pontius Pilate, 26 to 36 CE, (Matthew 27:24-26, Mark 15:15, Luke 23:24, John 19:15-16). It is further pointed that the ministry of Jesus begun after that of John the Baptist, son of Zechariah, (Matthew 3, Mark 1, Luke 3:1-2) to whom the "divine message" came in the desert in 15th year of the reign of the Roman Emperor Tiberius Caesar (14 to 37 CE). This infers that the ministry of John the Baptist started the year 29 CE and therefore the crucifixion must have happened between the year 29 and 36 CE. On the other hand, the ministry of Jesus recorded three consecutive Passovers (John 2:13, John 6:4, and John 11:55) and has lasted about three and a half years, thus excluding the years 29, 30 and 31 CE from being possible because they do not provide enough time. The year 36 CE must be eliminated for making the ministry of Jesus a six-year event. The final candidate years are, therefore, 32-35 CE.

It is also commonly agreed that hostility started in the morning of a Friday, just before a Sabbath, referred to as "Yoma da'Arubta," or "the day of preparation"—that is, the day on which Jews prepared for the Sabbath. There is also an agreement that this Friday coincided with the annual feast of Passover. The eight-day festival of Passover is celebrated in early spring, from the 15th day through the 22nd day of the Hebrew month of Nissan. Also, reference is made in the Bible to the Last Supper on Holy Thursday as a Passover meal. Using the photometric calendar presented earlier, we can go back and relate the 15th day of Nissan, counting from the first visible moon in Jerusalem for each Gregorian candidate year. We can also establish which day of the week "Yoma da-'Arubta" was in each candidate year. Finally, Matthew, Mark, and Luke stated that Jesus expired about "the ninth hour," which is what is today about 3:00 PM, i.e., about 1:00 PM UT.

The elements of the 15th day of Nissan in each year between 32 and 35 CE are given in Table 10.3-1. The only date that fits the above facts is April 1, 33 CE. Note that the moon phase is given at 13:00 UT. Figure 10.3-2 shows the solar and lunar photometric situation. The globe is illuminated exclusively by either the sun or the full moon with both terminators perfectly aligned so that sunset corresponds to moonrise and vice-versa. We coin this occurrence a state of ***orthogonal illumination***. In what follows, we study this state and derive the principal features of such an event.

Let us recall the transformation between the horizon and equatorial coordinate systems from Section 8.3.6. We have for a rising or setting sun an altitude $a_s=0$ and some azimuth As in the horizon coordinate system. These coordinates correspond to declination δ_s and an hour angle Hs in the equatorial coordinate system. These coordinates relate for an observer at a latitude ϕ through the following transformations:

$$\sin\delta_s = \sin a_s \sin\phi + \cos a_s \cos\phi \cos A_s = \cos\phi \cos A_s$$
$$\cos H_s = \frac{\sin a_s - \sin\phi \sin\delta_s}{\cos\phi\cos\delta_s} = -\tan\phi\tan\delta_s \quad . \qquad 10.3\text{-}1$$

Similarly, for a setting moon with zero altitude and an azimuth Am in the horizon coordinate system, and a corresponding declination δ_m and hour angle Hm, we have for the same latitude

$$\sin\delta_m = \sin a_m \sin\phi + \cos a_m \cos\phi \cos A_m = \cos\phi \cos A_m$$
$$\cos H_m = \frac{\sin a_m - \sin\phi \sin\delta_m}{\cos\phi\cos\delta_m} = -\tan\phi\tan\delta_m \quad . \qquad 10.3\text{-}2$$

Table 10.3-1 The 15th of the month of Nissan between 32 and 35 CE

Greogorian Date	Time, UT	Day of the week	Moon phase	Day of Nissan
Apr 12, 0032	13:00	Mon	0.99998	15
Apr 1, 0033	**13:00**	**Fri**	**0.99962**	**15**
Mar 21, 0034	13:00	Tue	0.99948	15
Apr 9, 0035	13:00	Mon	0.99872	15

In addition, using conversion from equatorial to ecliptic coordinates for the sun, we have

$$\tan\lambda_s = \frac{\sin\alpha_s \cos\varepsilon + \tan\delta_s \sin\varepsilon}{\cos\alpha_s} \quad . \qquad 10.3\text{-}3$$
$$\sin\beta_s = \cos\varepsilon\sin\delta_s - \sin\varepsilon\cos\delta_s\sin\alpha_s = 0$$

Note that we wrote that the latitude of the sun β_s is nil using the definition of the ecliptic coordinate system, λ_s is the ecliptic longitude of the sun and ε is the obliquity angle which is equal to about 23°.26'. For the moon, we also have similar conversions which we express as:

$$\tan\lambda_m = \frac{\sin\alpha_m \cos\varepsilon + \tan\delta_m \sin\varepsilon}{\cos\alpha_m} \quad . \qquad 10.3\text{-}4$$
$$\sin\beta_m = \cos\varepsilon\sin\delta_m - \sin\varepsilon\cos\delta_m\sin\alpha_m$$

The above equations are summarized in the following system of equations:

$$\begin{aligned}
&\sin\delta_s = \cos\phi\cos A_s,\ \cos H_s = -\tan\phi\tan\delta_s, \\
&\sin\delta_m = \cos\phi\cos A_m,\ \cos H_m = -\tan\phi\tan\delta_m, \\
&\tan\lambda_s = \frac{\sin\alpha_s\cos\varepsilon + \tan\delta_s\sin\varepsilon}{\cos\alpha_s}, \\
&\sin\beta_s = \cos\varepsilon\sin\delta_s - \sin\varepsilon\cos\delta_s\sin\alpha_s = 0, \\
&\tan\lambda_m = \frac{\sin\alpha_m\cos\varepsilon + \tan\delta_m\sin\varepsilon}{\cos\alpha_m}, \\
&\sin\beta_m = \cos\varepsilon\sin\delta_m - \sin\varepsilon\cos\delta_m\sin\alpha_m.
\end{aligned} \tag{10.3-5}$$

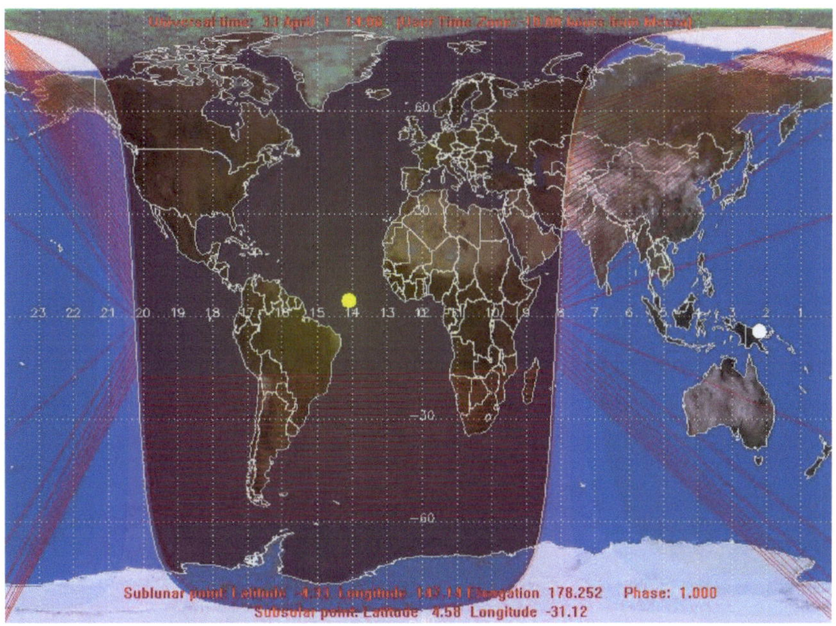

Figure 10.3-2 Photometric situation -Friday, April 1, of the year 33 CE, 14:00 UT (15th of Nissan)

Next, we express that the terminator at sunset coincides with the terminator at moonrise at the same time and at any geographic latitude. The time of sunset and its azimuth are given by:

$$LST_s = \frac{1}{15}\arccos(-\tan\varphi\tan\delta_s) + \alpha_s,\quad \mathrm{A}_s = 360 - \arccos\left(\frac{\sin\delta_s}{\cos\varphi}\right)$$

And the time of moonset and its Azimuth are given by

$$LST_m = 24 - \frac{1}{15}\arccos(-\tan\varphi\tan\delta_m) + \alpha_m, \quad A_m = -\arccos\left(\frac{\sin\delta_m}{\cos\varphi}\right), \text{ where}$$

A_s and A_r are given in degrees. $H_s = \frac{1}{15}\arccos(-\tan\varphi\tan\delta_s)$, $H_m = \frac{1}{15}\arccos(-\tan\varphi\tan\delta_m)$ are given in hours. If LST_s or LST_r is greater than 24hr, then subtract 24 hr. Setting $LST_s = LST_r$ for any latitude φ, we obtain

$$\delta_m = -\delta_s, \; \alpha_m = \alpha_s + 180^o, \text{ and } A_m = A_s - 180^o.$$

Because the declination of the moon is always close to that of the sun due to the small inclination of the moon's orbital plane with respect to the ecliptic, the moon and the sun declinations are equal and opposite only near the Equinoxes. This can be easily seen in a plot of the sun and moon declinations over the year, such as the one shown in Figure 9.3-4.

In summary, a state of *orthogonal illumination* is achieved when the moon and the sun coordinates are related using the following equation:

$$\begin{cases} \text{Equatorial coordinates:} \\ \delta_m = -\delta_s, \; \alpha_m \simeq \alpha_s - \pi \\ \text{Horizon coordinates (at Sunset/Moonrise):} \\ a_m = 0, \; a_s = 0, \; A_m \simeq A_s - \pi \end{cases} \qquad 10.3\text{-}6$$

We could have also established the condition for simultaneous sunrise and moonset events and obtained the same equations. Finally, using the solution from Equation 10.3-6, the ecliptic coordinates of the moon given in the last two Equations of 10.3-5 become

$$\begin{aligned} \tan\lambda_m &= \frac{-\sin\alpha_s\cos\varepsilon - \tan\delta_s\sin\varepsilon}{-\cos\alpha_s} = \tan\lambda_s \\ \sin\beta_m &= -\cos\varepsilon\sin\delta_s + \sin\varepsilon\cos\delta_s\sin\alpha_s = -\sin\beta_s = 0 \end{aligned} \qquad 10.3\text{-}7$$

This means that we have

$$\lambda_m = \lambda_s - \pi, \quad \beta_m = \beta_s = 0 \qquad 10.3\text{-}8$$

The first equation is the actual statement of the astronomical definition of a full moon, while the second equation means that the moon is at the ecliptic plane. Therefore, we can state that orthogonal illumination occurs only when the earth is in the vicinity of the Equinoxes and at the same time the moon is full at a lunar node (the lunar nodes are the orbital nodes of the moon, which are the points where the orbit of the moon crosses the ecliptic plane). We also know that lunar eclipses occur only when the longitude of a full moon happens to be within a little more than 11° of a lunar node.

Tracing back to the photometric event of April 1, of the year 33 CE requires that we account for the frictional tidal effects of the moon on the rotation of the earth. The spin of the earth has been slowing down over time while transferring momentum to the moon, which uses that energy to grow more and more distant from the earth. The reciprocal tidal effect of the earth on the moon, which as one would expect is much larger, has slowed its rotation about its axis over time and has brought it to a standstill. Today, the rotation of the moon has ceased, and only one side can be seen from the earth.

The reference unit of time used in tracking celestial objects has been the rotational period of the earth about its axis. The Universal Time (*UT*) also referred to as Greenwich Mean Time or GMT, a mean solar time with a geographic reference at Greenwich, England, was the standard adopted by early astronomers. The period of the earth's rotation is, however, not constant. The length of the day has been increasing by tiny amounts that add up over centuries to make the solar time fall behind. The main influences on earth's rotation about its axis are the internal relative motions between the planet's mantle and the core, and climate and sea-level fluctuations. While these effects are hard to quantify, historical accounts can provide a good estimate of such changes over the past three millennia. Today, astronomers are using the Terrestrial Dynamical Time (*TDT*) which uses stable vibration of atoms because of its independence from complex fluctuations in the periodicities of celestial objects. Events from the past few millennia can be predicted using ΔT, the earth's rotational clock error, which is basically the difference between *TDT* and *UT*, given by

$$\Delta T = TDT - UT$$

Earth's rotation over the past several millennia was studied by Stephenson and coworkers, e.g. (Stephenson, 1997) and (Morrison & Stephenson, 2004). They have compiled hundreds of eclipses and occultation observations from early European, Middle Eastern and Chinese records and proposed polynomial best-fits describing the variation of ΔT from 700 BCE to 1600 CE. Their results formed a cornerstone for the Five Millennium Canon of Lunar Eclipses from 1999 BCE to 3000 CE (Aspenak & Meeus, 2009). Abbreviated results from their work are shown in Table 10.3-2.

Table 10.3-2 Historical values of ΔT

Year	0	33	500	1000	1500	1970	1980	1990	2000	2007
ΔT (sec)	10534	10202	5681	1561	195	40.18	50.54	56.86	64	65.15

Back to the topic of Jesus crucifixion, data for the year 33 gives ΔT =10,202 s, that is 02 h 50m 2s, or a change in the longitude of 42.5 degrees. The time of maximum eclipse[22], which is predicted at 17:37:53 *TDT*, happened at 17:37:53 minus 02 h 50m 2s, or 14:47:51 *UT*. As one would expect, estimating the standard error in ΔT prior to 1600 CE is difficult and depends on several factors, mainly the accuracy of historical records and modeling the geophysical mechanisms behind the changes in the rate of rotation of the earth. Morrison and Stephenson (Morrison & Stephenson, 2004) give the following simple parabolic relation to estimating the standard error (σ), considered to be valid over 1000 BCE-1200 CE:

$$\sigma = 0.8 \times t^2 \; seconds,$$
$$\text{where } t = (year - 1820)/100.$$

For the year 33, we have an uncertainty σ=255 s (4m: 15s, or 1.06-degree longitude). The time of greatest eclipse was 17:37:53 *TDT* corresponds to 14:47:51 UT +/- 4 m: 15 s that is between 14:43:36 and 14:52:06 *UT* or between 16:43:36 and 16:52:06 Jerusalem time. Furthermore, the time between first and last external tangencies of the moon with the penumbral shadow lasted 332.8 min and the time between first and last external tangencies of the moon with the umbral shadow lasted 170.1 min. Therefore, umbral eclipse begun and ended over a period of 01:25:03, that is, it began sometime between 15:18:33 and 15:27:03 and ended between 18:08:39 and 18:17:09.

Reference to darkness in the middle of the day and an earthquake shaking Jerusalem the day of the crucifixion was reported (e.g., Matthew 27:51-54), with the crucifixion ending in the ninth hour. During those days, one hour is 1/12 of the time between sunset and sunrise, or 1/12 of the time between sunrise and sunset. The ninth hour is, therefore, 15:00 in the modern 24-hr system. This fits with the moon entering the Umbral zone sometime between 15:18:33 and 15:27:03, with the partial eclipse remaining until sometime between 18:08:39 and 18:17:09. The time of sunset in Jerusalem on April 01, 33 is calculated to have occurred at 18:00:31 Jerusalem time, and at the same time the moon has risen in the East, 180 degrees opposite to the point of sunset with a partial eclipse as it would not have exited the umbra until sometime between 18:08:39 and 18:17:09. One can imagine the scene of a darkened full moon rising for about 10 to 15 minutes during a day of turmoil. This does not necessarily provide an explanation for the reported darkness, but instead, suggests that another event may have happened to darken the sky, making the event of a rising moon with an eclipse not visible, otherwise it would have been widely reported.

---

[22] We shall carry the calculation up to the second, even though there is no need for such accuracy.

Orthogonal illumination can be regarded as a special event in an astronomical sense. There are 160 such events occurring near the Spring Equinox from 33 CE to 3000 CE, as documented in Table 11.5-1 in Section 11.5. The Table gives Julian and Gregorian dates, the time of the greatest eclipse (closest proximity to shadow axis), time (in *TDT* and ΔT) as given in (Aspenak & Meeus, 2009) as well as the coordinates of the sun and the moon in the Ecliptic and Equatorial coordinate systems. Events of orthogonal illumination amount to only 2.2 % of all eclipses. And the event itself accounts for a probability of occurrence less than 0.015 %.

There is also reference to earth-shaking violently the day of crucifixion. The Dead Sea Transform (DST) fault system, also known as the Dead Sea Rift, is a fault system running from the East of the Anatolian plate in Southeastern Turkey to the northern end of the Red Sea (just offshore of the southern tip of the Sinai Peninsula) as shown in Figure 10.3-3. The fault forms the transition between the African Plate and the Arabian Plate butting and moving in a northerly motion. Data compiled in (Ambraseys, 1994), mainly from the Ottoman Empire era of 49 earthquakes provided information to correlate with the lunar day of the month. The result of which is shown Figure 10.3-4, where the cycle is repeated over two months to illustrate the point that the probability of occurrence of an earthquake around any lunar day, +/- 2 days, peaks when the tide of the moon and the tide from the sun lineup. This produces one peak around a full moon and another peak at a new moon. The calculated equal probability of earthquake occurrence for any day +/- 2 days is 17 %. There is an obvious effect from the lunar tidal pull and from the compiled data, we can state that if an earthquake were to occur, a chance for it to happen around a full moon is twice that of any other normal day. Furthermore, at the writing of this book, ten out of the seventeen largest earthquakes with a magnitude above 8.5 ever recorded around the world in modern history happen either around a new moon or a full moon. This is probably not a chance, but instead must have a geophysical explanation. The first candidate must be a drop in gravitational forces on the earth plates, which in turn alleviates the friction between the plates, leading to slip and relative motion at the faults.

The modern Centennial Catalog (Engdah & Villaseño, 2002), also published by USGS, is a global catalog of about 13,500 instrumentally recorded earthquakes from 1900 to 2008. The catalog is used to draw a global picture of tidal effects. This catalog of earthquakes was assembled by combining preexisting catalogs, reducing all available magnitudes to a commonly corrected magnitude scale. For the years 1964 to present, a cut-off magnitude of 5.5 is used in the catalog while the cut-off is magnitude 6.5 prior to 1964. Also, between the 1930s and 1963, the catalog is complete to magnitude 6.5 and prior to that, the catalog is complete all the way to magnitude 7.0. Figure 10.3-5 clearly shows a jump in the probability of earthquake events during a full moon. The drop of probability around a new moon is an artifact however due to the lunar month averages to 29.5 days. This translates to the fact that the last day of the lunar month has a probability of occurrence of 0.5 and therefore the actual probability of earthquake occurrence is reduced by about 50 %. An accurate estimate of what the moon is responsible for can be established using the shift in probability computed from the area under the probability curve minus the area of equal probability. Using integration, it is shown that the lunar tidal effect is responsible for 14.4 % increase in probability of earthquake occurrence.

Figure 10.3-3 Dead Sea Transform fault system

Back to the day of crucifixion topic, the reported earthquake event is in line with the odds of occurrence when it comes to the effect of the moon, either in the Dead Sea Transform or around fault lines elsewhere.

Separately, a high-resolution Holocene seismic history of the DST from laminated sedimentary cores from the shores of the Dead Sea has been produced. Radiocarbon dating and annual lamina count showed agreement between disturbed sedimentary structures and the historical earthquake record, including the major earthquakes of 1927, 1837, 1212, 1033, and 749 CE. In addition, old earthquakes were correlated, including one in the year 33 CE and another major one in the year 31 BCE in a study led by a team from Germany.

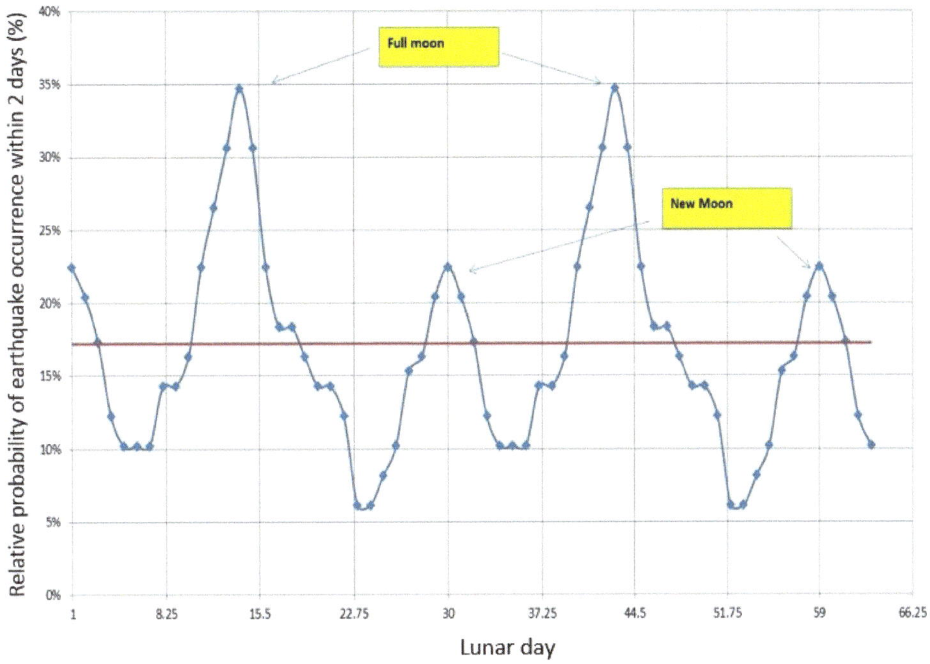

Figure 10.3-4 Relative probability of seismic events along the Dead Sea Transform, 363-1995 CE

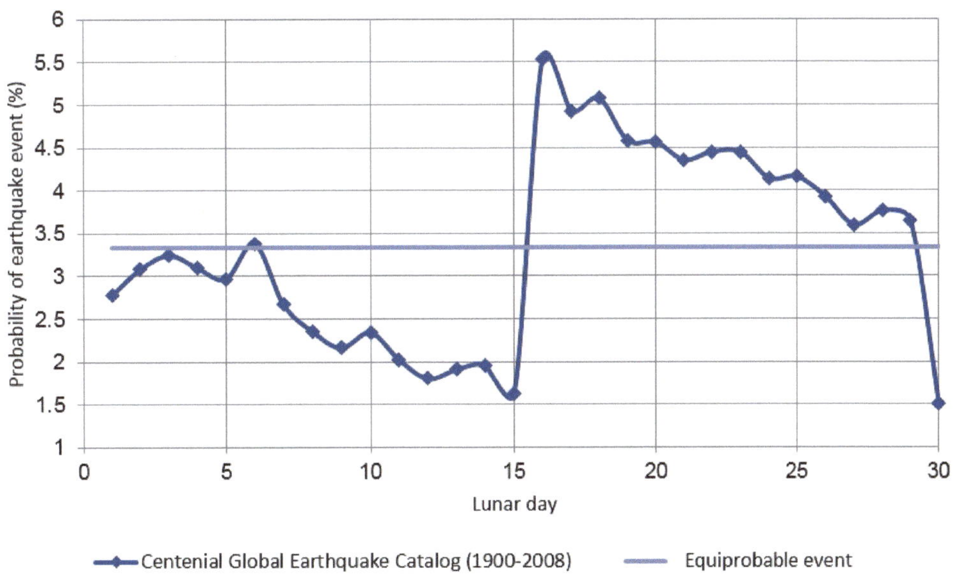

Figure 10.3-5 Relative probability of earthquake over a lunar month

10.4 Emergence of the Islamic calendar

As discussed in Section 2.7.4, the Islamic lunar calendar which is also known as the Hijra calendar was built on the Meccan pre-Islam lunisolar Arabic calendar during the time of the second Caliph, about 7 years after the death of the Prophet Mohammed. The Hijra era was then established, and the new Islamic calendar started being used by Muslims. The transition from the old Arabic calendar to the new Islamic calendar was prompted by the elimination of the intercalation month based on the interpretation of verses from the Qur'an and statements of the prophet during the Farewell Hajj, which took place a few months before his death.

In Section 10.4.1, we shall address what prompted the calendar change and its effects on the people of Arabia after the death of the Prophet Mohammed. In Section 10.4.2, the old Arabic calendar is reconstructed using date conversion, between the Gregorian, Islamic, and Hebrew calendars. The conversion is carried from the time of the Farewell Hajj, just a few months prior to the death of the prophet, until the time in which the Islamic calendar was established, i.e., about seven years later. Next, because the origin of the Islamic calendar is set retroactively to the time of the Prophet's migration from Mecca to Medina, i.e. seventeen years before the emergence of the Islamic calendar, it is often not clear whether the dates of historical events refer to the Islamic calendar or to its predecessor, especially both calendars use the same names for the months. Therefore, we continued the conversion back to the year of migration in Section 10.4.4, where technical aspects of the retroactive datum are discussed. Later, in Section 10.4.5, arguments for the merits of the time of Farewell Hajj as an alternative origin for the Islamic calendar, which the Prophet himself referred to as "origin" in the Farewell Hajj, are presented.

Finally, at the end of this Section, we shall discuss the original construct of the Arabic calendar from an etymological perspective. We show that the names of the months in the Arabic calendar point to the wrong season of the year, very much like in the Islamic calendar, even though the lunisolar calendar is supposed to be in step with the seasons. We shall propose that the origin of the Arabic calendar is solar instead of lunisolar, derived from the Egyptian-Alexandrian calendar, and analogous to the Julian, Coptic and Ethiopian calendars.

In pre-Islamic time, Arabia was without centralized administrative or clergy institutions. Driven by economics and logistics, the powerful Meccan traders shifted from a solar calendar to a lunisolar one. The shift not only correlated then with the Jewish calendar as it was maintained by the Jewish tribes which were spread over the Peninsula and who attended the fairs around the time of the Hajj but also brought a reliable approach to maintaining the calendar without a need for special institutions. The mechanism for intercalation in the Arabic Meccan calendar shall be discussed in detail in the next section.

10.4.1 Postponement of the sacred month

To understand what led to the dismantling of the lunisolar calendar in favor of a purely lunar calendar through abolishing of intercalation, we consider the two main references in Islam: The Qur'an and the Hadith. The main information on this topic comes from the interpretations of Qur'an: 9; 36 and 9; 37 verses, which translate from Arabic into:

The number of months
In the sight of Allah
Is twelve
So ordained by Him
The day He created
The heavens and the Earth;
Of them four are sacred;
That is the straight usage

(Qur'an: 9; 36)

Verily the postponing
(of a prohibited month)
Is an addition to Unbelief:
The Unbelievers are led
To wrong thereby: for they make
It lawful one year,
And forbidden another year,
Of months forbidden by Allah
And make such forbidden ones lawful

(Qur'an: 9; 37)

It is clearly indicated in Qur'an: 9;36 that a year should use 12 months, and not thirteen months as one may erroneously think intercalation does through the insertion of one additional month. Also, interpretation of Qur'an: 9; 37 from multiple sources indicate that the issue is with the postponement of the sacred month of Al-Muharram and not the lunisolar calendar: "*Postponement (of a sacred month) is only an excess of disbelief whereby those who disbelieve are misled.*" The classical reference *Tafseer Al-Tabary* [23] (items 16706-16717) provides over a dozen of opinions[24] and records on the interpretation of Qu'ran: 9; 37 verses. These records were documented over a span of two generations after the passing of the prophet. With minor differences in wording, it is mainly reported across these sources that

> *The month following the month of pilgrimage, i.e. the first month of the calendar AL-Muharram, is made sacred in one year and the subsequent month (i.e., Safar) is made sacred in another year.*

There was not a single reference in *Tafseer Al-Tabary* or other sources that addressed the concept of *intercalation*. Instead, all references addressed the sanctity of the month following Dhū al-Ḥijja, i.e., the month of the Hajj. Also, in these opinions, several accounts refer to the

---

[23] One of the main sources of Qur'an explanations - *The reader may find a concise electronic English version in www.Quran.ksu.edu.sa/tafseer/Tabary*

[24] The substantial number of opinions and interpretations indicates that the topic of intercalation was of concern to religious leaders and thinkers in the years following the passing of the prophet.

practice of using alternative names for the month AL-Muharram and to the number of months in a year in pre-Islam. And clearly, there is a general theme:

- *Thirteen months were in use on-and-off in a year,*
- *Sometimes, Arabs did not name the month following Dhū* al-Ḥijja, and
- *Mention of use of safarayn,* i.e., *two Safar months (following the month of Dhū al-Ḥijja).*

These reports clearly point to some on-and-off changes that occurred in the cyclic structure of the calendar from one year to another, which can only be interpreted as an implicit system of intercalation. Moreover, there is a unanimous agreement that Meccan leaders postponed the sanctity of the month Al-Muharram in some years and enforced it in other years. It must be noted that none of the references commented on the concept of intercalation, its purpose or mechanism. Instead, some references attributed the changes to leveraging advantages in tribal warfare. There is also mention in *Tafseer Al-Tabary* that for forty years prior to Islam, a man characterized as "of opinion" by the name *Jounadah Ibn Aouf Ibn Oumayah Al-Kinane* (Ibn means son of and al-Kinane means of the *Kinana* tribe), known as *Aba Thoumama*[25] *(Aba means father of)*, used to come out at the end of the yearly Pilgrimage and proclaims the sanctity of the month following Dhū al-Ḥijja. He is quoted:

"Aba Thoumama has no misdeeds or abnormalities - I proclaim Al-Muharram to be Sacred," and the following year he repeats the same except for stating *Safar to be Sacred."*

This is interpreted as either the first month following Dhū al-Ḥijja is named Al-Muharram and declared Sacred or the month is not given a name and the subsequent month of Safar becomes Al-Muharram and is considered sacred. In the latter case, there are reports that the first month is then called *Safar al-awwal* (i.e. Safar-I) and is followed by *Safar al-thaani* (that is Safar-II), thus the reference to the s*afarayn (i.e. the two safar months)*. The probable intention is that the knowledge of the first month of the Calendar, i.e., Al-Muharram to be the first or second month following Dhū al-Ḥijja fixes the upcoming calendar year. Technically, this practice was functional in setting up the Hajj and related Fairs for the upcoming year, as pilgrims are set to leave back to their home tribes which are spread over the large swath of the Arabian desert. The lack of structured religious institutions and reliable communication networks, which are capable of maintaining a calendar outside the city of Mecca, prompted a clever solution to maintaining the calendar. This is also the reason why intercalation takes place at the beginning of the calendar year, while it takes place at the end of the year in the

[25] The declaration of the sanctity of the month following Dhū al-Ḥijja was an honor passed to Aba Thoumama through a succession of at least six generations of ancestors, placing the usage of the Arabic calendar as early as the second half of the fourth century and around the time of the release of the Jewish Calendar by Hillel II, and just prior to the reported meeting of the masters of the Arab tribes to unify the names of the Calendar months in 412 CE.

Jewish calendar. The problem with this strategy is the sacred month of Al-Muharram is postponed in some years, leaving the first month open to aggression and score-settling. It is, however, well-established that pilgrims must be given safe passage as a fundamental of the Arabian society.

There is a consensus that *postponement* of the sacred month of Al-Muharram was the pivotal matter, at odds with Islam, that led to ending the use of the Arabic lunisolar calendar. In fact, the name *Al-Muharram* means "The sacred" and being sacred on-and-off was unacceptable to Islam. The Arabic word used for postponement (of a sacred month) in 9;37 is *"al-Nasi"* which means discarding, purging and forgetting. This is the same word used for the name of the short 13th month called *nasie*[26] of the Coptic solar calendar. And the use of the word "addition" in 9;37 does not relate to an "added" thirteenth month, like in the case of the Coptic calendar, and should not be misinterpreted. Intercalation is a process by which one month is discarded from the calendar count as the word *al-Nasi* bears the meaning. The verses identify *al-Nasi* as an *addition to unbelief* <u>for</u> postponing the forbidden months from one year to another and not for the practice of intercalation.

The twelve months in 9;36 refer to twelve synodic[27] months, which are used in lunar calendars because they give a day count closest to a solar year of 365.25 days (clearly there is an implicit reference the solar year). To be more precise, the number of lunations in a solar year is 12 and a little less than 11 days (365.25 divided by the average of 29.53 days in a lunar month gives 12.37 months. That is twelve lunations and 0.37 times 29.53 = 10.9 days). When rounding, the resulting number of months is 12. The use of twelve months is obviously acceptable in one solar year. But over a little less than three years, it becomes clear that an entire month is required to come out of the count. That is, to keep the calendar in par with the seasons, a lunation is omitted every 2.7 years (2.702703 x 0.37=1 month). The practice does not increase the number of months to 13 in a solar year. It makes the number of months actually 12, as it is shown from the simulation of two consecutive Metonic cycles in Table 10.4-1. The first column in the Table provides a counter of the solar year number, while the second column gives the cumulative number of days from the beginning of year 1 until the end of the current solar year. Also shown in the third column of the Table, is the cumulative numbers of lunar months (obtained by dividing the second column by 29.53). In the fourth column, the cumulative excess of the number of months per year above 12, calculated using the 3rd column entry minus 12 times the value in the first column. In the fifth column, it is shown that the average[28] number of months per year is constant throughout the years and equals to 12.37

---

26 See Section 2.7.1. Even the original Arabic calendar may have had such thirteenth month when we address its solar origin in Section10.6.

27 Other lunar periods (see Section 2.3) are not relevant to the topic of intercalation.

28 Divide 3rd column entry by the corresponding first column entry

Table 10.4-1 Number of months per solar year in pure lunar versus lunisolar calendars

Calendar							
Solar		Pure Lunar			Lunisolar		
Solar Year Number	Total number of days	Total number of months	Total Eccess months	Number of months per year	Intercalation month	Total number of months	Number of months per year
1	365.25	12.37	0.37	12.37	0	12.37	12.37
2	730.5	24.74	0.74	12.37	0	24.74	12.37
3	1095.75	37.11	1.11	12.37	1	36.11	12.04
4	1461	49.48	1.48	12.37	0	48.48	12.12
5	1826.25	61.84	1.84	12.37	0	60.84	12.17
6	2191.5	74.21	2.21	12.37	1	72.21	12.04
7	2556.75	86.58	2.58	12.37	0	84.58	12.08
8	2922	98.95	2.95	12.37	1	95.95	11.99
9	3287.25	111.32	3.32	12.37	0	108.32	12.04
10	3652.5	123.69	3.69	12.37	0	120.69	12.07
11	4017.75	136.06	4.06	12.37	1	132.06	12.01
12	4383	148.43	4.43	12.37	0	144.43	12.04
13	4748.25	160.79	4.79	12.37	0	156.79	12.06
14	5113.5	173.16	5.16	12.37	1	168.16	12.01
15	5478.75	185.53	5.53	12.37	0	180.53	12.04
16	5844	197.90	5.90	12.37	1	191.90	11.99
17	6209.25	210.27	6.27	12.37	0	204.27	12.02
18	6574.5	222.64	6.64	12.37	0	216.64	12.04
19	6939.75	235.01	7.01	12.37	1	228.01	12.00
20	7305	247.38	7.38	12.37	0	240.38	12.02
21	7670.25	259.74	7.74	12.37	0	252.74	12.04
22	8035.5	272.11	8.11	12.37	1	264.11	12.01
23	8400.75	284.48	8.48	12.37	0	276.48	12.02
24	8766	296.85	8.85	12.37	0	288.85	12.04
25	9131.25	309.22	9.22	12.37	1	300.22	12.01
26	9496.5	321.59	9.59	12.37	0	312.59	12.02
27	9861.75	333.96	9.96	12.37	1	323.96	12.00
28	10227	346.33	10.33	12.37	0	336.33	12.01
29	10592.25	358.69	10.69	12.37	0	348.69	12.02
30	10957.5	371.06	11.06	12.37	1	360.06	12.00
31	11322.75	383.43	11.43	12.37	0	372.43	12.01
32	11688	395.80	11.80	12.37	0	384.80	12.03
33	12053.25	408.17	12.17	12.37	1	396.17	12.01
34	12418.5	420.54	12.54	12.37	0	408.54	12.02
35	12783.75	432.91	12.91	12.37	1	419.91	12.00
36	13149	445.28	13.28	12.37	0	432.28	12.01
37	13514.25	457.64	13.64	12.37	0	444.64	12.02
38	13879.5	470.01	14.01	12.37	1	456.01	12.00

Finally, in the last section under Lunisolar Calendar, a schedule of intercalation is introduced at a rate of 2.7 solar years with a zero-entry in a regular year without intercalation and a unit value in the year in which intercalation takes place. The total number of months, with intercalation months discarded, is shown in column 7 by evaluating column 3 minus column 6. Finally, the number of months per year, accounting for intercalation, is shown in column 8. Its entries are calculated by dividing column 7 entries by the first column entries. After three years into the simulation, which are affected by the initial conditions of the simulation, the number of months per year remains consistently close to 12, while the pure lunar calendar has a constant 12.37 months per year. Therefore, it may be stated that the lunisolar calendar is actually more in-line with Qur'an 9; 36.

Addressing this very matter, the prophet mentioned in his "Farewell Sermon" during the Hajj, about three months before he passed away:

"...Time has cycled back to the state on the day that Allah created the heavens and the earth. The year is twelve months, four of which are sacred, three in sequence: Dhū al-Qiʿda, Dhu'l-Hijjah and Al-Muharram, and Rajab (of the tribe) of Mudar which comes between Jumādā and Shaʿbān."

The sermon addressed questions from the prophet's followers about the postponement of sacred months and the decree of the Qur'an: 9; 37. He reaffirmed that the number of months must be twelve, named the months that are sacred, and provided two additional elements.

First, the "year" is meant to be the *calendar year* and refers to the *solar year*. There are no other definitions for a year. This is the only cycle associated with the seasons.

Second, the lunisolar calendar has 12 months in a calendar year, while the pure lunar calendar has 12 months and 10.9 days.

Third, the Prophet stated that the sacred months are four[29], Dhū al-Qiʿda, Dhu'l-Hijjah and Al-Muharram must be in sequence[30], while the fourth sacred month, Rajab of the tribe of Mudar[31], must come between Jumādā and Shaʿbān. The purpose of the sanctity of these months is to guarantee the peace and safety of pilgrims, a practice that started long before

[29] This is in response to the tribe of Bani Aouf which used 8 sacred months instead. Arabs respected their tradition. They were never aggressed or aggressed anyone during these months.

[30] This is in response to postponement of Al-Muharram to Safar according to "Al-seerah al-Nabaweyah" by Ismail Ibn Omar Ibn Katheer, Vol I (in Arabic).

[31] The tribe of Mudar adopted the sacred month of Rajab as the month between Jumādā and Shaʿbān while the tribe of Rabiaa adopted the month of Rajab as the one between Sahabaane and Shawaal.

Islam. Specifying three months must be in sequence, and a fourth month must be between two named months indicates that some form of postponement was expected outside the above-mentioned sacred months. Otherwise, why would he emphasize that Dhū al-Qiʿda, Dhu'l-Hijjah, and Al-Muharram must be in sequence? Specifying that Rajab *must occur between* Jumādā and Shaʿbān is, in fact, an instruction not to execute intercalation either right before or after Rajab. This gives flexibility in choosing where intercalation may occur. Examining the months of the calendar, it seems that intercalation right after in the month of Al-Muharram could be a good possibility. For example, we would have Safar I and Safar II in a leap year to replace Safar.

Fourth, the statement that time has cycled back to the state of creation meant that the effect of prior postponements of sacred months has ended at the time of the Farewell Pilgrimage. This, in turn, means the date of Al-Muharram 1 of the 11th year of Hijra should have been used as the new origin of the Islamic calendar. We shall discuss the particularities of this datum in detail in Section 10.4.4. In addition, the mention of time has cycled back recognizes the Metonic cycle of 19 years, which is the only cycle for the moon.

In summary, there seems to be no reference made to canceling intercalation altogether, ending the leap year, or bringing the Hajj and Ramadan to occur during the varying season. The texts from the *Qur'an* and *Hadith* did not specifically instruct Muslims to abolish the lunisolar calendar. Instead, there is an indication that intercalation ought to occur, but no change can be made to the sanctity of the months.

10.4.2 Calendar transition timeline

To understand the consequences of the transition from the Arabic calendar to the Islamic calendar, we convert the 1st day of the lunar month between the two calendars and corresponding dates in the Gregorian (GR) and Jewish (JW) calendars. The latter was in use by the Jewish tribes of Medina and must have been the same as the Arabic calendar of Medina, as discussed earlier in Section 2.7.3. Conversion is initially carried over the first seven years following the death of the prophet Mohammed (632-638 CE), which are the years that preceded the transition to the Islamic calendar. Conversion of the dates of the first day of each lunar month for this period is shown in Table 10.4-3 with the age of the prophet given at the rightest column. It was reported that the prophet Mohammed died at the age of about 63, on 12 Rabīʿ I, 11 AH, shortly after his return from "the Farewell pilgrimage." This report, which is assumed to refer to the Islamic calendar, should refer to the Arabic calendar instead because the Islamic calendar did not exist when he passed away.

Over the following six years after the death of the Prophet, Arabia was in turmoil. The exact timing for the trip to the Hajj was a key issue for the followers. Pilgrims who were used to come to the main fair to buy and sell goods and pay their dues to their pagan gods in the month of Dhū al-Ḥijja (per the Arabian calendar) were probably told they were too late for the *Hajj*, and they could instead perform *Omra*[32]. Many pilgrims must have arrived at the wrong "new" time and were off by as much as a month. They no longer had their old gods and were told of a new timing all together for the season, which did not fit the economics of the desert. The talk of the day was about the civil war that was raging between the new Islamic power structure which swiftly ascended to dominate Arabia and those who wanted nothing to do with the new religion after the passing of the prophet. Doubts and confusion must have settled in the tribes of Arabia because the difference between the Arabic and Islamic calendars was not felt over the first three years.

One must keep in mind that information was communicated from word of mouth in those days, and it must have taken time for information about the calendar change to be passed correctly over large swaths of land. This was a time when no written records were available, and not everyone was literate. Therefore, keeping track of when events happened was probably hampered by the calendar change.

While the study of historical events is outside the scope of this work, it is important to note that detailed technical work is needed to clear the confusion about the chronology of events associated with the rise of Islam. The confusion is mainly due to the calendric shift from the Arabian calendar to the Islamic calendar and made even worse by the latter having a retroactive origin. Other errors, due to the shift from the Julian to the Gregorian calendar, have hampered conversions prior to the advent of computers and from the use of the accurate concept of Julian Day.

To establish the timeline associated with the emergence of Islam and the introduction of the new Islamic calendar, we first focus on the conversion of the first of the month of Dhū al-Ḥijja and the intercalation schedule around this date in the old Arabic and Jewish calendars. Subsequently, we convert the first of each lunar month in the four calendars using the following assumptions:

1. We assume that old Arabian and Jewish lunisolar calendars must have had the same intercalation schedule. The former is marked by the addition of the month (Veadar) at the end of the year, while the latter is achieved by not naming the first month of the year. The month following Dhū al-Ḥijja is shown as "No Name" in Table 10.4-3. Evidently, the time for Hajj was well established from seasonal pointers associated with the start of the spring[33] season that could not be missed; Table 10.4-2 illustrates the transition from winter to spring in Dhū al-Ḥijja (February-March time frame). In addition, advance knowledge of the time of Hajj was

---

[32] This is not a required trip to Mecca to perform certain rituals in Islam

[33] We refer here to spring in the northern hemisphere. Note that Arabs called the fall season spring.

made available to pilgrims and everyone attending the related Fairs during the prior season. In pre-Islam, the fair was visited by Christians, Jews, and Pagans from as far as Palestine and Syria in the North, and from the ancient Christian kingdom of Habasha, which extended to today's Ethiopia in the south. Such long travel required advance knowledge of the start of spring, which was common knowledge from clues related to aspects of the Horizon Calendar and other environmental indications. Postponement in the Arabic calendar is designed so that pilgrims and fair visitors depart the Hajj with knowledge about whether Al-Muharram is about to start or is postponed to the following moon. This ensures that they return home knowing if Al-Muharram is about to start or will start at the subsequent new moon and therefore, the timing for the Hajj in the upcoming year has been already decided. The sequence of the months and their names remained the same from year to year irrespective of the leap/common type. The suggestion that the caretaker of the Arabic calendar from the Kinana tribe had advance knowledge of whether the year is leap comes from the Jewish calendar one month ahead could be true but is certainly not needed and no reference backs such claim (Ioh, 2014). Intercalation was common practice since the time of Babylon, and the Horizon calendar provided the necessary indications for when to carry them. The accumulated drift caused by the lunar month manifests itself clearly in the seasonal pointers of rising and setting celestial bodies.

2. We shall artificially take reference of the yearly lunar cycle at 1 Dhū al-Ḥijja. This was selected because the last month of Dhū al-Ḥijja preceding the death of the prophet is common to both Arabic and Islamic calendars. This is also the most important month in pre-and post-Islam from religious, social, and economic standpoints. We also note that Dhū al-Ḥijja corresponds to either the month Nissan or Veadar in the Jewish calendar (JW), depending on whether intercalation is taking place. The first row in

Table 10.4-3 starts with the 1st of Dhū al-Ḥijja, the year of the prophet's farewell pilgrimage and the last row corresponds to the year in which the Islamic calendar was officially adopted, that is in the month of Rabīʿ al-Awwal (3rd month of the year 17 AH which starts with the month of AL-Muharram in the Islamic calendar). This time-reference could not have been made in the old Arabic calendar for obvious reasons. The same procedure is followed in Tables 10.4-4 and 10.4-5 in which the Islamic calendar is extended artificially beyond its defined origin to check on whether historical references were made in the Arabic calendar or in the Islamic calendar.

3. We use an algorithmic methodology to determine the start of lunations, and supplement the results using lunar-solar photometry to verify the date of historical events. The computation of the pure lunar Islamic calendar is based on a 30 year cycle of nineteen 354-day common years: 2, 3, 4, 6, 8, 9, 11, 12, 14, 15, 17, 19, 20, 22, 23, 25, 27, 28, 30 and eleven 355-day leap years: 1, 5, 7, 10, 13, 16, 18, 21, 24, 26, and 29. Each year has 12 months of

alternating 29 and 30 days with the last 29-day month in a leap year accounts for 30-days. For the Jewish calendar, years are either normal or leap. Leap years are 3, 6, 8, 11, 14, 17, and 19 in a 19-year cycle with an extra month of 29 days, called "Veadar," placed at the end of the year. Years are also classified as either deficient with 353 days, regular with 354 days or complete with 385 days in a common year and have 383, 384, or 385 days when the year is a leap.

Table 10.4-2 Typical seasonal environmental data during Arabic calendar months

Climate data for Mecca												
Month	Mar	Apr	May	Jun	Jul	Aug	Sep	Oct	Nov	Dec	Jan	Feb
Average high °C	35	39	42	44	43	43	43	40	35	32	31	32
Daily mean °C	27	31	34	36	36	36	35	32	28	26	24	25
Average low °C	21	25	28	29	29	30	29	26	23	20	19	19
Average precipitation mm	6	10	1	0	1	5	5	15	23	22	21	3
Average precipitation days	2	2	1	0	0	2	2	2	4	4	4	1
Mean daily sunshine hours	9	9	10	11	10	10	9	10	9	8	8	9

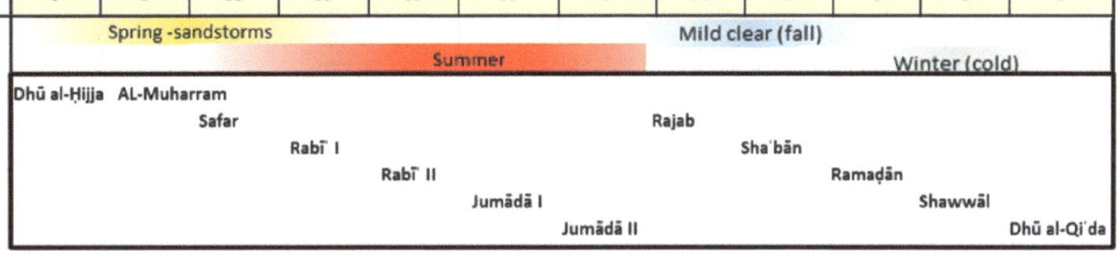

The first row of Table 10.4-3 gives the date of the 1st of Dhū al-Ḥijja, 10 AH in the Islamic calendar together with its conversion to the Jewish, Gregorian and Arabic calendars. This is the start of the month in which the landmark pilgrimage "Farewell Hajj" of the prophet had taken place three months before he died. The last month of the year 10 AH, i.e., Dhū al-Ḥijja, in the Arabic and Islamic calendars must coincide. Because it is a leap year, Dhū al-Ḥijja corresponds to the Jewish month of Veadar and at the start of the year 11 AH, the Arabic calendar introduces postponement by not naming the month. But then the new Islamic calendar drops the intercalation process altogether as seen starting in the second row. The first of Dhū al-Ḥijja, 10 AH corresponds to March 2, 632 CE from an algorithmic standpoint as tabulated under the Gregorian heading and as widely reported in the literature. However, refined analysis shows that from a Photometric standpoint, the New moon must have been clearly seen during twilight on March 1, 632 CE around the city of Medina. And because the day in the Islamic, Arabic and Jewish calendars goes from sunset to a subsequent sunset and does not start at midnight, as in the case for the Gregorian calendar, it may be stated that the first of the month of Dhū al-Ḥijja was actually March 1, 632 in Medina in both Islamic and Arabic calendars.

The rightest column shows the age of the prophet Mohammed. In Tables 10.4-3-10.4-6, the age refers to the old Arabic calendar instead of the Islamic calendar, which properly fits the narration of his history. Unfortunately, most references do not specify whether the Arabic or the Islamic calendar is being used when giving Hijra dates. And even though the first year of the new Islamic calendar was made retroactive back to the year of the Hijra of the prophet, we shall artificially take the conversions of the first of the month of Dhū al-Ḥijja all the way back to the beginning of the Islamic prophecy, to the years 609/610 CE in Table 10.4-5. We shall address the topic of the death of the prophet in Section 10.4.3, and give some details to show why such precision is required so that proper historical conclusions are drawn.

The Table shows that while the Arabic and Islamic calendars started in synch in the month of Dhū al-Ḥijja in early March 632 CE, they were out of phase by 2 months in the Gregorian calendar and by 3 lunar months the year of initiation of the Islamic calendar in seven years after the death of the prophet due to postponement (see the end of the Table).

Table 10.4-3 Post "Farwell Hajj" date conversion of the first of the lunar month (632-638 CE)

First of Lunar Month									Age
Jewish		Gregorian			Arabic		Islamic		
Y	M	Y	M	D	M	Y	M	Y	
4392	Veadar	632	Mar	2	Dhū al-Ḥijja	10	Dhū al-Ḥijja	10	62
4392	Nissan	632	Mar	31	Noname	11	AL-Muharram	11	62
4392	Lyyar	632	Apr	30	AL-Muharram	11	Safar	11	62
4392	Sivan	632	May	29	Safar	11	Rabīʿ I	11	62
4392	Tammuz	632	Jun	28	**Rabīʿ I**	11	Rabīʿ II	11	**63**
4392	Av	632	Jul	27	Rabīʿ II	11	Jumādā I	11	-
4392	Elul	632	Aug	26	Jumādā I	11	Jumādā II	11	-
4393	Teshrei	632	Sep	24	Jumādā II	11	Rajab	11	-
4393	Heshvan	632	Oct	24	Rajab	11	Shaʿbān	11	-
4393	Kislev	632	Nov	22	Shaʿbān	11	Ramaḍān	11	-
4393	Teveth	632	Dec	21	Ramaḍān	11	Shawwāl	11	-
4393	Shevat	633	Jan	19	Shawwāl	11	Dhū al-Qiʿda	11	-
4393	Adar	633	Feb	18	Dhū al-Qiʿda	11	Dhū al-Ḥijja	11	-
4393	Nissan	633	Mar	19	Dhū al-Ḥijja	11	AL-Muharram	12	-
4393	Lyyar	633	Apr	18	AL-Muharram	12	Safar	12	-
4393	Sivan	633	May	17	Safar	12	Rabīʿ I	12	-
4393	Tammuz	633	Jun	16	Rabīʿ I	12	Rabīʿ II	12	-
4393	Av	633	Jul	15	Rabīʿ II	12	Jumādā I	12	-
4393	Elul	633	Aug	14	Jumādā I	12	Jumādā II	12	-

First of Lunar Month									Age
Jewish		Gregorian			Arabic		Islamic		
Y	M	Y	M	D	M	Y	M	Y	
4394	Teshrei	633	Sep	12	Jumādā II	12	Rajab	12	-
4394	Heshvan	633	Oct	12	Rajab	12	Shaʿbān	12	-
4394	Kislev	633	nov	11	Shaʿbān	12	Ramaḍān	12	-
4394	Teveth	633	Dec	11	Ramaḍān	12	Shawwāl	12	-
4394	Shevat	634	Jan	9	Shawwāl	12	Dhū al-Qiʿda	12	-
4394	Adar	634	Feb	8	Dhū al-Qiʿda	12	Dhū al-Ḥijja	12	-
4394	Nissan	634	Mar	9	Dhū al-Ḥijja	12	AL-Muharram	13	-
4394	Lyyar	634	Apr	8	AL-Muharram	13	Safar	13	-
4394	Sivan	634	May	7	Safar	13	Rabīʿ I	13	-
4394	Tammuz	634	Jun	6	Rabīʿ I	13	Rabīʿ II	13	-
4394	Av	634	Jul	5	Rabīʿ II	13	Jumādā I	13	-
4394	Elul	634	Aug	4	Jumādā I	13	Jumādā II	13	-
4395	Teshrei	634	Sep	2	Jumādā II	13	Rajab	13	-
4395	Heshvan	634	Oct	2	Rajab	13	Shaʿbān	13	-
4395	Kislev	634	Oct	31	Shaʿbān	13	Ramaḍān	13	-
4395	Teveth	634	Nov	30	Ramaḍān	13	Shawwāl	13	-
4395	Shevat	634	Dec	29	Shawwāl	13	Dhū al-Qiʿda	13	-
4395	Adar	635	Jan	28	Dhū al-Qiʿda	13	Dhū al-Ḥijja	13	-
4395	Veadar	635	Feb	27	Dhū al-Ḥijja	13	AL-Muharram	14	-
4395	Nissan	635	Mar	28	Noname	14	Safar	14	-
4395	Lyyar	635	Apr	27	AL-Muharram	14	Rabīʿ I	14	-
4395	Sivan	635	May	26	Safar	14	Rabīʿ II	14	-
4395	Tammuz	635	Jun	25	Rabīʿ I	14	Jumādā I	14	-
4395	Av	635	Jul	24	Rabīʿ II	14	Jumādā II	14	-
4395	Elul	635	Aug	23	Jumādā I	14	Rajab	14	-
4396	Teshrei	635	Sep	21	Jumādā II	14	Shaʿbān	14	-
4396	Heshvan	635	Oct	21	Rajab	14	Ramaḍān	14	-
4396	Kislev	635	Nov	20	Shaʿbān	14	Shawwāl	14	-
4396	Teveth	635	Dec	20	Ramaḍān	14	Dhū al-Qiʿda	14	-
4396	Shevat	636	Jan	18	Shawwāl	14	Dhū al-Ḥijja	14	-
4396	Adar	636	Feb	17	Dhū al-Qiʿda	14	AL-Muharram	15	-
4396	Nissan	636	Mar	17	Dhū al-Ḥijja	14	Safar	15	-

First of Lunar Month									Age
Jewish		Gregorian			Arabic		Islamic		
Y	M	Y	M	D	M	Y	M	Y	
4396	Lyyar	636	Apr	16	AL-Muharram	15	Rabīʿ I	15	-
4396	Sivan	636	May	15	Safar	15	Rabīʿ II	15	-
4396	Tammuz	636	Jun	14	Rabīʿ I	15	Jumādā I	15	-
4396	Av	636	Jul	13	Rabīʿ II	15	Jumādā II	15	-
4396	Elul	636	Aug	12	Jumādā I	15	Rajab	15	-
4397	Teshrei	636	Sep	10	Jumādā II	15	Shaʿbān	15	-
4397	Heshvan	636	Oct	10	Rajab	15	Ramaḍān	15	-
4397	Kislev	636	Nov	8	Shaʿbān	15	Shawwāl	15	-
4397	Teveth	636	Dec	7	Ramaḍān	15	Dhū al-Qiʿda	15	-
4397	Shevat	637	Jan	5	Shawwāl	15	Dhū al-Ḥijja	15	-
4397	Adar	637	Feb	4	Dhū al-Qiʿda	15	AL-Muharram	16	-
4397	Veadar	637	Mar	6	Dhū al-Ḥijja	15	Safar	16	-
4397	Nissan	637	Apr	4	Noname	16	Rabīʿ I	16	-
4397	Lyyar	637	May	4	AL-Muharram	16	Rabīʿ II	16	-
4397	Sivan	637	Jun	2	Safar	16	Jumādā I	16	-
4397	Tammuz	637	Jul	2	Rabīʿ I	16	Jumādā II	16	-
4397	Av	637	Jul	31	Rabīʿ II	16	Rajab	16	-
4397	Elul	637	Aug	30	Jumādā I	16	Shaʿbān	16	-
4398	Teshrei	637	Sep	28	Jumādā II	16	Ramaḍān	16	-
4398	Heshvan	637	Oct	28	Rajab	16	Shawwāl	16	-
4398	Kislev	637	Nov	26	Shaʿbān	16	Dhū al-Qiʿda	16	-
4398	Teveth	637	Dec	26	Ramaḍān	16	Dhū al-Ḥijja	16	-
4398	Shevat	638	Jan	24	Shawwāl	16	AL-Muharram	17	-
4398	Adar	638	Feb	23	Dhū al-Qiʿda	16	Safar	17	-
4398	Nissan	638	Mar	24	Dhū al-Ḥijja	16	Rabīʿ I	17	-

10.4.3 Death and burial of the Prophet

It is well established that the prophet passed away at the age of 63 on 12 Rabīʿ I, 11 AH. It has been always assumed that this date is in the Islamic calendar, and corresponds to June 10, 632 CE in the Gregorian calendar (June 7, 632 in the Julian calendar). Analysis of the photometry at sunset shows that the day should be June 9, 632 CE, as the new crescent moon of 1 Rabīʿ I, 11 AH would have been visible in Medina on May 29. Intricate details of the unfolding of the event of the Prophet's death are documented in several references which state that the day was a Monday and the prophet was buried the following day, Tuesday evening. But, calculations show that June 10, 632 CE was a Sunday and June 9 was a Saturday, and not a Monday as reported. Note that Muslims track the day of the week accurately because of the Friday prayer, which must take place in a congregation. This, in turn, makes the day of the week a reliable reference. The Islamic calendar did not exist at the time of the death of the Prophet, and the Arabic calendar was the one in use.

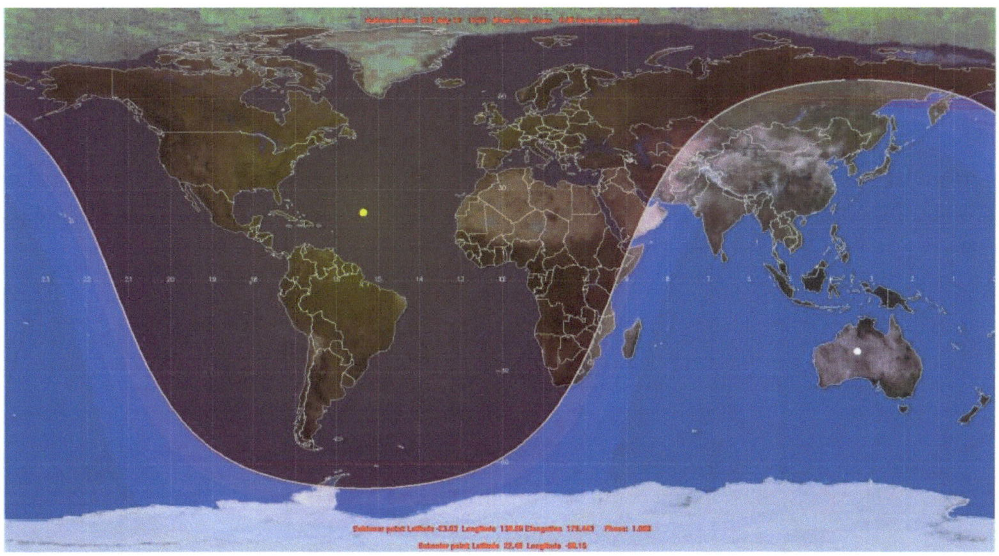

Figure 10.4-1 Lunar Eclipse the night of burial of the prophet Mohammed, 10 July, 632 CE

From Table 10.4-3, the 1st of Rabīʿ I in the Arabic calendar corresponds to June 28, 632 CE and 12 Rabīʿ I corresponds to Monday, July 9, 632 CE. The date in the Arabic calendar fits the narration on the sequence of days leading to his death. Tuesday, July 10 was the night the prophet was lowered in his tomb which was said to be in his room edging his mosque in Medina, right where he died. That night was the start of 14th of Rabīʿ I in the Arabic calendar (remember the day starts at sunset). According to astronomical calculations, an eclipsed full moon happened to be over Medina, like the one during the crucifixion of Jesus which was covered in Section 10.2. The eclipse occurred at 15:22 TD, i.e., 18:22 local time with the lunar and solar terminators co-located, as shown in Figure 10.4-1. It must be noted that the occurrence of this eclipse is not referred to in the narrative of the passing of the prophet. Either because the prophet made it clear that eclipses are not to associate with births and deaths, or it could simply be that this eclipse was not visible in Medina due to atmospheric conditions. For reference, the Prophet was born on either the 9th or the 12th of Rabīʿ I (the second date is popular, but the first is considered more reliable. It is also said that he was born in "the year of the Elephant", the year Abraha, an Abyssinian, Christian ruler of Yemen, who was at one time a subject to the Kingdom of Aksum of Ethiopia before revolting, attacked Mecca using a powerful army supported by elephants seeking the destruction of the Kaaba in 570 CE.

There is a well-documented event of a solar eclipse which occurred the day the toddler son of the prophet Ibrahim (Abraham) died. It was reported from multiple sources that the prophet led an "eclipse" prayer on that day and stated that eclipses are not signs associated with someone's death or birth. Using the coordinates of Medina of 24.4° N and 39.7° E, the only solar eclipse visible during the reported era was the one which occurred Monday morning Jan 30, 632 CE (i.e., Jan 27 in the Julian calendar and 29 Shawwal, 10 AH). Calculations show that the start and end of the partial eclipse as seen from Medina corresponded to when the Alt-Az coordinates of the sun were (1°, 110°) and (32°, 133°), respectively while accounting for Terrestrial Dynamic Time. This timing fits in the Shawwal month of the second row from the end of Table 10.4-4. Note that the Arabic and the Islamic calendars coincide on this date.

Table 10.4-4 Calendar date conversion post Prophet migration to Medina, 623-632 CE.

First of Lunar Month									Age
Jewish		Gregorian			Arabic		Islamic		
Y	M	Y	M	D	M	Y	M	Y	
4383	Nissan	623	Mar	11	Dhū al-Ḥijja	1	Ramaḍān	1	53
4383	Lyyar	623	Apr	10	AL-Muharram	2	Shawwāl	1	53
4383	Sivan	623	May	9	Safar	2	Dhū al-Qiʿda	1	53
4383	Tammuz	623	Jun	8	Rabīʿ I	2	Dhū al-Ḥijja	1	54
4383	Av	623	Jul	7	Rabīʿ II	2	AL-Muharram	2	54
4383	Elul	623	Aug	6	Jumādā I	2	Safar	2	54
4384	Teshrei	623	Sep	4	Jumādā II	2	Rabīʿ I	2	54
4384	Heshvan	623	Oct	4	Rajab	2	Rabīʿ II	2	54
4384	Kislev	623	nov	3	Shaʿbān	2	Jumādā I	2	54
4384	Teveth	623	Dec	3	Ramaḍān	2	Jumādā II	2	54
4384	Shevat	624	Jan	1	Shawwāl	2	Rajab	2	54
4384	Adar	624	Jan	31	Dhū al-Qiʿda	2	Shaʿbān	2	54
4384	Veadar	624	Mar	1	Dhū al-Ḥijja	2	Ramaḍān	2	54
4384	Nissan	624	Mar	30	Noname	3	Shawwāl	2	54
4384	Lyyar	624	Apr	29	AL-Muharram	3	Dhū al-Qiʿda	2	54
4384	Sivan	624	May	28	Safar	3	Dhū al-Ḥijja	2	54
4384	Tammuz	624	Jun	27	Rabīʿ I	3	AL-Muharram	3	55

First of Lunar Month									Age
Jewish		Gregorian			Arabic		Islamic		
Y	M	Y	M	D	M	Y	M	Y	
4384	Av	624	Jul	26	Rabīʿ II	3	Safar	3	55
4384	Elul	624	Aug	25	Jumādā I	3	Rabīʿ I	3	55
4385	Teshrei	624	Sep	23	Jumādā II	3	Rabīʿ II	3	55
4385	Heshvan	624	Oct	23	Rajab	3	Jumādā I	3	55
4385	Kislev	624	nov	21	Shaʿbān	3	Jumādā II	3	55
4385	Teveth	624	Dec	21	Ramaḍān	3	Rajab	3	55
4385	Shevat	625	Jan	19	Shawwāl	3	Shaʿbān	3	55
4385	Adar	625	Feb	18	Dhū al-Qiʿda	3	Ramaḍān	3	55
4385	Nissan	625	Mar	19	Dhū al-Ḥijja	3	Shawwāl	3	55
4385	Lyyar	625	Apr	18	AL-Muharram	4	Dhū al-Qiʿda	3	55
4385	Sivan	625	May	17	Safar	4	Dhū al-Ḥijja	3	55
4385	Tammuz	625	Jun	16	Rabīʿ I	4	AL-Muharram	4	56
4385	Av	625	Jul	15	Rabīʿ II	4	Safar	4	56
4385	Elul	625	Aug	14	Jumādā I	4	Rabīʿ I	4	56
4386	Teshrei	625	Sep	12	Jumādā II	4	Rabīʿ II	4	56
4386	Heshvan	625	Oct	12	Rajab	4	Jumādā I	4	56
4386	Kislev	625	nov	10	Shaʿbān	4	Jumādā II	4	56
4386	Teveth	625	Dec	9	Ramaḍān	4	Rajab	4	56
4386	Shevat	626	Jan	7	Shawwāl	4	Shaʿbān	4	56
4386	Adar	626	Feb	6	Dhū al-Qiʿda	4	Ramaḍān	4	56
4386	Nissan	626	Mar	7	Dhū al-Ḥijja	4	Shawwāl	4	56
4386	Lyyar	626	Apr	6	AL-Muharram	5	Dhū al-Qiʿda	4	56
4386	Sivan	626	May	5	Safar	5	Dhū al-Ḥijja	4	56
4386	Tammuz	626	Jun	4	Rabīʿ I	5	AL-Muharram	5	57
4386	Av	626	Jul	3	Rabīʿ II	5	Safar	5	57
4386	Elul	626	Aug	2	Jumādā I	5	Rabīʿ I	5	57
4387	Teshrei	626	Aug	31	Jumādā II	5	Rabīʿ II	5	57
4387	Heshvan	626	Sep	30	Rajab	5	Jumādā I	5	57
4387	Kislev	626	Oct	30	Shaʿbān	5	Jumādā II	5	57
4387	Teveth	626	nov	29	Ramaḍān	5	Rajab	5	57
4387	Shevat	626	Dec	28	Shawwāl	5	Shaʿbān	5	57
4387	Adar	627	Jan	27	Dhū al-Qiʿda	5	Ramaḍān	5	57

First of Lunar Month									Age
Jewish		Gregorian			Arabic		Islamic		
Y	M	Y	M	D	M	Y	M	Y	
4387	Veadar	627	Feb	26	Dhū al-Ḥijja	5	Shawwāl	5	57
4387	Nissan	627	Mar	27	Noname	6	Dhū al-Qiʿda	5	57
4387	Lyyar	627	Apr	26	AL-Muharram	6	Dhū al-Ḥijja	5	57
4387	Sivan	627	May	25	Safar	6	AL-Muharram	6	57
4387	Tammuz	627	Jun	24	Rabīʿ I	6	Safar	6	58
4387	Av	627	Jul	23	Rabīʿ II	6	Rabīʿ I	6	58
4387	Elul	627	Aug	22	Jumādā I	6	Rabīʿ II	6	58
4388	Teshrei	627	Sep	20	Jumādā II	6	Jumādā I	6	58
4388	Heshvan	627	Oct	20	Rajab	6	Jumādā II	6	58
4388	Kislev	627	nov	18	Shaʿbān	6	Rajab	6	58
4388	Teveth	627	Dec	18	Ramaḍān	6	Shaʿbān	6	58
4388	Shevat	628	Jan	16	Shawwāl	6	Ramaḍān	6	58
4388	Adar	628	Feb	15	Dhū al-Qiʿda	6	Shawwāl	6	58
4388	Nissan	628	Mar	15	Dhū al-Ḥijja	6	Dhū al-Qiʿda	6	58
4388	Lyyar	628	Apr	14	AL-Muharram	7	Dhū al-Ḥijja	6	58
4388	Sivan	628	May	13	Safar	7	AL-Muharram	7	58
4388	Tammuz	628	Jun	12	Rabīʿ I	7	Safar	7	59
4388	Av	628	Jul	11	Rabīʿ II	7	Rabīʿ I	7	59
4388	Elul	628	Aug	10	Jumādā I	7	Rabīʿ II	7	59
4389	Teshrei	628	Sep	8	Jumādā II	7	Jumādā I	7	59
4389	Heshvan	628	Oct	8	Rajab	7	Jumādā II	7	59
4389	Kislev	628	nov	6	Shaʿbān	7	Rajab	7	59
4389	Teveth	628	Dec	5	Ramaḍān	7	Shaʿbān	7	59
4389	Shevat	629	Jan	3	Shawwāl	7	Ramaḍān	7	59
4389	Adar	629	Feb	2	Dhū al-Qiʿda	7	Shawwāl	7	59
4389	Veadar	629	Mar	4	Dhū al-Ḥijja	7	Dhū al-Qiʿda	7	59
4389	Nissan	629	Apr	2	Noname	8	Dhū al-Ḥijja	7	59
4389	Lyyar	629	May	2	AL-Muharram	8	AL-Muharram	8	59
4389	Sivan	629	May	31	Safar	8	Safar	8	59
4389	Tammuz	629	Jun	30	Rabīʿ I	8	Rabīʿ I	8	60
4389	Av	629	Jul	29	Rabīʿ II	8	Rabīʿ II	8	60
4389	Elul	629	Aug	28	Jumādā I	8	Jumādā I	8	60

First of Lunar Month									Age
Jewish		Gregorian			Arabic		Islamic		
Y	M	Y	M	D	M	Y	M	Y	
4390	Teshrei	629	Sep	26	Jumādā II	8	Jumādā II	8	60
4390	Heshvan	629	Oct	26	Rajab	8	Rajab	8	60
4390	Kislev	629	nov	25	Sha'bān	8	Sha'bān	8	60
4390	Teveth	629	Dec	25	Ramaḍān	8	Ramaḍān	8	60
4390	Shevat	630	Jan	23	Shawwāl	8	Shawwāl	8	60
4390	Adar	630	Feb	22	Dhū al-Qi'da	8	Dhū al-Qi'da	8	60
4390	Nissan	630	Mar	23	Dhū al-Ḥijja	8	Dhū al-Ḥijja	8	60
4390	Lyyar	630	Apr	22	AL-Muharram	9	AL-Muharram	9	60
4390	Sivan	630	May	21	Safar	9	Safar	9	60
4390	Tammuz	630	Jun	20	Rabī' I	9	Rabī' I	9	61
4390	Av	630	Jul	19	Rabī' II	9	Rabī' II	9	61
4390	Elul	630	Aug	18	Jumādā I	9	Jumādā I	9	61
4391	Teshrei	630	Sep	16	Jumādā II	9	Jumādā II	9	61
4391	Heshvan	630	Oct	16	Rajab	9	Rajab	9	61
4391	Kislev	630	nov	14	Sha'bān	9	Sha'bān	9	61
4391	Teveth	630	Dec	14	Ramaḍān	9	Ramaḍān	9	61
4391	Shevat	631	Jan	12	Shawwāl	9	Shawwāl	9	61
4391	Adar	631	Feb	11	Dhū al-Qi'da	9	Dhū al-Qi'da	9	61
4391	Nissan	631	Mar	12	Dhū al-Ḥijja	9	Dhū al-Ḥijja	9	61
4391	Lyyar	631	Apr	11	AL-Muharram	10	AL-Muharram	10	61
4391	Sivan	631	May	10	Safar	10	Safar	10	61
4391	Tammuz	631	Jun	9	Rabī' I	10	Rabī' I	10	62
4391	Av	631	Jul	8	Rabī' II	10	Rabī' II	10	62
4391	Elul	631	Aug	7	Jumādā I	10	Jumādā I	10	62
4392	Teshrei	631	Sep	5	Jumādā II	10	Jumādā II	10	62
4392	Heshvan	631	Oct	5	Rajab	10	Rajab	10	62
4392	Kislev	631	nov	4	Sha'bān	10	Sha'bān	10	62
4392	Teveth	631	Dec	4	Ramaḍān	10	Ramaḍān	10	62
4392	Shevat	632	Jan	2	Shawwāl	10	Shawwāl	10	62
4392	Adar	632	Feb	1	Dhū al-Qi'da	10	Dhū al-Qi'da	10	62

10.4.4 Retroactive time-origin of the Islamic Calendar

Six years after the death of the prophet Mohammed, the civil war between the followers of Islam and those resisted the new religion has come to an end with Islam spreading to a large swath of land that extended far outside Arabia. Those in charge of the new nation needed clarity in communication and precision in time reckoning from the Calif. In addition, the timing for two pillars of Islam, i.e., fasting in the month Ramadan and the Hajj to Mecca needed proper definition. Abu Musa Ashaari, the governor of Basra, Iraq, wrote the Caliph *Umar* ibn Al-Khattāb (583-644 CE), asking for clarification about dates mentioned in communication he received. Back in Arabia, the time for pre-Islam Hajj has arrived again (end of March-early April lunation). Travel was possible due to a new peace that Arabia never had before and people from the vast desert headed for the fare around Mecca, where they have traditionally exchanged goods for generations and run by the Kaaba to seek help from their favorite gods. But upon arrival to Mecca, they were told they were late by three moons; the new time for the Hajj was in the cold winter season (early January). The visitors could, however, perform the non-required rituals of the *Omra* instead. Arabs continued with the fare at the time of the old Hajj for many years and ultimately ditched pagan traditions and converted to Islam.

The understanding was that no more intercalation could take place. The Caliph, in consultation with his aids, realized the need for a calendar and initiated the Islamic lunar calendar, retroactive to 1st of AL-Muharram, 1 (AH). The task was technically simple for the calendar-savvy new Muslims who were part of the inner circle of the Caliph because there was no intercalation to account for. The starting a new calendar was a courageous step outside the *Sunna* (tradition of the prophet) because devoted Muslims followed "the way" of the prophet, and never engaged in something he did not do. Additions beyond what the prophet did were viewed as "bid'a," i.e., deviation from the righteous path.

The retroactive origin of the 1st of AL-Muharram, the year 1 (AH) made the Islamic Calendar weakly defined. Mathematically, a datum must be properly and uniquely defined. This is not the case for the Islamic Calendar because it is not clear if the chosen origin is referencing the old Arabian calendar or the new Islamic calendar. Even though the designer(s) of the calendar probably knew which of the two they were referring to, there are issues with both alternatives. Should the Calendar references the 1st of AL-Muharram of the year of the Hijra in the old Arabic calendar, then the reference is:

(a) Unacceptable and erroneous by the reason the new calendar was introduced in the first place. The chronology referred to uses postponement of the month of Al-Muharram, which was viewed to be in contradiction with Qur'an 12:35 and therefore could not be used.

(b) Inconsistent with the event of the Hijra (being referenced) and other events which occurred between the Hijra and the death of the prophet. In the Arabic calendar, the 1st of AL-

Muharram came in April while it occurred three months later in the month of July according to the Islamic calendar (see Table 10.4-5). Furthermore, the first of the month of AL-Muharram in the Arabic calendar occurs in the year before the Hijra.

Alternatively, if 1st of AL-Muharram, 1 (AH) is chosen retroactively per the new calendar system, free of intercalation, then the reference point in time which defines the Hijra is not chronologically correct and therefore invalidates the basis for the new calendar altogether. In addition, all historical events referenced before the Farewell Hajj must pertain to the old Arabian calendar.

The year of Hijra of the prophet Mohammed is said to be a key milestone in Islam and was therefore chosen as the starting datum for the Islamic calendar. The names and sequence of the months were kept the same as those of the Arabic calendar with AL-Muharram as the first month. Other dates could have been candidates for such datum, e.g., the year of revelation or the year of birth of the prophet, or his death. One can argue that they are equally or even more important. However, to maintain some continuity in the dating system, a retroactive origin must at least go to a common origin in the two calendars. Starting with Table 10.4-3, the first of Al-Muharram following the Farewell Hajj presents a good origin to start the new calendar. Not only because the prophet performed the Hajj, which signals the month of the Hajj was correct, but also because he stated that *``time having cycled back to the state of creation."*

The alternative datum of the 1st of Al-Muharram following the Farewell Hajj would have lifted the definition problems, as the concordance of the month of Dhū al-Ḥijja between the Islamic and Arabian calendars at the time of the sermon has been implicitly stated, and no confusion about historical events would have occurred. Because the prophet died only three months into the New Year, the first year would be the year when the prophet died[34]. We shall discuss the subject of such alternative origin in the following Section.

10.4.5 Alternative time-origin for the Islamic Calendar

Historically, the month Dhū al-Ḥijja, 10 (AH) is quite important for several reasons. First, the verses from the Qur'an and Hadith addressing the topic of the sanctity of the months and their numbers were revealed during the Farewell Hajj in the month Dhū al-Ḥijja. And it was the interpretation of those verses and Hadith by the followers of the Prophet that led to the design of the Islamic calendar seven years after his death. Second, the month of Dhū al-Ḥijja, in the year 10 (AH) was characterized by the prophet as a time when the state of the heavens is identical to that of the time of creation, stating that the time has cycled back to the origin. Third, because the month Dhū al-Ḥijja, 10 (AH) is common to both Islamic and Arabic

[34] Muammar Gaddafi adopted the death of the Prophet as datum in his short-lived and unsuccessful calendar during his reign over Libya fourteen centuries later.

calendars, any reconciliation needs to have Dhū al-Ḥijja, 10 (AH) the same month in the Arabic calendar, which is what we did in the reconstruction of the conversion Tables.

The next question: is there anything special about Dhū al-Ḥijja, 10 (AH) from a photometric perspective? We shall look at the astronomical positions of the moon and the sun and see if we can recover some particularities about this origin.

Table 10.4-5 Conversion of the First Day of the lunar month from the start of revelation to the Prophet Mohammed's migration to Medina (609-623 CE)

First of Lunar Month									Age
Jewish		Gregorian			Arabic		Islamic		
Y	M	Y	M	D	M	Y	M	Y	
4369	Nisan	609	Mar	16	Dhū al-Ḥijja	-14	Rabīʿ II	-14	39
4369	Lyyar	609	Apr	15	AL-Muharram	-13	Jumādā I	-14	39
4369	Sivan	609	May	14	Safar	-13	Jumādā II	-14	39
4369	Tammuz	609	Jun	13	Rabīʿ I	-13	Rajab	-14	40
4369	Av	609	Jul	12	Rabīʿ II	-13	Shaʿbān	-14	40
4369	Elul	609	Aug	11	Jumādā I	-13	Ramaḍān	-14	40
4370	Teshrei	609	Sep	9	Jumādā II	-13	Shawwāl	-14	40
4370	Heshvan	609	Oct	9	Rajab	-13	Dhū al-Qiʿda	-14	40
4370	Kislev	609	Nov	7	Shaʿbān	-13	Dhū al-Ḥijja	-14	40
4370	Teveth	609	Dec	6	Ramaḍān	-13	AL-Muharram	-13	40
4370	Shevat	610	Jan	4	Shawwāl	-13	Safar	-13	40
4370	Adar	610	Feb	3	Dhū al-Qiʿda	-13	Rabīʿ I	-13	40
4370	Veadar	610	Mar	5	Dhū al-Ḥijja	-13	Rabīʿ II	-13	40
4370	Nisan	610	Apr	3	Noname	-12	Jumādā I	-13	40
4370	Lyyar	610	May	3	AL-Muharram	-12	Jumādā II	-13	40
4370	Sivan	610	June	1	Safar	-12	Rajab	-13	40
4370	Tammuz	610	Jul	1	Rabīʿ I	-12	Shaʿbān	-13	41
4370	Av	610	Jul	30	Rabīʿ II	-12	Ramaḍān	-13	41
4370	Elul	610	Aug	29	Jumādā I	-12	Shawwāl	-13	41
4371	Teshrei	610	Sep	27	Jumādā II	-12	Dhū al-Qiʿda	-13	41
4371	Heshvan	610	oct	27	Rajab	-12	Dhū al-Ḥijja	-13	41
4371	Kislev	610	Nov	25	Shaʿbān	-12	AL-Muharram	-12	41
4371	Teveth	610	Dec	25	Ramaḍān	-12	Safar	-12	41

First of Lunar Month									Age
Jewish		Gregorian			Arabic		Islamic		
Y	M	Y	M	D	M	Y	M	Y	
4371	Shevat	611	Jan	23	Shawwāl	-12	Rabīʿ I	-12	41
4371	Adar	611	feb	22	Dhū al-Qiʿda	-12	Rabīʿ II	-12	41
4371	Nisan	611	Mar	23	Dhū al-Ḥijja	-12	Jumādā I	-12	41
4371	Lyyar	611	Apr	22	AL-Muharram	-11	Jumādā II	-12	41
4371	Sivan	611	May	21	Safar	-11	Rajab	-12	41
4371	Tammuz	611	Jun	20	Rabīʿ I	-11	Shaʿbān	-12	42
4371	Av	611	Jul	19	Rabīʿ II	-11	Ramaḍān	-12	42
4371	Elul	611	Aug	18	Jumādā I	-11	Shawwāl	-12	42
4372	Teshrei	611	Sep	16	Jumādā II	-11	Dhū al-Qiʿda	-12	42
4372	Heshvan	611	Oct	16	Rajab	-11	Dhū al-Ḥijja	-12	42
4372	Kislev	611	Nov	15	Shaʿbān	-11	AL-Muharram	-11	42
4372	Teveth	611	Dec	15	Ramaḍān	-11	Safar	-11	42
4372	Shevat	612	Jan	13	Shawwāl	-11	Rabīʿ I	-11	42
4372	Adar	612	Feb	12	Dhū al-Qiʿda	-11	Rabīʿ II	-11	42
4372	Nisan	612	Mar	12	Dhū al-Ḥijja	-11	Jumādā I	-11	42
4372	Lyyar	612	Apr	11	AL-Muharram	-10	Jumādā II	-11	42
4372	Sivan	612	may	10	Safar	-10	Rajab	-11	42
4372	Tammuz	612	Jun	9	Rabīʿ I	-10	Shaʿbān	-11	43
4372	Av	612	Jul	8	Rabīʿ II	-10	Ramaḍān	-11	43
4372	Elul	612	Aug	7	Jumādā I	-10	Shawwāl	-11	43
4373	Teshrei	612	Sep	5	Jumādā II	-10	Dhū al-Qiʿda	-11	43
4373	Heshvan	612	oct	5	Rajab	-10	Dhū al-Ḥijja	-11	43
4373	Kislev	612	Nov	3	Shaʿbān	-10	AL-Muharram	-10	43
4373	Teveth	612	Dec	2	Ramaḍān	-10	Safar	-10	43
4373	Shevat	612	Dec	31	Shawwāl	-10	Rabīʿ I	-10	43
4373	Adar	613	Jan	30	Dhū al-Qiʿda	-10	Rabīʿ II	-10	43
4373	Veadar	613	Mar	1	Dhū al-Ḥijja	-10	Jumādā I	-10	43
4373	Nisan	613	Mar	30	Noname	-9	Jumādā II	-10	43
4373	Lyyar	613	Apr	29	AL-Muharram	-9	Rajab	-10	43
4373	Sivan	613	May	28	Safar	-9	Shaʿbān	-10	43
4373	Tammuz	613	Jun	27	Rabīʿ I	-9	Ramaḍān	-10	44

First of Lunar Month									Age
Jewish		Gregorian			Arabic		Islamic		
Y	M	Y	M	D	M	Y	M	Y	
4373	Av	613	Jul	26	Rabīʿ II	-9	Shawwāl	-10	44
4373	Elul	613	Aug	25	Jumādā I	-9	Dhū al-Qiʿda	-10	44
4374	Teshrei	613	Sep	23	Jumādā II	-9	Dhū al-Ḥijja	-10	44
4374	Heshvan	613	Oct	23	Rajab	-9	AL-Muharram	-9	44
4374	Kislev	613	Nov	22	Shaʿbān	-9	Safar	-9	44
4374	Teveth	613	Dec	22	Ramaḍān	-9	Rabīʿ I	-9	44
4374	Shevat	614	Jan	20	Shawwāl	-9	Rabīʿ II	-9	44
4374	Adar	614	Feb	19	Dhū al-Qiʿda	-9	Jumādā I	-9	44
4374	Nisan	614	Mar	20	Dhū al-Ḥijja	-9	Jumādā II	-9	44
4374	Lyyar	614	Apr	19	AL-Muharram	-8	Rajab	-9	44
4374	Sivan	614	May	18	Safar	-8	Shaʿbān	-9	44
4374	Tammuz	614	Jun	17	Rabīʿ I	-8	Ramaḍān	-9	45
4374	Av	614	Jul	16	Rabīʿ II	-8	Shawwāl	-9	45
4374	Elul	614	Aug	15	Jumādā I	-8	Dhū al-Qiʿda	-9	45
4375	Teshrei	614	Sep	13	Jumādā II	-8	Dhū al-Ḥijja	-9	45
4375	Heshvan	614	Oct	13	Rajab	-8	AL-Muharram	-8	45
4375	Kislev	614	Nov	11	Shaʿbān	-8	Safar	-8	45
4375	Teveth	614	Dec	11	Ramaḍān	-8	Rabīʿ I	-8	45
4375	Shevat	615	Jan	9	Shawwāl	-8	Rabīʿ II	-8	45
4375	Adar	615	Feb	8	Dhū al-Qiʿda	-8	Jumādā I	-8	45
4375	Nisan	615	Mar	9	Dhū al-Ḥijja	-8	Jumādā II	-8	45
4375	Lyyar	615	Apr	8	AL-Muharram	-7	Rajab	-8	45
4375	Sivan	615	May	7	Safar	-7	Shaʿbān	-8	45
4375	Tammuz	615	Jun	6	Rabīʿ I	-7	Ramaḍān	-8	46
4375	Av	615	Jul	5	Rabīʿ II	-7	Shawwāl	-8	46
4375	Elul	615	Aug	4	Jumādā I	-7	Dhū al-Qiʿda	-8	46
4376	Teshrei	615	Sep	2	Jumādā II	-7	Dhū al-Ḥijja	-8	46
4376	Heshvan	615	Oct	2	Rajab	-7	AL-Muharram	-7	46
4376	Kislev	615	Nov	1	Shaʿbān	-7	Safar	-7	46
4376	Teveth	615	Dec	1	Ramaḍān	-7	Rabīʿ I	-7	46
4376	Shevat	615	dec	30	Shawwāl	-7	Rabīʿ II	-7	46

First of Lunar Month									Age
Jewish		Gregorian			Arabic		Islamic		
Y	M	Y	M	D	M	Y	M	Y	
4376	Adar	616	Jan	29	Dhū al-Qiʿda	-7	Jumādā I	-7	46
4376	Veadar	616	Feb	28	Dhū al-Ḥijja	-7	Jumādā II	-7	46
4376	Nisan	616	Mar	28	Noname	-6	Rajab	-7	46
4376	Lyyar	616	Apr	27	AL-Muharram	-6	Shaʿbān	-7	46
4376	Sivan	616	may	26	Safar	-6	Ramaḍān	-7	46
4376	Tammuz	616	Jun	25	Rabīʿ I	-6	Shawwāl	-7	47
4376	Av	616	Jul	24	Rabīʿ II	-6	Dhū al-Qiʿda	-7	47
4376	Elul	616	Aug	23	Jumādā I	-6	Dhū al-Ḥijja	-7	47
4377	Teshrei	616	Sep	21	Jumādā II	-6	AL-Muharram	-6	47
4377	Heshvan	616	oct	21	Rajab	-6	Safar	-6	47
4377	Kislev	616	Nov	19	Shaʿbān	-6	Rabīʿ I	-6	47
4377	Teveth	616	Dec	18	Ramaḍān	-6	Rabīʿ II	-6	47
4377	Shevat	617	Jan	16	Shawwāl	-6	Jumādā I	-6	47
4377	Adar	617	Feb	15	Dhū al-Qiʿda	-6	Jumādā II	-6	47
4377	Nisan	617	Mar	16	Dhū al-Ḥijja	-6	Rajab	-6	47
4377	Lyyar	617	Apr	15	AL-Muharram	-5	Shaʿbān	-6	47
4377	Sivan	617	May	14	Safar	-5	Ramaḍān	-6	47
4377	Tammuz	617	Jun	13	Rabīʿ I	-5	Shawwāl	-6	48
4377	Av	617	Jul	12	Rabīʿ II	-5	Dhū al-Qiʿda	-6	48
4377	Elul	617	Aug	11	Jumādā I	-5	Dhū al-Ḥijja	-6	48
4378	Teshrei	617	Sep	9	Jumādā II	-5	AL-Muharram	-5	48
4378	Heshvan	617	Oct	9	Rajab	-5	Safar	-5	48
4378	Kislev	617	Nov	7	Shaʿbān	-5	Rabīʿ I	-5	48
4378	Teveth	617	Dec	7	Ramaḍān	-5	Rabīʿ II	-5	48
4378	Shevat	618	Jan	5	Shawwāl	-5	Jumādā I	-5	48
4378	Adar	618	Feb	4	Dhū al-Qiʿda	-5	Jumādā II	-5	48
4378	Veadar	618	Mar	6	Dhū al-Ḥijja	-5	Rajab	-5	48
4378	Nisan	618	Apr	4	Noname	-4	Shaʿbān	-5	48
4378	Lyyar	618	May	4	AL-Muharram	-4	Ramaḍān	-5	48
4378	Sivan	618	Jun	2	Safar	-4	Shawwāl	-5	48
4378	Tammuz	618	Jul	2	Rabīʿ I	-4	Dhū al-Qiʿda	-5	49

First of Lunar Month									Age
Jewish		Gregorian			Arabic		Islamic		
Y	M	Y	M	D	M	Y	M	Y	
4378	Av	618	Jul	31	Rabīʿ II	-4	Dhū al-Ḥijja	-5	49
4378	Elul	618	Aug	30	Jumādā I	-4	AL-Muharram	-4	49
4379	Teshrei	618	Sep	28	Jumādā II	-4	Safar	-4	49
4379	Heshvan	618	Oct	28	Rajab	-4	Rabīʿ I	-4	49
4379	Kislev	618	Nov	27	Shaʿbān	-4	Rabīʿ II	-4	49
4379	Teveth	618	Dec	27	Ramaḍān	-4	Jumādā I	-4	49
4379	Shevat	619	Jan	25	Shawwāl	-4	Jumādā II	-4	49
4379	Adar	619	Feb	24	Dhū al-Qiʿda	-4	Rajab	-4	49
4379	Nisan	619	Mar	25	Dhū al-Ḥijja	-4	Shaʿbān	-4	49
4379	Lyyar	619	Apr	24	AL-Muharram	-3	Ramaḍān	-4	49
4379	Sivan	619	may	23	Safar	-3	Shawwāl	-4	49
4379	Tammuz	619	Jun	22	Rabīʿ I	-3	Dhū al-Qiʿda	-4	50
4379	Av	619	Jul	21	Rabīʿ II	-3	Dhū al-Ḥijja	-4	50
4379	Elul	619	Aug	20	Jumādā I	-3	AL-Muharram	-3	50
4380	Teshrei	619	Sep	18	Jumādā II	-3	Safar	-3	50
4380	Heshvan	619	Oct	18	Rajab	-3	Rabīʿ I	-3	50
4380	Kislev	619	Nov	17	Shaʿbān	-3	Rabīʿ II	-3	50
4380	Teveth	619	Dec	17	Ramaḍān	-3	Jumādā I	-3	50
4380	Shevat	620	Jan	15	Shawwāl	-3	Jumādā II	-3	50
4380	Adar	620	Feb	14	Dhū al-Qiʿda	-3	Rajab	-3	50
4380	Nisan	620	Mar	14	Dhū al-Ḥijja	-3	Shaʿbān	-3	50
4380	Lyyar	620	Apr	13	AL-Muharram	-2	Ramaḍān	-3	50
4380	Sivan	620	May	12	Safar	-2	Shawwāl	-3	50
4380	Tammuz	620	Jun	11	Rabīʿ I	-2	Dhū al-Qiʿda	-3	51
4380	Av	620	Jul	10	Rabīʿ II	-2	Dhū al-Ḥijja	-3	51
4380	Elul	620	Aug	9	Jumādā I	-2	AL-Muharram	-2	51
4381	Teshrei	620	Sep	7	Jumādā II	-2	Safar	-2	51
4381	Heshvan	620	Oct	7	Rajab	-2	Rabīʿ I	-2	51
4381	Kislev	620	Nov	5	Shaʿbān	-2	Rabīʿ II	-2	51
4381	Teveth	620	Dec	4	Ramaḍān	-2	Jumādā I	-2	51
4381	Shevat	621	Jan	2	Shawwāl	-2	Jumādā II	-2	51

First of Lunar Month									Age
Jewish		Gregorian			Arabic		Islamic		
Y	M	Y	M	D	M	Y	M	Y	
4381	Adar	621	Feb	1	Dhū al-Qiʿda	-2	Rajab	-2	51
4381	Veadar	621	Mar	3	Dhū al-Ḥijja	-2	Shaʿbān	-2	51
4381	Nisan	621	Apr	1	Noname	-1	Ramaḍān	-2	51
4381	Lyyar	621	May	1	AL-Muharram	-1	Shawwāl	-2	51
4381	Sivan	621	May	30	Safar	-1	Dhū al-Qiʿda	-2	51
4381	Tammuz	621	Jun	29	Rabīʿ I	-1	Dhū al-Ḥijja	-2	52
4381	Av	621	Jul	28	Rabīʿ II	-1	AL-Muharram	-1	52
4381	Elul	621	Aug	27	Jumādā I	-1	Safar	-1	52
4382	Teshrei	621	Sep	25	Jumādā II	-1	Rabīʿ I	-1	52
4382	Heshvan	621	oct	25	Rajab	-1	Rabīʿ II	-1	52
4382	Kislev	621	Nov	23	Shaʿbān	-1	Jumādā I	-1	52
4382	Teveth	621	Dec	23	Ramaḍān	-1	Jumādā II	-1	52
4382	Shevat	622	Jan	21	Shawwāl	-1	Rajab	-1	52
4382	Adar	622	feb	20	Dhū al-Qiʿda	-1	Shaʿbān	-1	52
4382	Nisan	622	Mar	21	Dhū al-Ḥijja	-1	Ramaḍān	-1	52
4382	Lyyar	622	Apr	20	AL-Muharram	1	Shawwāl	-1	52
4382	Sivan	622	may	19	Safar	1	Dhū al-Qiʿda	-1	52
4382	Tammuz	622	Jun	18	Rabīʿ I	1	Dhū al-Ḥijja	-1	53
4382	Av	622	Jul	17	Rabīʿ II	1	AL-Muharram	1	53
4382	Elul	622	Aug	16	Jumādā I	1	Safar	1	53
4383	Teshrei	622	Sep	14	Jumādā II	1	Rabīʿ I	1	53
4383	Heshvan	622	Oct	14	Rajab	1	Rabīʿ II	1	53
4383	Kislev	622	Nov	13	Shaʿbān	1	Jumādā I	1	53
4383	Teveth	622	dec	13	Ramaḍān	1	Jumādā II	1	53
4383	Shevat	623	Jan	11	Shawwāl	1	Rajab	1	53
4383	Adar	623	Feb	10	Dhū al-Qiʿda	1	Shaʿbān	1	53
4383	Nisan	623	Mar	11	Dhū al-Ḥijja	1	Ramaḍān	1	53

(1) 23 Ramaḍān (-13 AH), 19 Aug 610 (Julian) - 22 Aug 610 (GR) and 24 Av 4370 (JW)

(2) 23 Ramaḍān (-13 AH), 25 Dec 609 (Julian) - 28 Dec 609 (GR) and 23 Teveth 4370 (JW)

The astronomical setting of **Dhū al-Ḥijja,** 10 (AH) is shown in Table 10.4-6 with the position of the moon and the sun at three important times: (a) the beginning of the photometric lunar month, (b) right at the middle of the Hajj when the Farewell sermon was given, and at (c) the day of Vernal Spring Equinox. The sublunar point of the new moon of **Dhū al-Ḥijja** is shown to be right on the equator in Figure 10.4-2. Also shown are the vertical lines of moonrise and moonset. First light of the month occurred over the Atlantic Ocean, following sunset on Feb 29, 632 CE 20:21 UT. Most of the northern latitudes started seeing the new moonlight, and by the time the sunset over the Mecca, the entire globe could see the young moon. The month of February in the year CE 632 had 29 days (a leap year). After midnight, universal time, 1-**Dhū al-Ḥijja**, corresponds to March 1, 632 CE, even though it is calculated to be March 2 algorithmically[35]. New moon visibility must have taken place the night of March 1 over Arabia. We call this time the V***ernal Lunar Equinox*** (i.e., Lunar Spring, paralleling the nomenclature used for Solar Equinoxes), a time when the lunar terminator is perpendicular to the equator. The sublunar point crosses the equator twice a year. Once in the fall in the month of September moving southward and once in the Spring, moving northward in March, within days from when the sun does (see Figure 9.3-4).

Table 10.4-6 Position of the moon and the sun at New moon, the time of Haj and Spring Equinox during the month of Dhū al-Ḥijja, 10 AH, 632 CE

		Ecliptic Coordinates (deg)				Equatorial Coordinates (deg)			
Gregorian Date	Time	Sun[Long,Lat]		Moon[Long,Lat]		Sun[ra,dec]		Moon[ra,dec]	
Wed Feb 29, 632	20:21:00	340.29499	-0.060006	351.02558	3.868630	341.83254	-7.7633812	350.23353	**-0.000438**
Sun Mar 11, 632	16:24:00	351.00428	-0.027818	123.21755	**-0.002591**	351.74686	-3.5914809	125.51533	19.43502
Tue Mar 20, 632	15:20:00	359.80385	-0.000601	251.85076	-4.006391	359.82027	**-0.0785775**	249.74199	-26.17829

As time passed, and right in the middle of the Hajj, the moon crossed the ecliptic plane. This has occurred March 11, 632 CE, as illustrated in Table 10.4-6. The same Table shows the Vernal Equinox occurred on March 20, 632 CE. At the same time, the ecliptic latitude of the moon is almost equal and opposite to its latitude at the beginning of the month. Were these events special from an astronomical standpoint? The answer is, obviously "yes." The Farwell Hajj of 10 (AH) occurred at the onset of a lunar node[36], midway between two springs: the "moon spring" and the "sun spring." The photometry of the two events is illustrated in Figure 10.4-2 and Figure 10.4-3, respectively.

---

[35] This case shows the mathematical algorithm of the Islamic calendar can be off by a full day in predicting exact dates when it comes to the photometry of the month; nevertheless, the algorithm is accurate over long periods of time.

[36] A lunar node is a point where the lunar orbit crosses the plane of the ecliptic.

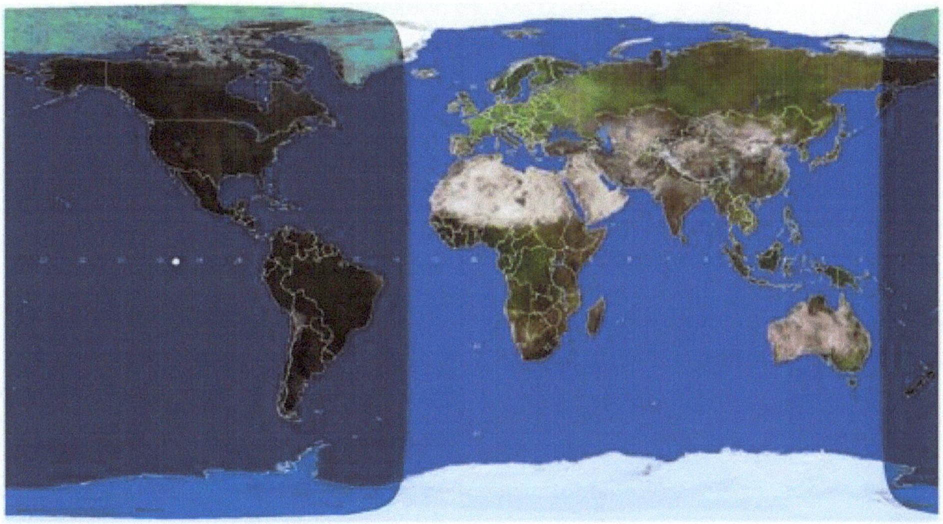

Figure 10.4-2 Time of first new moon visibility on the first of Dhū al-Ḥijja in the year 632 CE, the month of the Farewell Hajj.

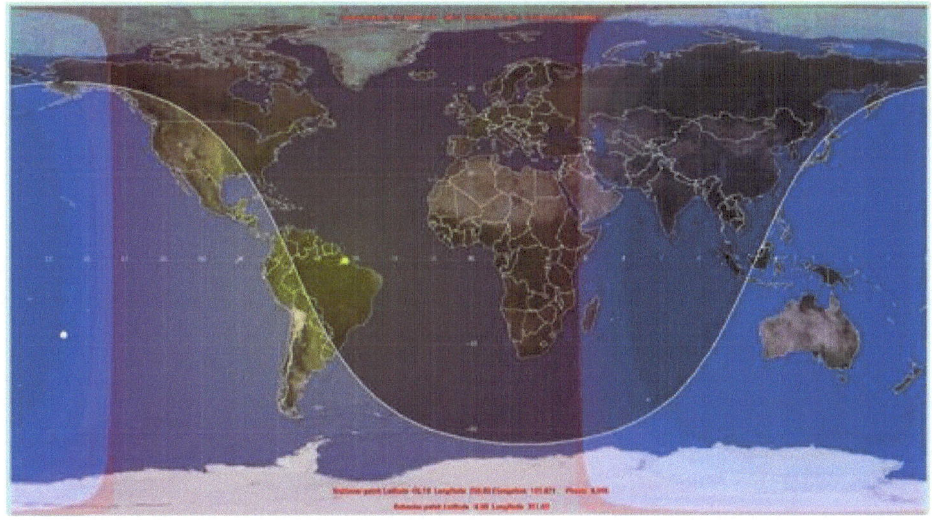

Figure 10.4-3 The photometry of the Spring Equinox, in the year 632 CE. Note the vertical solar terminator being parallel to earth's rotation axis.

10.5 Christmas and the night of Qur'an Revelation

The holiest night in Islam is called *Laylatul Qadr* (*Laylatul* is Arabic for "night of"). Its name means the *Night of Power*, or the *Night of Decree* and some refer to it as the *Night of Destiny*. It was on a similar night that the first verses of the Quran were first revealed to the prophet Mohammed while he was meditating in the cave of *Hera*, just a few miles outside Mecca. He was then about forty years old. Each year, on the anniversary of the event, Islam teaches that the *angels* and *the Spirit* descend onto the earth, in peace, to take care of certain matters. It is explained that it is a night of limitless blessings, forgiveness, and mercy. The following translation of the five-verse Qur'an chapter of *Al-Qadr* stresses the importance of the anniversary of *Laylatul Qadr*:

"We have indeed revealed this (Message) in the night of Qadr:
And what will explain to thee what the Night of Qadr is?
The night of Qadr is better than a thousand months.
Therein come down the Angels and the Spirit by God's permission, on every errand:
Peace! This until the rise of dawn!"

(Qur'an: 97; 1-5)

The specific date for this night is not known and is, in fact, a mystery. But several clues are provided in the Qur'an and Hadith about when the anniversary might be. The following verse is explicit about *Laylatul Qadr* occurring *i*n the month of Ramadan:

"Ramadan is (the month) in which was revealed the Qur'an, as a guide to mankind, also clear (signs) for guidance and judgment (between right and wrong)."

(Qur'an: 2; 185)

Furthermore, this night is one of the last ten nights of the month of Ramadan, as established from several reliable references from *Hadith*. For example, Ibn Umar is reported to have quoted the prophet: *"He who is anxious to seek it (Lailat-ul-Qadr) should seek it in the last ten (nights of Ramadan),"* [Sahih Muslim 006-2622]. Other reports confirm its occurrence in the last 10 nights of Ramadan, specifying odd night numbers or in one case the night of the 27th of Ramadan (see, e.g., Sahih Muslim 006-2630 and 2634). There is however a consensus that *Laylatul Qadr* is one of the last ten nights of the month of Ramadan.

Before addressing which one of the last ten nights of Ramadan, *Laylatul Qadr* might be, let us discuss the interpretations of verses 97; 1-5. First, the agreed-upon interpretation of the thousand months is the equivalency to a human's life span, averaging little over 80 solar years. An accepted prayer during this night (referred to as blessed in another verse in the Qur'an) is a lifetime worth and therefore, the spiritual importance of this night is clear. Second, Verse 97; 4 mentions the Angels and the Spirit descending onto the earth. Except for some minor details, the concept of Angels in the Qur'an is like what is found in other Biblical religions with only minor differences. Angels are non-physical, honorable, obedient, and modest creations of God to whom are assigned all sorts of divine duties. The Spirit (Arabic: الروح *al-Ruh*) is, however, the high-ranking powerful Angel Gabriel (Arabic *Jibrīl*), who is entrusted with delivering the Qur'an from God to Mohammed. Islam teaches that Gabriel also supported other prophets before Mohammed since the time of the creation of Adam. It is important to note that reference to the "Holy Spirit" is made in several Qur'an verses. The common opinion is that it is about Gabriel who was sent to Maryam (Arabic for Mary), mother of Jesus (Isaa in Arabic) in a human form:

"And mention, [O Muhammad], in the Book [the story of] Maryam (Mary), when she withdrew from her family to a place toward the east. And she took, in seclusion from them, a screen. Then We sent to her Our Angel [i.e., *Jibrīl*], and he represented himself to her as a well-proportioned man. She said, "Indeed, I seek refuge in the Most Merciful from you, [so leave me] if you should be fearing of Allah." He said, "I am only the messenger of your Lord to give you [news of] a pure boy [i.e., Son].'"

[Qur'an: 19; 16-19]

It should also be noted that multiple Qur'an verses refer to Christ as the Spirit of God. For example, it is stated in verse 4; 171 that The Messiah, Jesus son of Mary, was a Spirit from God. And the prophet Mohammed refers to Jesus as "the word of Allah and Spirit" in more than one occasion. Also, the Qur'an and Muslims use the phrase "Peace upon him" when mentioning the name of *al Masih* (the Messiah) Isaa (Jesus). For Muslims, the proper address to Jesus in Arabic translates to "The Messiah Jesus Peace upon him." In verse 19; 33 of the Qur'an when Jesus was speaking from the cradle to the crowd surrounding the virgin Mariam (Marry), he said:

"Peace be upon me, the day I was born, and the day I die, and the day I am raised up alive!"

(Qur'an: 19; 33)

Gabriel is also mentioned by Muslims in the same way as Jesus (i.e., the proper reference is "Gabriel Peace upon him"). And when in his role of supporting prophets, Gabriel was referred to as the *Holy Spirit*. Both Jesus and Gabriel are referred to as the: "Spirit" (of God). Finally, because Gabriel has an exclusive supporting role to prophets in Islamic teaching[37], and because Mohammed is the last prophet, why would Gabriel descend onto earth on the night of Qadr, if there are no more prophets to assist or give messages to after Mohammed dies? The only plausible reason[38] is the outstanding return of Jesus who was brought back to the heavens during the event of crucifixion (again, per Islam teaching). Based on the above, the night of Qadr, the anniversary of the revelation of the Qur'an to Mohammed might have something to do with Jesus as we will find next.

The timing of Qur'an revelation and therefore, the start of the message of the prophet Mohammed may be determined with an appropriate level of accuracy. Table 10.4-5 shows

[37] This made some scholars wonder if the virgin Mary ought to be considered as prophet due to Gabriel's manifestation

[38] From an Islamic perspective

the conversion of the first day of the lunar month between 609 CE (the year of the Qur'an revelation and the beginning of prophecy) and the year 615 CE. The Qur'an clearly states that the beginning of revelation occurred in Ramadan. And based on Hadith, it is placed sometime in one of the last 10 days of the month when the prophet turned 40 years old. It happens that the 23rd of Ramadan corresponds to December 25, 609 in the Julian calendar and to December 28, 609 CE in the Gregorian calendar which had yet to be adopted at the time but is given here for reference. The date of Coptic Christmas is observed on December 25 in the Julian calendar, a date that corresponds to January 7 in the Gregorian calendar today, which is also the observed date of Christmas in Eastern Orthodox countries, such as Greece and Russia. The 25th of December happens to be the 20th of Ramadan. This places Ramadan 23 (-13 AH) on the Orthodox Christmas Eve and Ramadan 20 on the Gregorian Christmas eve because the Islamic day starts at sunset. Even though the exact date of *Laylatul Qadr* could be in any evening between Ramadan 20 and Ramadan 29 (or 30), the actual timing of the event did coincide with Christmas. In fact, when the prophet was asked by his followers regarding the fasting on a Monday[39], he said: "on this day I was born, I received the prophecy and divine revelation, and on this day, I fled (referring to Hijra)." This makes only two possible candidates for *Laylatul Qadr:* either the evening of the 25th of December or the evening of January 1st, i.e., either Christmas Eve or the Gregorian New Year's Eve. It should be noted that the Prophet Mohammed died on a Monday, as we saw in Section 10.4.3. Also shown in Table 10.4-5 is the Islamic calendar extending back in time to the year of revelation for reference. It can be immediately seen that as the months of the Islamic calendar were set to float and circle around the year in its 33-year period, concordance of Christmas and *Laylatul Qadr* was lost.

The 25th of March was the first day of the year in the medieval Julian calendar and the date of the nominal vernal equinox at the time when the Julian calendar was designed. This is also the anniversary of Creation in Christian belief. Considering that Jesus Christ was conceived on that date, known as the Feast of the *Annunciation*, nine months later was argued to be the date of celebration of his birth, i.e., Christmas, on December 25. The Nativity of Jesus was instated by Hippolytus of Rome, who was one of the most important theologians of the third century CE, eight days before the month of January in the forty-second year of the reign of Augustus. Earlier, in the second century CE, the Bishop of Caesarea Theophilus stated that Christians needed to celebrate the birthday of Christ on the 25th of December. By 367 CE, the 25th of December became accepted in the Christian world. Prior to that time, the Eastern Church had January 6 as the Nativity.

Muslims consider Jesus to be the Messiah and believe in his resurrection and return to lead the believers. From the above information, the holiest Islamic evening of the year could be Christmas Eve. Also, could it be that Christmas not being celebrated the same day by

---

[39] This question is not related to our topic.

Orthodox, Catholics and other Christians, together with the non-committal of Islam to a specific day around Christmas, represents biblical discord that needs clarification? Clearing this discord opens the door to common celebrations. There is a precedence to cross-faith observances. For example, when the prophet Mohammed first arrived in medina the year of Hijra, it is said that he found the Jews of Medina fasting on the day of *Ashurah* (derived from the word *ashara* which means ten). This is the fasting day of Kippur; see (Al-Biruni, 1879). Ashurah is an Arabized word of the 10th day of Tishri. When the Prophet arrived in Medina, he found that the Jews were observing this day in remembrance of the escape of Moses and the Israelites from the Pharaoh of Egypt. He said, "we have a nearer claim to Moses than they," and initiated fasting on that day.

10.6 Origin and names of the Arabic calendar months

Conversion between the four calendars in Tables 10.4-3 - 10.4-5 indicate an outstanding puzzle about the names of the months of the Arabic calendar. Many of these months were named[40] to indicate the season in which they occur or social behavior, yet they are flagrantly out of phase with the seasons. For example, in the year of Farewell Hajj, the month of Jumādā I, which means the first month of Freeze, came in summer while Ramadan which refers to heat came in winter, as shown in Table 10.4-4. The discrepancy is summarized in Table 10.4-2, where the correspondence between the months and the seasons in Arabia is highlighted through the monthly average weather conditions.

The rich and powerful Arabic literature is nearly void of information about the calendar. In fact, the few references to the calendar are pertaining to the Hajj, and there is no explanation to why the original Arabic calendar has the names of the months pointing to the wrong seasons. In fact, it is very strange that it comes from a culture whose people are defined by the power of their language and profound expressions. According to the 14th century North African scholar Ibn Khaldun (Ibn-Khaldun, c. 1350), the word Arabs is derived from the Arabic word "A'araba" which means to articulate. And the people of the peninsula are referred to as Arabs because of their superior eloquence and capacity to express and articulate. It is hard to believe that they have messed up their calendar to the point that they refer to seasons by the wrong names as proposed by Al-Biruni, see translation (Al-Biruni, 1879) who was not too fond of the Arab Muslim occupants.

To put the history of the Arabic calendar in its context, we go back in time to the pre-Islam era. The Himyarite kingdom of the South Arabian Peninsula was the default nominal force of the region, which started a six- century-long dynasty in 110 BCE. It is well established that they used to be a pagan society and have adopted a Julian type of calendar with an epoch of 110

[40] The names of the months of the Arabic and Islamic calendar are discussed in Section 2.7.3.

BCE. Four centuries later, right after the neighboring and rival Ethiopian kingdom of Axum converted to Christianity in 328 CE, the ruler of Himyar who had reigned over the nearby regions Saba' (Sheba), Qataban, and Haḍramaut which they the Himyarite occupied in c. 25 BCE, c. 200 CE and c. 300 CE, respectively, converted his own people to Judaism. This change of religion came after a trip North into the Arabian heartland in which he lost his son to his recent acquisition of Yathrib (Medina) in 390 CE. Their use of the solar calendar ceased altogether with the worship of pagan gods, to be replaced by the monotheistic Jewish religion as taught by the people of Yathrib. With the new religion came the lunisolar calendar, of which a new version was derived. And within two decades, in 412 CE, the meeting of leaders of Arab tribes which led to the unification of the names of the months and their sequence as mentioned in Section 2.7.3 occurred, shifting from a solar calendar to a lunisolar one.

There are several indications which point to the origin of the Arabic calendar being a solar one. Recall the similarities observed in the names of the months of the Ethiopian and Coptic solar calendars, which are both based on the Egyptian Alexandrian calendar, as discussed earlier in Section 2.7.1. Comparing these two calendars with the Arabic calendar, we make the following observations:

(1) The first month in the Arabic calendar AL-Muharram and the name of the first month of the Ethiopian calendar, Meskerem, are phonetically similar. In fact, one of the meanings of Meskerem is "spring." The spring season runs between September and November in the horn of Africa (located south of Arabia), as discussed in Section 2.7.1.

(2) The twelfth month, *Nashi* in the Ethiopian calendar is the same as that of the thirteenth month of the Coptic calendar, *Nasie,* and is the same word used to refer to the act of postponement of Al-Muharram.

(3) Aligning a precursor solar Arabic calendar to the Coptic and Ethiopian calendars in the sequence shown in Table 10.6-1 brings the names of the months in line with the seasons. It must be noted that Arabs also called the fall season spring like the rest of the people of the Horn of Africa. The sequence of names is the same as that of the lunisolar Arabic calendar, and the similarity with the Ethiopian and Coptic cannot be ignored. It should be noted that the Christian presence was prevalent in Arabia with monasteries everywhere. In fact, the Prophet Mohammed's wife Khadija had a first cousin who was a Christian monk.

(4) There is a report on Arabs having used a solar calendar at one time. This report is, however, not well-founded.

An Arabic solar calendar was derived from the Egyptian Alexandrian calendar and emerged in parallel with the Ethiopian, Coptic and Julian calendars. It was probably maintained by Arab

Christian monks in a dominant pagan Arabia. The question is then when and how this calendar was altered to become a lunisolar one? The naming sequence of the months of this Arabic solar calendar, together with a rough conversion to the Gregorian calendar, the Ethiopian and Coptic calendars can be seen in Table 10.6-1. The calendar had 12 regular months and a short month called Al-Nasi.

Before the start of the fourth century CE, in addition, to be the center of pagan worship, Mecca became a destination for major fairs. People were showing up either one month early or one month late for a trip in which they combine trade with spirituality and entertainment. To ensure that visitors show up at the same time, the leaders of Mecca set up a clever way which works well for self-governing tribal societies extending beyond the vast and dangerous Arabian Desert, to reach Yemen and the kingdom of Aksum in the south and Syria in the north. They decided to adopt a lunisolar calendar which can be managed in Mecca through intercalation, once every two to three years. The leap or no-leap year is announced at Hajj closure every year. This is easy, yet a risky solution for the pilgrims in years when postponement takes place, as they lose the protection of the sanctity of the month. It could have been argued that this is not that risky, as pilgrims are going to travel in large caravans as everyone is leaving Mecca at the same time and therefore are not vulnerable to aggression by desert bandits. Postponement being announced at the end of the Hajj gives enough time for pilgrims to return home and disseminate information on whether the new moon or the following one is that of Al-Muharram, thus pinpointing the datum for the New Year and therefore the time of the subsequent Hajj. The new calendar kept the names of the months in the same sequence as its solar precursor. Because there are two seasons for trips to Mecca, one in the fall and one in spring, with the latter being the most important one from a trade perspective, they shifted the start of the year from the Fall Equinox (September-October) to the Spring Equinox (March-April)[41]. The new calendar made the short month of Al-Nasi the intercalation month. The shift was initiated either in 412 CE when the Arabic calendar was unified or slightly before. The new calendar used an intercalation system like the Jewish one which was maintained by local Jewish tribes who were active in trade but had nothing to do with Pagan rituals. The Jewish tribes were also present in Arabia with their celebrations during both fall and spring seasons. They also had pilgrimage requirements, but not to Mecca, and their calendar also starts at the Fall Equinox.

Based on the above, it is concluded that Al-Muharram used to occur at the start at the Fall Equinox as illustrated in **Table 10.6-2**. Its shift to the Spring Equinox not only fits well with the economics of the region of the time but was also easy to implement in a society with no central governance. Mecca traded with the kingdom of Aksum (Habasha) and the southern tribes of Yemen in the fall season (which Arabs called spring) and traded with the north, i.e., Al-Shaam

[41] Therefore, travel in the month of Rajab is also protected through sanctity

(today Iraq, Syria and even further north) in the spring and summer seasons. They always had two events one in the Fall Equinox, accommodating business with the south and one in spring doing business with the north. No doubt the shift was triggered by needs for trade. Gerald de Gaury (De Gaury, 1991) wrote: "the pilgrimage time being customarily in the autumn when the nomads were obliged to send their caravans to the Markets to restock ready for the winter migration in search for grazing."

One may ask the question, why this was never addressed. The answer is simple. With the advent of Islam, the Islamic calendar has its months go around the solar year in a cycle of about 33-years, and therefore the months are expected to float over the year. Despite the importance of the months in Islam, the dilemma of the names of the months not pointing to the right season seems to have not bothered thinkers and writers for the past 14 centuries. Reference to the seasons in the names of the months of the Islamic calendar is erroneously said to indicate the lunisolar nature of the original Arabic calendar from which it resulted. It has been simply assumed that the perpetual shift in the months across the seasons is due to the transition from a lunisolar to the pure lunar calendar construct. And because that happened following what was interpreted as divine instruction, the matter never received a critical assessment by Muslim chronologists.

Table 10.6-1 Solar Arabic calendar names correlate with Ethiopian and Coptic calendars

	Solar calendars			
#	Greg.	Arabic Solar	Ethiopian	Coptic
1.	Sept	AL-Muharram	Meskerem	Thuout
2.	Oct	Safar	Tikemet	Paopi
3.	Nov	Rabīʿ al-Awwal	Hidar	Athor
4.	Dec	Rabīʿ ath-Thānī	Tahsas	Khoiak
5.	jan	Jumādā al-Ūlā	Tarr	Tobi
6.	Feb	Jumādā ath-Thāniya	Yekatit	Meshir
7.	Mar	Rajab	Makwit	Baramhat
8.	Apr	Shaʿbān	Miyaziah	Baramouda
9.	may	Ramaḍān	Genbot	Pashons
10	Jun	Shawwāl	Sanne	Paony
11	Jul	Dhū al-Qiʿda	Hamli	Epep
12	Aug	Dhū al-Ḥijja	Nashi	Mesori
13	Sep	no name (nasi)	Quaggimi	Nasie

Table 10.6-2 Arabic calendar names correlate perfectly with the seasons, assuming a solar calendar origin

	Climate data for Mecca											
Month	Mar	Apr	May	Jun	Jul	Aug	Sep	Oct	Nov	Dec	Jan	Feb
Average high °C	35	39	42	44	43	43	43	40	35	32	31	32
Daily mean °C	27	31	34	36	36	36	35	32	28	26	24	25
Average low °C	21	25	28	29	29	30	29	26	23	20	19	19
Average precipitation mm	6	10	1	0	1	5	5	15	23	22	21	3
Average precipitation days												
Mean daily sunshine hours												
Original solar	Rajab	Sha'bān	Ramaḍān	Shawwāl	Dhū al-Qi'da	Dhū al-Ḥijja	AL-Muharram	Safar	Rabī' I	Rabī' II	Jumādā I	Jumādā II
Ensuing lunisolar	Dhū al-Ḥijja	AL-Muharram	Safar	Rabī' I	Rabī' II	Jumādā I	Jumādā II	Rajab	Sha'bān	Ramaḍān	Shawwāl	Dhū al-Qi'da

The matter of Arabic names in the Arabic calendar is not as simple as one may think, despite the above findings, one must be careful making final claims on the matter. In addition to the existence of other calendars and other names of calendar months, depending on the tribe being referred to, it is also well established that the people of central Arabia have long relied on the Horizon calendar in which the position of the stars relative to the horizon at sunset or sunrise provided the time and the season of the year. Arabs divided the ecliptic into 28 equal parts corresponding to the location of the moon in each successive month. The names of the months in that calendar have continued in the Islamic calendar.

10.7 Historical events of early Islam

Gregorian dates of historical events during the emergence of Islam are marred by uncertainties and lack of precision. Establishing Gregorian dates of events to relate seasonal contexts or parallel events in other cultures can be a tough challenge due to the shift from the Arabic to the Islamic calendar, which occurred two and a half years into the succession of the second Caliph, *Omar ibn al-Khattab*, in the year 17 AH. There are several reasons behind the discrepancies and lack of accuracy: First, none of the dates mentioned in the quotes and collections of narrations specify which calendar of the two is intended. Probably, most of the people of the time were unaware of the difference between the two calendars, and therefore, there was no reason for them to point to a specific one. Second, the retroactive origin of the Islamic calendar adds a virtual timeline out-of-phase with the events that took place during the decade of the Hijra while the prophet lived in Medina, as discussed in Section 10.4.4. Third, the lack of central governance and lack of written records provide reliance on what someone said. Ask anyone which day of the month is today, there certainly be multiple answers, despite support from technology. Let alone during a time in which information is communicated through word of mouth.

Table 10.7-1 presents revised Gregorian dates of events which occurred mainly at the time of the Emergence of Islam. In this revision, we use the conversion discussed earlier in Section 10.4. The first column in the Table gives the age of the prophet. The cell is highlighted if the Prophet was physically present. The second column gives a brief description of the event. The Gregorian date or approximate time of the event is given in the third column followed by the Hijra year in the fourth column. Finally, the rightest column gives more details about the event. References to rain in the last column are highlighted for the reader to compare with the seasons from time information, given in column 3, and precipitation information from Table 10.6-2. There is an excellent correlation between the time of events in which it rained and the seasonal precipitation which occurs mainly in the months of November through January. Another anchor is the well-established date for the death of the son of the prophet Ibrahim, which occurred in a day with a solar eclipse.

The conversion presented earlier in Tables 10.4-3-10.4-5 and used herein fits several historical events. For example, during one of the last ten nights of Ramadan, the prophet engaged in prayers late at night with some of his followers, as it may be *Laylatul Qadr*. It is reported that it rained heavily to a point where the roof of the Prophet's mosque in Medina, which was made of reinforced palms, was letting water through. Ramadan occurred either at the end of November, over the month of December, or early January, making the last ten days occur sometime around the end of December: a time of peak rain in Arabia.

While historical dates can ultimately get sorted out through detailed forensic evaluation of events and environmental information, the crucial point is in the matter of intercalation. From our effort herein, the religious texts -as translated in the English language- appear not to conclusively instruct Muslims to abolish the lunisolar calendar altogether. The issue is in shifting of the sanctity of the months as discussed earlier.

Table 10.7-1 Revised dates of events during the Emergence of Islam

Age	Event	C.E. date	A.H. year	Note
	Emergence of the Himyarite Kingdom	**110 BCE**		The Himyarite kingdom was the nominal force in Arabia (tip western corner of today's Yemen). They started a six-century long dynasty. The Himyarite used a Julian-type Calendar with an epoch of 110 BCE.

Age	Event	C.E. date	A.H. year	Note
	The Ethiopian Kingdom of Axum converts to Christianity	328		
	Theodosius I dismantled the Sanhedrin, the head of which, Hillel II released the first Jewish calendar	358		
	Himyarite King converts to Judaism	390		
	Unification of Arabic calendar names	412		Meeting of leaders of Arab tribes led to the unification of the names of the months and their sequence
	An attempt to resurrect a clandestine Sanhedrin	425		Resulted in the execution of the last supreme Rabbi Gamaliel VI by Theodosius II
	The Aksumite takeover of the Himyarite Kingdom	520		End of Himyarite dominance over Arabia
	Massive volcanic eruption in Iceland (named worst year to be alive)	536		Icelandic volcano erupts Summer temperature drops 1.5 to 2.5 deg C dimming the sun for 18 months and the coldest decade in 2000 years, failing crops around most of the world including Europe, Mesopotamia and China
	Second volcanic eruption in Iceland	540-541		Temperature drops again 1.4-2.7 deg C
	Justinian Bubonic plague kills 35% to 55 % of the population	541-543		
0.2	Ethiopian General Abraha, who revolted against Aksum and took over a large swath of Yemen, attacks Mecca	Apr 570	-53	15 Al-Muharram
0	**Prophet Mohammed is born**	Jul 2, 570	-53	12 Rabīʿ al-Awwal (Monday)
2 or3	Mohammed's Grandfather Abdul-Muttalib went to King Saif Bin dhi Yazn	572 or 573	-50 or -51	King Saif Bin dhi Yazn ascended to power when Mohammed was 2 or 3 years old. He led a revolt against Masrūq b. Abraha. The Himyarite. Saif was denied help from the Byzantines, but then received support from Khosrow I, Vahrēz of Persia. Saif killed Masrūq and the Ethiopians, and was installed as ruler of Yemen.
	Prophet Mohammed was returned to the custody of his mother from his caring nurse	574		Famine and drought around Mecca
	Traveled with his mother to Medina to visit his uncles; in the way back, his mother died	576		

Age	Event	C.E. date	A.H. year	Note
	Saif Bin dhi Yazn dies	578		Reference to being stabbed to death by Ethiopian servants
	Grandfather Abdul-Muttalib dies	Sep 578		He was over 80 years old
	Byzantines Emperor Justin II died on Oct 5, 578 and was succeeded by Tiberius II Constantine	Oct 5, 578		In 572 his refusal to pay tribute to the Persians in combination with overtures to the Turks led to a war with the Sassanid Empire. After two disastrous campaigns, in which the Persians under Khosrow I overran Syria and captured the strategically important fortress of Dara, Justin reportedly lost his mind. In 574, Justin had a mental breakdown, forcing Empress Sophia to turn to Tiberius to manage the empire, which was fighting the Persians to the east and dealing with the internal crisis of the plague.
	Persia's Khosrow I (also known as Kasra, the twenty-second Shahanshah of the Sasanian Empire, died after 48 years of reign 531 –Jan 31, 579), known as the Just, learned philosophy, a builder of roads, cities and walls, succeeded by his son Hormizd IV	Jan 31, 579	-44	Shawwāl -44 AH, he led multiple wars with Byzantines
12	Mohammed traveled to Shaam with uncle Abu-Taleb for trade	May-Jul, 582	-41	Traveled at the age of 12; Prophet would travel later for future wife Khadijah's business and stops in Basra.
	Byzantines Emperor Tiberius II Constantine died and succeeded by Emperor Maurice	Aug 14, 582		
	Fujjar War	Nov-Dec 589		Occurred in Sha'bān
20	Alfoudhul Treaty	Feb-Mar 590	-33	Occurred Dhū al-Qi'da, 20 years before the prophecy
25	Married Khadijah	May-Jul 595	-28	At the age of 25, after coming back from Al-Shaam. Prophet married Khadijah with whom he had 3 boys and four girls (Khadijah could not be 40 years old at the time of marriage. There is a reference she was 35)
	Byzantine's Emperor Maurice died and succeeded by Phocas	Nov 27, 602		Phocas took over the throne, having Maurice and his six sons executed. He purged the elite and placed family members in charge of the Empire. This event would prove a disaster for the Empire, sparking a twenty-six-year war with Sassanid Persia which would leave both empires devastated prior to the emergence of Islam.
35	Prophet Mohammed helped resolve the conflict of reconstruction of the Kaaba	605	-38	He was 35 years of age.

Age	Event	C.E. date	A.H. year	Note
40	**Prophecy begins**	Dec 609	-13	The evening of the 25th of December or the evening of January 1st, a Monday (strong Hadith). The former is prevalent
	Emperor Phocas killed by Heraclius to Reign over the Byzantine Empire for the next 30 years.	Oct 5, 610	-12	Because of Phocas' incompetence and brutality, the Exarch of Carthage, Heraclius the Elder, rebelled against him. Heraclius the Elder's son, Heraclius, succeeded in taking Constantinople on Oct. 5, 610. He executed Phocas on the same day before declaring himself Emperor.
44	Hijra of followers to Habashah	Oct-Nov, 613	-9	Rajab
44	Jerusalem Fell to Persia/ surat al-Room	Apr-May 614	-9	Roman information / corroborates prophet's hadeeth
46	Hamza + Omar accept Islam	Mar 616	-7	Dhū al-Ḥijja, year 6 of prophecy
49	Abou Taleb (prophet's uncle) dies	Oct-Dec 619	-3	Rajab or Ramaḍān, year 10 of prophecy
49	Khadijah (prophet's wife) dies	Dec-Jan, 619	-3	Ramaḍān, year 10 of prophecy
53	**Mohammed's migration -Hijra**	Jun 17, 622	1	Initiated during the night, the 1st of Rabīʿ al-Awwal (Monday)
53	Mohammed arrives in Medina	Jun 28, 622	1	12 Rabīʿ al-Awwal, entering Medina on a Friday (for sure)
54	Expedition of Hamza ibn 'Abdul-Muttalib	Dec 622 -Jan 623	1	Seef al-Bahr in Ramaḍān
54	Prophet married Aisha	Jan 623	1	Shawwāl
54	Expedition of Ubaydah ibn al-Harith	Jan - Feb 623	1	Shawwāl, they skirmished w Abu Soufian
54	Al-Kharrar expedition	Feb 623	1	Dhū al-Qiʿda, Saad ibn Abil Wakkas
54	Patrol of Waddan	May 623	2	Safar
55	Patrol of Buwat	Jun 622	2	Rabi' al'awal
55	First Expedition to Badr	Jun 622	2	Rabi' al'awal
55	Patrol of Zul Al-Ushairah	End of Aug - Sep 623	2	Jumādi al-Ūlā and Jumādā ath-Thāniya
55	Nakhla Raid	Oct 623	2	Rajab not supposed to have occurred (sacred month -Mohamed paid the jizya)
	The direction of prayer switched from Jerusalem to Mecca	Nov 623	2	Shaʿbān
55	Battle of Badr	Dec 20, 623	2	17 Ramaḍān, it **rained** the night of the battle

Age	Event	C.E. date	A.H. year	Note
55	The killing of Asma bint Marwan	Nov 27, 623	2	24 Sha'bān, 19 months after Hijra
55	The killing of Abu Afak	Jan 21, 624	2	Shawwāl 20 months after Hijra
55	Invasion of Banu Qaynuqa	Jan 3-18, 624	2	Shawwāl 3-18 (probably started on a Saturday, first half of Shawwāl)
55	Invasion of Sawiq	Mar 624	2	Dhū al-Ḥijja
	-			
56	Al Kudr Invasion	May 624	3	
56	Dhu 'Amarr	Jul 8, 624	3	Occurred 12 Rabīʿ al-Awwal
56	The killing of Ka'b ibn al-Ashraf	Jul 10, 624	3	Occurred 14 Rabīʿ al-Awwal
56	Invasion of Buhran	Aug 624	3	Rabīʿ ath-Thānī or Jumādā al-Ūlā
56	Al-Qarada raid	Sep-Oct 624	3	Jumādā ath-Thāniya, there is also reference to 28 months after Hijra
56	Battle of Uhud	Feb 5, 625	3	Middle of Shawwāl, a Saturday -probably Shawwal 18.
56	Battle of Hamra al-Asad	Feb 6, 625	3	8 Shawwāl, Sunday (following day from Uhud)
56	Expedition of Qatan	Apr 18-25, 625	4	Beginning of Al-Muharram, 35 months after Hijra may indicate the intercalation was in effect, also a reference to one month after Uhud strengthen the date for Uhud itself
56	Expedition of Abdullah Ibn Unais	Apr-May 625	4	Al-Muharram
56	Expedition of Bir Maona	May-Jun 625	4	
56	Invasion of Banu Nadir	Jun-Jul 625	4	Rabīʿ al-Awwal
56	Expedition of Dhat al-Riqa	Aug-Sep	4	
56	Expedition of Badr al-Maw'id	Nov-Dec 625	4	Sha'bān
56	Expedition of Dumat al-Jandal	June 626	5	Rabīʿ al-Awwal
57	Expedition of al-Muraysi'	Nov 626	5	Sha'bān
57	Battle of the Trench (or Al-Ahzab)	Dec 626-Jan 627	5	Shawwāl – Dhu al-Qi'dah (extreme cold and hunger) lasted 1 month a strong cold wind blew tents. It was also the start of the forbidden month; the invaders had to break.

Age	Event	C.E. date	A.H. year	Note
57	Invasion of Banu Qurayza	End Jan to End of Feb. 627	5	Started at the end of Al-Ahzab siege and lasted 25 days siege in Dhū al-Qiʿda and start of Dhū al-Ḥijja
57	Expedition of Muhammad ibn Maslamah	Apr-May 627	6	Al-Muharram
57	Invasion of Banu Lahyan	End of Aug-Early Sep 627	6	Jumādā al-Ūlā
57	Expedition of Dhu Qarad	End of Sep 627	6	Muhammad stayed only a few nights in Medina when this happened
58	Expedition of Abdu-Rahman bin Auf	Nov/Dec 627	6	
58	Expedition of Kurz bin Jabir Al-Fihri	Jan/Feb 628	6	
58	Treaty of Hudaybiyyah	March 628	6	Well-referenced date of Dhu al-Qi'dah, 6 AH - **rained** at night
58	Battle of Khyber	end Apr-early May 628	7	Back from treaty stayed 20 days (or part of Dhū al-Ḥijja and part of Al-Muharram) then left to Khyber in Al-Muharram back to medina in Safar
58	Surrender of Fidak	May 628	7	2nd month of 7 AH
58	Third Expedition of Wadi al Qura	May 628	7	While back from Khaibar, heading to Medina (Prophet was poisoned and married Safiya)
58	Expedition of Umar ibn al-Khattab to Turbah	Jun-Jul 628	7	3rd month (weak reference)
58	Expedition of Abu Bakr As-Siddiq (to Nejd -Banii Fazara)	Jun-Jul 628	7	3rd month -weak
58	Expedition of Bashir ibn Sa'd al-Ansari (Fadak)	Nov-Dec 628	7	Shaʿbān; many Muslims killed -It **rained**
59	Expedition of Ghalib ibn Abdullah al-Laithi (Mayfah)	Dec 628	7	Ramaḍān
59	Expedition of Bashir ibn Sa'd al-Ansari	Jan 628	7	Expedition to Yemen
	Umrat al-Qada	Feb 629	7	Prophet Returned to mecca in fulfillment of Hudaybiyyah pact, married Maymuna
59	Expedition of Ibn Abi Al-Awja Al-Sulami to Banu Sulaym	Mar-Apr 629	7	1 month after returning from Mecca, Al-Awja, hurt, made it back to medina Safar 1
	Jerusalem back under Byzantine	21 Mar 629	7	Heraclius restored the "True Cross" to Jerusalem in a majestic ceremony
59	Expedition of Ghalib ibn Abdullah al-Laithi (Fadak)	Dec 629	7	
59	Expedition of Shuja ibn Wahb al-Asadi	Jun 629	8	Not certain

Age	Event	C.E. date	A.H. year	Note
	Acceptance of Islam by 'Amr Ibn Al-'As, Khalid Ibn Al-Walid, and 'Uthman Ibn Talha Ibn Abu Talib	Jun 629	8	Safar 8 AH
59	Expedition of Ka'b ibn 'Umair al-Ghifari	Jul 629	8	date not certain
59	Battle of Mu'tah	Aug 28 629	8	1 Jumādā al-Ūlā
59	Expedition of Amr ibn al-As	Sep 629	8	Reference to the 6th month
59	Expedition of Abu Ubaidah ibn al-Jarrah (the expedition of Fish)	Oct 629	8	Reference to the 7th month
59	Expedition of Abu Qatadah ibn Rab'i al-Ansari (Khadirah)	Dec 629	8	
59	Expedition of Abu Qatadah ibn Rab'i al-Ansari (Batn Edam)	Dec 629	8	
	Abu-Sufyan pleads with Muslims to reinforce the truce	Dec	8	The month is probable; Mohammed decided to march on Mecca immediately (saw a **cloud** in the sky)
60	**Mohammed takes over Mecca**	Jan 12, 630	8	Occurred in Ramaḍān, 18
60	Expedition of Khalid ibn al-Walid (Nakhla)	Jan 630	8	Occurred in Ramaḍān
60	Raid of Amr ibn al-As	Jan 630	8	
60	Raid of Sa'd ibn Zaid al-Ashhali	Jan 630	8	
60	Expedition of Khalid ibn al-Walid (Banu Jadhimah)	Jan 630	8	
60	Battle of Hunayn	Feb 3, 630	8	Occurred Wednesday night, Shawwāl 10
60	Battle of Autas	Feb, Mar 630	8	
60	Siege of Ta'if	Mar 630 back in Medina 16 Apr 630	8	Prophet involved in a long siege with no result. After 30 days, he was done with Taif matters, returned to Medina six days prior to the end of Dhū al-Ḥijja
60	Battle of Tabuk	Start: Aug-Sep End: Oct-Nov	9	The trip lasted about 50 days. Prophet left Jumādā al-Ūlā - Jumādā ath-Thāniya; stayed in Tabuk 20 days and was back in Rajab 9 (AH), i.e. (October/Nov) 630. It was very hot, but it ultimately **rained**. Also rained in Medina (or at Banu Murra) when he was back
	Najashi Died	Oct-Nov	9	
61	Expedition of Khalid ibn Al-Walid (Najran)	Jul 631	10	
61	Expedition of Ali ibn Abi Talib (Mudhij)	Dec 631	10	

Age	Event	C.E. date	A.H. year	Note
	Prophet's child Ibrahim dies	**Jan 30, 632**	**10**	**An eclipse** reported; it occurred on 29 Shawwāl
	Farewell Hajj	**Mar 9, 632**	**10**	
61	Expedition of Usama bin Zayd	**End of June**	**11**	25 Safar, it was also reported Al-Muharram, as followers were set to prepare for the expedition, the prophet fell sick
62	**Prophet Mohammed dies**	**Jul 10, 632**	**11**	12 Rabīʿ al-Awwal
	Abou Bakr dies	**Aug 24, 634**	**13**	21 Jumādā ath-Thāniya
	Omar became khalif	**Aug 26, 634**	**13**	23 Jumādā ath-Thāniya
	Byzantine forces defeated by Arabs	**Aug 15-20, 636**		The Byzantines suffered crushing defeat by Arabs at the Battle of Yarmouk
	Start of two decades of economic recovery in the Mediterranean triggered by the rise of the Islamic empire	**640**	**20**	Following a decline in mid 500s, a surge in the atmospheric lead signal increase in silver mining
	Heraclius dies, and Constantine III ruled as Co-Emperor for four months in 641 and Heraklonas in Feb/May 641. Succeeded by Constans II	**Feb 11, 641**		Constans II ruled until Sep 15, 668
	Omar Ibn Al-Khattab assassinated	**Nov 10, 644**	**24**	1 AL-Muharram (Islamic calendar date)
	Othman ibn Affan is Calif	**Nov 10, 644**	**24**	2 AL-Muharram (Islamic calendar date)
	Othman ibn Affan is assassinated	**Jun 19, 656**	**35**	17 Dhū al-Ḥijja (Islamic calendar date)

Chapter 11 Calendric data

"It is a capital mistake to theorize before one has data."

— Sherlock Holmes

In this Chapter, we provide data related to topics which are addressed throughout the book. Section 11.1 provides the Ephemeris for the 2010 -USNO astronomical lunar calendar phases, as well as an example of the daily lunar and solar rise and set times of the moon and the sun for the same year. Calendar data of the Jewish New Year based on photometry for the years 2014-2045 is presented in Section 11.2. Section 11.3 and Section 11.4 present the start time of conjunction and crescent first-visibility defining the start and end of Ramadan respectively for a century. And finally, Section 11.5 shows the dates of Jesus Crucifixion under the same conditions of orthogonal illumination which took place in the year 33, for 3000 years.

11.1 Ephemeris for 2010 -USNO data

Table 11.1-1 Lunar Ephemeris for the year 2010 (USNO)

	New Moon				First Quarter				Full Moon				Last Quarter		
	d	h	m		d	h	m		d	h	m		d	h	m
												Jan	7	10	39
Jan	15	7	11	Jan	23	10	53	Jan	30	6	18	Feb	5	23	48
Feb	14	2	51	Feb	22	0	42	Feb	28	16	38	Mar	7	15	42
Mar	15	21	1	Mar	23	11	0	Mar	30	2	25	Apr	6	9	37
Apr	14	12	29	Apr	21	18	20	Apr	28	12	18	May	6	4	15
May	14	1	4	May	20	23	43	May	27	23	7	Jun	4	22	13
Jun	12	11	15	Jun	19	4	29	Jun	26	11	30	Jul	4	14	35
Jul	11	19	40	Jul	18	10	11	Jul	26	1	37	Aug	3	4	59
Aug	10	3	8	Aug	16	18	14	Aug	24	17	5	Sep	1	17	22
Sep	8	10	30	Sep	15	5	50	Sep	23	9	17	Oct	1	3	52
Oct	7	18	44	Oct	14	21	27	Oct	23	1	36	Oct	30	12	46
Nov	6	4	52	Nov	13	16	39	Nov	21	17	27	Nov	28	20	36
Dec	5	17	36	Dec	13	13	59	Dec	21	8	13	Dec	28	4	18

Table 11.1-2 Moonrise and moonset for the year 2010 (USNO)

Moon Rise and Set for 2010

Location: Riyadh area E46° 46', N24° 38'
Blank: no rise/set

	Jan				Feb				Mar			
	Rise		Set		Rise		Set		Rise		Set	
Day	h	m	h	m	h	m	h	m	h	m	h	m
1	18	6	7	3	20	12	7	54	18	53	6	24
2	19	16	7	55	21	15	8	33	19	57	7	3
3	20	23	8	41	22	17	9	11	21	1	7	43
4	21	27	9	22	23	18	9	50	22	4	8	25
5	22	28	10	1			10	31	23	5	9	9
6	23	28	10	37	0	18	11	15			9	56
7			11	14	1	16	12	2	0	4	10	46
8	0	27	11	52	2	12	12	52	0	59	11	39
9	1	26	12	33	3	5	13	45	1	49	12	33
10	2	25	13	17	3	53	14	39	2	34	13	27
11	3	22	14	5	4	36	15	32	3	14	14	20
12	4	17	14	56	5	15	16	26	3	51	15	13
13	5	8	15	49	5	51	17	18	4	25	16	4
14	5	55	16	44	6	23	18	9	4	56	16	55
15	6	37	17	38	6	54	18	59	5	27	17	45
16	7	15	18	30	7	25	19	50	5	58	18	37
17	7	49	19	22	7	56	20	41	6	30	19	30
18	8	21	20	13	8	28	21	34	7	4	20	25
19	8	52	21	3	9	2	22	29	7	42	21	22
20	9	22	21	53	9	41	23	27	8	25	22	21
21	9	53	22	45	10	25			9	13	23	20
22	10	26	23	40	11	16	0	27	10	7		
23	11	2			12	13	1	28	11	6	0	19
24	11	44	0	37	13	17	2	27	12	10	1	14
25	12	31	1	37	14	24	3	23	13	15	2	5
26	13	27	2	40	15	33	4	14	14	21	2	52
27	14	30	3	43	16	41	5	1	15	27	3	35
28	15	38	4	43	17	48	5	43	16	31	4	15
29	16	49	5	38					17	35	4	54
30	17	59	6	28					18	39	5	34
31	19	6	7	13					19	44	6	15

	Apr				May				Jun			
	Rise		Set		Rise		Set		Rise		Set	
Day	h	m	h	m	h	m	h	m	h	m	h	m
1	20	47	6	59	21	29	7	17	22	19	8	47
2	21	49	7	46	22	20	8	12	22	54	9	41
3	22	47	8	36	23	5	9	8	23	26	10	32
4	23	41	9	29	23	46	10	4	23	57	11	23
5			10	24			10	57			12	13
6	0	29	11	19	0	22	11	50	0	28	13	4
7	1	11	12	13	0	56	12	41	1	0	13	56
8	1	50	13	6	1	27	13	31	1	34	14	51
9	2	24	13	58	1	58	14	22	2	12	15	49
10	2	57	14	48	2	29	15	14	2	55	16	49
11	3	28	15	39	3	2	16	8	3	44	17	51
12	3	59	16	30	3	38	17	4	4	40	18	52
13	4	31	17	23	4	18	18	3	5	43	19	49
14	5	5	18	18	5	4	19	5	6	48	20	42
15	5	42	19	15	5	56	20	5	7	55	21	29
16	6	23	20	14	6	54	21	4	9	0	22	11
17	7	10	21	15	7	56	21	58	10	4	22	50
18	8	3	22	14	9	1	22	47	11	6	23	28
19	9	1	23	10	10	5	23	31	12	6		
20	10	3			11	9			13	7	0	5
21	11	7	0	2	12	10	0	11	14	7	0	44
22	12	11	0	49	13	11	0	49	15	8	1	25
23	13	15	1	31	14	12	1	27	16	9	2	10
24	14	17	2	11	15	13	2	4	17	7	2	59
25	15	19	2	50	16	15	2	44	18	2	3	51
26	16	22	3	28	17	17	3	27	18	52	4	46
27	17	25	4	7	18	18	4	14	19	37	5	42
28	18	28	4	49	19	16	5	6	20	17	6	38
29	19	31	5	35	20	10	6	0	20	53	7	32
30	20	32	6	24	20	58	6	56	21	26	8	24
31					21	41	7	52				

	Jul				Aug				Sep			
	Rise		Set		Rise		Set		Rise		Set	
Day	h	m	h	m	h	m	h	m	h	m	h	m
1	21	57	9	15	22	5	10	31	22	56	12	6
2	22	28	10	5	22	42	11	24	23	53	13	4
3	22	58	10	55	23	24	12	20			14	0
4	23	31	11	46			13	18	0	54	14	53
5			12	39	0	12	14	18	2	0	15	43
6	0	6	13	34	1	6	15	17	3	6	16	28
7	0	46	14	33	2	7	16	14	4	14	17	11
8	1	32	15	33	3	13	17	7	5	20	17	53
9	2	24	16	35	4	22	17	55	6	26	18	34
10	3	24	17	34	5	30	18	40	7	32	19	16
11	4	29	18	30	6	37	19	21	8	38	20	0
12	5	36	19	20	7	43	20	1	9	42	20	47
13	6	44	20	6	8	47	20	41	10	45	21	37
14	7	51	20	48	9	51	21	23	11	45	22	31
15	8	56	21	27	10	54	22	6	12	40	23	26
16	9	58	22	5	11	56	22	53	13	30		
17	11	0	22	44	12	56	23	44	14	14	0	21
18	12	2	23	25	13	53			14	54	1	16
19	13	3			14	45	0	36	15	30	2	9
20	14	3	0	9	15	33	1	31	16	3	3	1
21	15	2	0	56	16	15	2	26	16	34	3	52
22	15	57	1	47	16	53	3	21	17	5	4	42
23	16	48	2	40	17	28	4	14	17	36	5	32
24	17	34	3	36	18	1	5	5	18	9	6	23
25	18	16	4	31	18	32	5	56	18	44	7	15
26	18	53	5	25	19	2	6	46	19	22	8	8
27	19	27	6	18	19	34	7	36	20	4	9	4
28	19	58	7	10	20	6	8	26	20	52	10	0
29	20	29	8	0	20	42	9	19	21	46	10	57
30	21	0	8	50	21	21	10	13	22	44	11	52
31	21	31	9	40	22	6	11	9				

	Oct				Nov				Dec			
	Rise		Set		Rise		Set		Rise		Set	
Day	h	m	h	m	h	m	h	m	h	m	h	m
1	23	46	12	45	0	46	13	38	1	43	13	34
2			13	34	1	48	14	17	2	45	14	16
3	0	49	14	19	2	51	14	57	3	49	15	1
4	1	54	15	2	3	55	15	38	4	53	15	52
5	2	59	15	43	5	0	16	23	5	55	16	46
6	4	4	16	24	6	5	17	12	6	55	17	43
7	5	9	17	5	7	10	18	5	7	49	18	42
8	6	15	17	49	8	12	19	1	8	37	19	40
9	7	21	18	36	9	10	19	59	9	20	20	36
10	8	27	19	26	10	1	20	57	9	58	21	29
11	9	30	20	20	10	46	21	53	10	33	22	21
12	10	29	21	16	11	26	22	47	11	5	23	11
13	11	22	22	13	12	2	23	39	11	36		
14	12	10	23	9	12	35			12	7	0	1
15	12	52			13	6	0	30	12	39	0	51
16	13	29	0	3	13	37	1	20	13	14	1	43
17	14	3	0	56	14	9	2	10	13	53	2	36
18	14	35	1	47	14	42	3	1	14	36	3	32
19	15	6	2	37	15	19	3	54	15	25	4	30
20	15	37	3	27	15	59	4	49	16	20	5	29
21	16	10	4	17	16	45	5	46	17	20	6	26
22	16	44	5	9	17	36	6	44	18	23	7	20
23	17	22	6	3	18	33	7	42	19	28	8	10
24	18	3	6	58	19	33	8	37	20	32	8	56
25	18	50	7	55	20	35	9	28	21	34	9	38
26	19	42	8	52	21	37	10	15	22	36	10	17
27	20	39	9	48	22	39	10	58	23	37	10	55
28	21	39	10	41	23	41	11	38			11	34
29	22	41	11	30			12	16	0	38	12	14
30	23	43	12	16	0	41	12	54	1	40	12	58
31			12	58					2	42	13	45

Table 11.1-3 Sunrise and sunset for the year 2010 (USNO)

Sun Rise and Set for 2010

Location: Riyadh area W46° 46', N24° 38'

Blank: no rise/set

	Jan				Feb				Mar				Apr				May				Jun			
	Rise		Set		Rise		Set		Rise		Set		Rise		Set		Rise		Set		Rise		Set	
Day	h	m	h	m	h	m	h	m	h	m	h	m	h	m	h	m	h	m	h	m	h	m	h	m
1	6	37	17	16	6	35	17	38	6	16	17	55	5	45	18	9	5	18	18	22	5	4	18	38
2	6	37	17	16	6	35	17	39	6	15	17	56	5	44	18	9	5	17	18	23	5	4	18	38
3	6	38	17	17	6	34	17	40	6	14	17	56	5	43	18	10	5	17	18	23	5	4	18	39
4	6	38	17	18	6	34	17	40	6	13	17	57	5	42	18	10	5	16	18	24	5	4	18	39
5	6	38	17	18	6	33	17	41	6	12	17	57	5	41	18	11	5	15	18	24	5	3	18	39
6	6	38	17	19	6	33	17	42	6	11	17	58	5	40	18	11	5	15	18	25	5	3	18	40
7	6	39	17	20	6	32	17	42	6	10	17	58	5	39	18	12	5	14	18	25	5	3	18	40
8	6	39	17	21	6	32	17	43	6	9	17	59	5	38	18	12	5	13	18	26	5	3	18	41
9	6	39	17	21	6	31	17	44	6	8	17	59	5	37	18	12	5	13	18	26	5	3	18	41
10	6	39	17	22	6	30	17	44	6	7	18	0	5	36	18	13	5	12	18	27	5	3	18	41
11	6	39	17	23	6	30	17	45	6	6	18	0	5	35	18	13	5	12	18	27	5	3	18	42
12	6	39	17	23	6	29	17	45	6	5	18	0	5	34	18	14	5	11	18	28	5	3	18	42
13	6	39	17	24	6	28	17	46	6	4	18	1	5	33	18	14	5	10	18	28	5	4	18	42
14	6	39	17	25	6	28	17	47	6	3	18	1	5	32	18	15	5	10	18	29	5	4	18	43
15	6	39	17	26	6	27	17	47	6	2	18	2	5	31	18	15	5	9	18	29	5	4	18	43
16	6	39	17	26	6	26	17	48	6	1	18	2	5	30	18	15	5	9	18	30	5	4	18	43
17	6	39	17	27	6	26	17	49	6	0	18	3	5	30	18	16	5	8	18	30	5	4	18	44
18	6	39	17	28	6	25	17	49	5	59	18	3	5	29	18	16	5	8	18	31	5	4	18	44
19	6	39	17	29	6	24	17	50	5	58	18	4	5	28	18	17	5	8	18	31	5	4	18	44
20	6	39	17	29	6	23	17	50	5	57	18	4	5	27	18	17	5	7	18	32	5	4	18	44
21	6	38	17	30	6	23	17	51	5	56	18	4	5	26	18	18	5	7	18	32	5	5	18	45
22	6	38	17	31	6	22	17	51	5	55	18	5	5	25	18	18	5	6	18	33	5	5	18	45
23	6	38	17	32	6	21	17	52	5	54	18	5	5	24	18	19	5	6	18	33	5	5	18	45
24	6	38	17	32	6	20	17	53	5	53	18	6	5	23	18	19	5	6	18	34	5	5	18	45
25	6	38	17	33	6	19	17	53	5	52	18	6	5	23	18	20	5	5	18	34	5	6	18	45
26	6	37	17	34	6	18	17	54	5	51	18	7	5	22	18	20	5	5	18	35	5	6	18	46
27	6	37	17	35	6	18	17	54	5	50	18	7	5	21	18	20	5	5	18	35	5	6	18	46
28	6	37	17	35	6	17	17	55	5	49	18	7	5	20	18	21	5	5	18	36	5	7	18	46
29	6	36	17	36					5	48	18	8	5	20	18	21	5	4	18	36	5	7	18	46
30	6	36	17	37					5	47	18	8	5	19	18	22	5	4	18	37	5	7	18	46
31	6	36	17	37					5	46	18	9					5	4	18	37				

	Jul				Aug				Sep				Oct				Nov				Dec			
	Rise		Set		Rise		Set		Rise		Set		Rise		Set		Rise		Set		Rise		Set	
Day	h	m	h	m	h	m	h	m	h	m	h	m	h	m	h	m	h	m	h	m	h	m	h	m
1	5	7	18	46	5	21	18	37	5	34	18	12	5	45	17	40	6	0	17	13	6	20	17	4
2	5	8	18	46	5	21	18	37	5	34	18	11	5	45	17	39	6	0	17	12	6	21	17	4
3	5	8	18	46	5	22	18	36	5	35	18	10	5	46	17	38	6	1	17	12	6	21	17	4
4	5	9	18	46	5	22	18	35	5	35	18	9	5	46	17	37	6	2	17	11	6	22	17	4
5	5	9	18	46	5	23	18	35	5	35	18	8	5	46	17	36	6	2	17	11	6	23	17	4
6	5	9	18	46	5	23	18	34	5	36	18	7	5	47	17	35	6	3	17	10	6	23	17	4
7	5	10	18	46	5	24	18	33	5	36	18	6	5	47	17	34	6	3	17	10	6	24	17	4
8	5	10	18	46	5	24	18	33	5	36	18	5	5	48	17	33	6	4	17	9	6	25	17	5
9	5	11	18	46	5	25	18	32	5	37	18	3	5	48	17	32	6	5	17	9	6	25	17	5
10	5	11	18	45	5	25	18	31	5	37	18	2	5	48	17	31	6	5	17	8	6	26	17	5
11	5	11	18	45	5	26	18	30	5	38	18	1	5	49	17	30	6	6	17	8	6	27	17	5
12	5	12	18	45	5	26	18	30	5	38	18	0	5	49	17	29	6	7	17	7	6	27	17	6
13	5	12	18	45	5	26	18	29	5	38	17	59	5	50	17	28	6	7	17	7	6	28	17	6
14	5	13	18	45	5	27	18	28	5	39	17	58	5	50	17	27	6	8	17	6	6	29	17	6
15	5	13	18	44	5	27	18	27	5	39	17	57	5	51	17	26	6	9	17	6	6	29	17	7
16	5	14	18	44	5	28	18	27	5	39	17	56	5	51	17	26	6	9	17	6	6	30	17	7
17	5	14	18	44	5	28	18	26	5	40	17	55	5	52	17	25	6	10	17	5	6	30	17	7
18	5	14	18	44	5	29	18	25	5	40	17	54	5	52	17	24	6	11	17	5	6	31	17	8
19	5	15	18	43	5	29	18	24	5	40	17	53	5	53	17	23	6	11	17	5	6	32	17	8
20	5	15	18	43	5	29	18	23	5	41	17	52	5	53	17	22	6	12	17	5	6	32	17	9
21	5	16	18	43	5	30	18	22	5	41	17	51	5	54	17	21	6	13	17	4	6	33	17	9
22	5	16	18	42	5	30	18	21	5	41	17	50	5	54	17	20	6	14	17	4	6	33	17	10
23	5	17	18	42	5	31	18	20	5	42	17	49	5	55	17	20	6	14	17	4	6	34	17	10
24	5	17	18	41	5	31	18	19	5	42	17	48	5	55	17	19	6	15	17	4	6	34	17	11
25	5	18	18	41	5	31	18	18	5	43	17	46	5	56	17	18	6	16	17	4	6	34	17	11
26	5	18	18	40	5	32	18	18	5	43	17	45	5	56	17	17	6	16	17	4	6	35	17	12
27	5	19	18	40	5	32	18	17	5	43	17	44	5	57	17	17	6	17	17	4	6	35	17	12
28	5	19	18	39	5	32	18	16	5	44	17	43	5	57	17	16	6	18	17	4	6	36	17	13
29	5	20	18	39	5	33	18	15	5	44	17	42	5	58	17	15	6	19	17	4	6	36	17	14
30	5	20	18	38	5	33	18	14	5	44	17	41	5	59	17	14	6	19	17	4	6	36	17	14
31	5	21	18	38	5	34	18	13					5	59	17	14					6	37	17	15

11.2 Jewish New Year calendar

Table 11.2-1 Universal Molad for the Month of Tishrei

Jewish New Year						GSFC Conjunction					Molad for Month of Tishrei								
Year Number	Year length (days)	Leap year	Day of the week	Month	day	year	Day of the week	Month	day	Time (UTC)	year	Day of the week	Month	day	Time (UTC)	Ccr	Lat (deg)	Long (deg)	Time elapsed from Conjunction
5775	354		Thu	Sep	25	2014	Wed	Sep	24	6:14	2014	Wed	Sep	24	19:40:04	6.35	-32.0	-26.9	13:27
5776	385	Yes	Mon	Sep	14	2015	Sun	Sep	13	6:41	2015	Sun	Sep	13	20:44:52	6.35	-25.9	-44.3	14:03
5777	353		Mon	Oct	3	2016	Sat	Oct	1	0:12	2016	Sat	Oct	1	18:05:12	6.37	39.9	-7.2	17:53
5778	354		Thu	Sep	21	2017	Wed	Sep	20	5:30	2017	Wed	Sep	20	18:14:53	6.47	-24.0	-6.1	12:45
5779	385	Yes	Mon	Sep	10	2018	Sun	Sep	9	18:01	2018	Mon	Sep	10	2:56:22	6.35	14.0	-133.9	8:55
5780	355		Mon	Sep	30	2019	Sat	Sep	28	18:26	2019	Sun	Sep	29	1:04:23	6.35	28.0	-110.0	6:38
5781	353		Sat	Sep	19	2020	Thu	Sep	17	11:00	2020	Thu	Sep	17	18:16:35	6.37	12.0	-5.5	7:16
5782	384	Yes	Tue	Sep	7	2021	Tue	Sep	7	0:52	2021	Tue	Sep	7	18:06:47	6.35	-45.7	-8.5	17:15
5783	355		Mon	Sep	26	2022	Sun	Sep	25	21:54	2022	Mon	Sep	26	9:04:28	6.35	4.0	131.3	11:10
5784	383	Yes	Sat	Sep	16	2023	fri	Sep	15	1:40	2023	fri	Sep	15	18:13:41	6.48	-41.9	-7.6	16:34
5785	355		Thu	Oct	3	2024	Wed	Oct	2	18:49	2024	Thu	Oct	3	8:47:05	6.35	-31.9	137.8	13:58
5786	354		Tue	Sep	23	2025	Sun	Sep	21	19:54	2025	Mon	Sep	22	8:56:14	6.35	-38.0	133.7	13:03
5787	385	Yes	Sat	Sep	12	2026	fri	Sep	11	3:27	2026	fri	Sep	11	18:16:35	6.43	-2.0	-5.5	14:50
5788	355		Sat	Oct	2	2027	Thu	Sep	30	2:36	2027	Thu	Sep	30	10:46:20	6.35	-59.9	110.5	8:10
5789	354		Thu	Sep	21	2028	Mon	Sep	18	18:24	2028	Tue	Sep	19	1:45:02	6.35	-60.0	-120.4	7:22
5790	383	Yes	Mon	Sep	10	2029	Sat	Sep	8	10:44	2029	Sat	Sep	8	18:12:38	6.35	-57.6	-12.6	7:28
5791	355		Sat	Sep	28	2030	fri	Sep	27	9:55	2030	fri	Sep	27	18:38:40	6.35	-54.0	-9.7	8:44
5792	354		Thu	Sep	18	2031	Tue	Sep	16	18:47	2031	Wed	Sep	17	18:19:40	6.62	16.0	-6.0	23:32
5793	383	Yes	Mon	Sep	6	2032	Sat	Sep	4	20:57	2032	Sun	Sep	5	8:39:46	6.35	-55.5	129.8	11:43
5794	355		Sat	Sep	24	2033	fri	Sep	23	13:40	2033	Sat	Sep	24	3:41:46	6.35	-28.0	-147.5	14:02
5795	385	Yes	Thu	Sep	14	2034	Tue	Sep	12	16:14	2034	Wed	Sep	13	5:17:56	6.35	-20.0	-172.2	13:04
5796	354		Thu	Oct	4	2035	Mon	Oct	1	13:07	2035	Mon	Oct	1	23:41:45	6.35	8.0	-88.9	10:34
5797	353		Mon	Sep	22	2036	Sat	Sep	20	1:51	2036	Sat	Sep	20	10:46:14	6.35	12.0	106.6	8:55
5798	385	Yes	Thu	Sep	10	2037	Wed	Sep	9	18:25	2037	Thu	Sep	10	2:27:03	6.35	17.9	-126.3	8:02
5799	354		Thu	Sep	30	2038	Tue	Sep	28	18:57	2038	Wed	Sep	29	1:39:42	6.35	28.0	-119.0	6:42
5800	355		Mon	Sep	19	2039	Sun	Sep	18	8:23	2039	Sun	Sep	18	18:14:05	6.45	-6.0	-5.5	9:51
5801	383	Yes	Sat	Sep	8	2040	Thu	Sep	6	15:13	2040	fri	Sep	7	0:20:04	6.35	23.9	-93.3	9:06
5802	354		Thu	Sep	26	2041	Wed	Sep	25	8:41	2041	Wed	Sep	25	21:34:50	6.35	-2.0	-56.2	12:53
5803	385	Yes	Mon	Sep	15	2042	Sun	Sep	14	8:50	2042	Sun	Sep	14	22:10:24	6.35	-6.0	-64.4	13:20
5804	353		Mon	Oct	5	2043	Sat	Sep	3	13:17	2043	Sun	Sep	4	1:45:34	6.35	-11.9	-118.5	12:28
5805	355		Thu	Sep	22	2044	Wed	Sep	21	11:03	2044	Wed	Sep	21	22:16:34	6.35	-44.0	-66.5	11:13
5806	384	Yes	Tue	Sep	12	2045	Mon	Sep	11	1:27	2045	Mon	Sep	11	11:13:21	6.35	-49.8	95.3	9:46

11.3 Centurial Ramaḍan Conjunction Calendar

Table 11.3-1: Conjunction date, time (UTC) and "location" for the month of Ramaḍan

New Moon									
NASA Goddard				The Witness					
Y	M	D	Time	Y	M	D	Time	Lat	Long
2015	Jun	16	14:05	2015	Jun	16	14:05	-66.7	-26.3
2015	Jul	16	1:24	2015	Jul	16	1:24	-68.6	-174.4
2016	Jun	5	3:00	2016	Jun	5	3:00	-67.4	151.1
2016	Jul	4	11:01	2016	Jul	4	11:01	-67.2	29.5
2017	May	25	19:44	2017	May	25	19:45	-68.9	-90.5
2017	Jun	24	2:31	2017	Jun	24	2:31	-66.6	146.0
2018	May	15	11:48	2018	May	15	11:48	-71.1	39.5
2018	Jun	13	19:43	2018	Jun	13	19:43	-66.8	-108.6
2019	May	4	22:45	2019	May	4	22:45	-73.9	-114.5
2019	Jun	3	10:02	2019	Jun	3	10:01	-67.7	47.8
2020	Apr	23	2:26	2020	Apr	23	2:25	-77.4	-157.9
2020	May	22	17:39	2020	May	22	17:38	-69.4	-56.0
2021	Apr	12	2:31	2021	Apr	12	2:30	-81.3	-148.1
2021	May	11	19:00	2021	May	11	18:59	-71.9	-65.3
2022	Apr	1	6:24	2022	Apr	1	6:24	-85.4	164.2
2022	Apr	30	20:28	2022	April	30	20:27	-75.0	-76.2
2023	Mar	21	17:23	2023	Mar	21	17:23	-89.7	9.6
2023	Apr	20	4:12	2023	Apr	20	4:12	-78.6	178.5
2024	Mar	10	9:00	2024	Mar	10	9:00	93.8	146.2
2024	Apr	8	18:21	2024	Apr	8	18:20	-82.4	12.2
2025	Feb	28	0:45	2025	Feb	28	0:45	97.9	-79.3
2025	Mar	29	10:58	2025	Mar	29	10:58	-86.4	114.5
2026	Feb	17	12:01	2026	Feb	17	12:01	101.9	122.0
2026	Mar	19	1:23	2026	Mar	19	1:24	90.6	-110.4
2027	Feb	6	15:56	2027	Feb	6	15:55	105.5	74.6
2027	Mar	8	9:29	2027	Mar	8	9:29	94.9	118.7
2028	Jan	26	15:12	2028	Jan	26	15:12	108.7	-6.3
2028	Feb	25	10:37	2028	Feb	25	10:37	99.2	91.8
2029	Jan	14	17:24	2029	Jan	14	17:23	111.2	-52.2
2029	Feb	13	10:31	2029	Feb	13	10:31	103.2	82.7
2030	Jan	4	2:49	2030	Jan	4	2:49	112.7	153.8
2030	Feb	2	16:07	2030	Feb	2	16:07	106.7	-12.6
2030	Dec	24	17:32	2030	Dec	24	17:32	113.4	-80.5
2031	Jan	23	4:31	2031	Jan	23	4:30	109.5	150.4
2031	Dec	14	9:06	2031	Dec	14	9:05	113.2	50.4
2032	Jan	12	20:07	2032	Jan	12	20:06	111.6	-96.2
2032	Dec	2	20:53	2032	Dec	2	20:53	112.1	-116.3
2033	Jan	1	10:17	2033	Jan	1	10:17	113.0	38.6
2033	Nov	22	1:39	2033	Nov	22	1:39	110.2	-176.1

2033	Dec	21	18:47	2033	Dec	21	18:46	113.4	-102.5
2034	Nov	11	1:16	2034	Nov	11	1:16	107.4	-159.3
2034	Dec	10	20:14	2034	Dec	10	20:14	113.0	-114.0
2035	Oct	31	2:59	2035	Oct	31	2:59	104.0	-174.0
2035	Nov	29	19:38	2035	Nov	29	19:38	111.5	-93.7
2036	Oct	19	11:50	2036	Oct	19	11:49	100.3	63.6
2036	Nov	18	0:14	2036	Nov	18	0:14	109.3	-151.1
2037	Oct	9	2:34	2037	Oct	9	2:34	96.4	-146.7
2037	Nov	7	12:03	2037	Nov	7	12:03	106.5	41.8
2038	Sep	28	18:57	2038	Sep	28	18:57	92.3	-22.6
2038	Oct	28	3:53	2038	Oct	28	3:52	103.1	175.2
2039	Sep	18	8:23	2039	Sep	18	8:22	-88.1	147.0
2039	Oct	17	19:09	2039	Oct	17	19:08	99.4	-44.0
2040	Sep	6	15:13	2040	Sep	6	15:14	-84.0	54.6
2040	Oct	6	5:26	2040	Oct	6	5:26	95.4	172.9
2041	Aug	26	16:16	2041	Aug	26	16:15	-79.9	50.3
2041	Sep	25	8:41	2041	Sep	25	8:40	91.1	135.0
2042	Aug	15	18:01	2042	Aug	15	18:01	-76.2	34.8
2042	Sep	14	8:50	2042	Sep	14	8:50	-86.7	143.7
2043	Aug	5	2:23	2043	Aug	5	2:22	-73.0	-79.2
2043	Sep	3	13:17	2043	Sep	3	13:17	-82.5	87.8
2044	Jul	24	17:10	2044	Jul	24	17:10	-70.4	68.6
2044	Aug	23	1:06	2044	Aug	23	1:05	-78.7	-78.2
2045	Jul	14	10:28	2045	Jul	14	10:28	-68.4	179.8
2045	Aug	12	17:39	2045	Aug	12	17:39	-75.3	43.8
2046	Jul	4	1:39	2046	Jul	4	1:38	-67.1	-36.7
2046	Aug	2	10:25	2046	Aug	2	10:25	-72.4	67.4
2047	Jun	23	10:36	2047	Jun	23	10:35	-66.6	-160.9
2047	Jul	22	22:49	2047	Jul	22	22:49	-69.9	-129.3
2048	Jun	11	12:50	2048	Jun	11	12:50	-66.9	157.5
2048	Jul	11	4:04	2048	Jul	11	4:04	-68.0	141.2
2049	May	31	14:00	2049	May	31	14:00	-68.0	-12.9
2049	Jun	30	4:50	2049	Jun	30	4:50	-66.9	117.8
2050	May	20	20:51	2050	May	20	20:51	-69.8	-102.5
2050	Jun	19	8:22	2050	Jun	19	8:22	-66.6	56.5
2051	May	10	10:29	2051	May	10	10:29	-72.3	64.3
2051	Jun	8	18:56	2051	Jun	8	18:56	-67.1	-91.9
2052	Apr	29	3:20	2052	Apr	29	3:21	-75.3	-178.0
2052	May	28	10:50	2052	May	28	10:50	-68.4	40.4
2053	Apr	18	18:48	2053	Apr	18	18:48	-78.8	-40.2
2053	May	18	3:43	2053	May	18	3:43	-70.4	157.8
2054	Apr	8	4:32	2054	Apr	8	4:32	-82.7	-174.9
2054	May	7	17:00	2054	May	7	17:00	-73.0	-31.5
2055	Mar	28	7:01	2055	Mar	28	7:00	-87.0	159.0
2055	Apr	26	23:17	2055	Apr	26	23:16	-76.2	-115.1
2056	Mar	16	6:52	2056	Mar	16	6:51	91.4	172.2
2056	Apr	14	23:50	2056	Apr	14	23:50	-80.0	-112.4
2057	Mar	5	11:25	2057	Mar	5	11:24	95.8	114.6

2057	Apr	4	1:31	2057	Apr	4	1:31	-84.2	-125.7
2058	Feb	22	22:56	2058	Feb	22	22:57	99.8	-48.4
2058	Mar	24	9:50	2058	Mar	24	9:50	-88.4	119.9
2059	Feb	12	14:27	2059	Feb	12	14:27	103.6	89.9
2059	Mar	14	0:05	2059	Mar	14	0:05	92.6	-82.9
2060	Feb	2	5:22	2060	Feb	2	5:22	106.9	-123.1
2060	Mar	2	16:11	2060	Mar	2	16:11	96.7	45.2
2061	Jan	21	15:16	2061	Jan	21	15:15	109.7	99.0
2061	Feb	20	5:31	2061	Feb	20	5:31	100.7	-143.7
2062	Jan	10	17:52	2062	Jan	10	17:52	111.8	70.7
2062	Feb	9	12:11	2062	Feb	9	12:10	104.5	126.9
2062	Dec	30	16:57	2062	Dec	30	16:57	113.1	95.6
2063	Jan	29	12:23	2063	Jan	29	12:23	107.8	134.9
2063	Dec	19	20:04	2063	Dec	19	20:03	113.4	55.3
2064	Jan	18	12:37	2064	Jan	18	12:36	110.5	142.5
2064	Dec	8	6:28	2064	Dec	8	6:28	112.8	-113.6
2064	Nov	8	19:45	2064	Nov	8	19:45	107.0	13.6
2065	Nov	27	21:40	2065	Nov	27	21:39	111.3	5.5
2065	Dec	27	8:27	2065	Dec	27	8:27	113.3	-133.7
2066	Nov	17	13:06	2066	Nov	17	13:06	109.1	122.3
2066	Dec	17	0:17	2066	Dec	17	0:17	113.3	-10.3
2067	Nov	7	0:14	2067	Nov	7	0:14	106.2	-55.6
2067	Dec	6	14:05	2067	Dec	6	14:05	112.5	-17.9
2068	Oct	26	4:17	2068	Oct	26	4:16	102.7	-127.1
2068	Nov	24	21:42	2068	Nov	24	21:42	110.8	-121.1
2069	Oct	15	4:03	2069	Oct	15	4:02	98.7	-133.9
2069	Nov	13	22:38	2069	Nov	13	22:37	108.3	-124.0
2070	Oct	4	7:01	2070	Oct	4	7:00	94.5	172.0
2070	Nov	2	22:42	2070	Nov	2	22:42	105.1	-114.3
2071	Sep	23	17:21	2071	Sep	23	17:20	90.3	6.3
2071	Oct	23	4:49	2071	Oct	23	4:49	101.4	165.6
2072	Sep	12	9:07	2072	Sep	12	9:06	-86.2	140.6
2072	Oct	11	17:55	2072	Oct	11	17:55	97.5	-20.8
2073	Sep	2	1:52	2073	Sep	2	1:52	-82.3	-100.2
2073	Oct	1	10:21	2073	Oct	1	10:21	93.5	103.4
2074	Aug	22	14:59	2074	Aug	22	14:59	-78.5	73.1
2074	Sep	21	1:28	2074	Sep	21	1:28	-89.4	-112.6
2075	Aug	11	21:10	2075	Aug	11	21:11	-75.0	-9.3
2075	Sep	10	11:02	2075	Sep	10	11:02	-85.2	114.1
2076	Jul	30	22:05	2076	Jul	30	22:06	-71.9	-12.1
2076	Aug	29	13:43	2076	Aug	29	13:43	-81.1	84.7
2077	Jul	20	0:41	2077	Jul	20	0:41	-69.4	-39.1
2077	Aug	18	14:17	2077	Aug	18	14:18	-77.3	87.1
2078	Jul	9	10:08	2078	Jul	9	10:08	-67.7	-170.5
2078	Aug	7	19:52	2078	Aug	7	19:52	-73.9	14.2
2079	Jun	29	1:31	2079	Jun	29	1:31	-66.8	-30.3
2079	Jul	28	8:33	2079	Jul	28	8:33	-71.1	-164.9
2080	Jun	17	18:40	2080	Jun	17	18:40	-66.6	76.6

2080	Jul	17	1:21	2080	Jul	17	1:21	-69.0	-46.3
2081	Jun	7	9:01	2081	Jun	7	9:01	-67.2	-149.8
2081	Jul	6	17:44	2081	Jul	6	17:44	-67.5	77.7
2082	May	27	16:47	2082	May	27	16:46	-68.6	81.8
2082	Jun	26	5:16	2082	Jun	26	5:15	-66.7	-84.1
2083	May	16	18:14	2083	May	16	18:13	-70.7	48.6
2083	Jun	15	9:37	2083	Jun	15	9:36	-66.7	-150.6
2084	May	4	19:32	2084	May	4	19:32	-73.6	17.8
2084	Jun	3	10:13	2084	Jun	3	10:13	-67.6	-172.0
2085	Apr	24	2:59	2085	Apr	24	2:59	-76.9	-103.3
2085	May	23	14:23	2085	May	23	14:23	-69.2	113.7
2086	Apr	13	16:53	2086	Apr	13	16:52	-80.6	38.4
2086	May	13	1:41	2086	May	13	1:41	-71.5	-66.0
2087	Apr	3	9:26	2087	Apr	3	9:25	-84.5	141.5
2087	May	2	17:51	2087	May	2	17:50	-74.4	40.6
2088	Mar	23	0:00	2088	Mar	23	0:00	-88.7	-85.6
2088	Apr	21	10:25	2088	Apr	21	10:24	-77.7	142.5
2089	Mar	12	8:24	2089	Mar	12	8:24	93.0	138.8
2089	Apr	10	22:45	2089	Apr	10	22:46	-81.5	-92.3
2090	Mar	1	9:46	2090	Mar	1	9:46	97.3	108.6
2090	Mar	31	3:48	2090	Mar	31	3:48	-85.7	-156.5
2091	Feb	18	9:38	2091	Feb	18	9:37	101.5	100.3
2091	Mar	20	3:45	2091	Mar	20	8:47	-90.0	139.3
2092	Feb	7	15:03	2092	Feb	7	15:02	105.2	7.8
2092	Mar	8	5:56	2092	Mar	8	5:55	94.5	-166.4
2093	Jan	27	3:22	2093	Jan	27	3:22	108.3	-88.0
2093	Feb	25	15:05	2093	Feb	25	15:05	98.7	66.5
2094	Jan	16	19:05	2094	Jan	16	19:06	110.7	45.3
2094	Feb	15	5:43	2094	Feb	15	5:43	102.5	-142.2
2095	Jan	6	9:33	2095	Jan	6	9:34	112.4	-160.5
2095	Feb	4	21:28	2095	Feb	4	21:28	105.9	-8.8
2095	Dec	26	18:24	2095	Dec	26	18:24	113.3	77.3
2096	Jan	25	9:45	2096	Jan	25	9:44	108.9	178.4
2096	Dec	14	20:05	2096	Dec	14	20:05	113.3	49.5
2097	Jan	13	15:00	2097	Jan	13	14:59	111.3	110.2
2097	Dec	3	19:26	2097	Dec	3	19:26	112.3	45.7
2098	Jan	2	14:32	2098	Jan	2	14:32	112.8	128.2
2098	Nov	22	23:51	2098	Nov	22	23:50	110.4	-33.3
2098	Dec	22	15:24	2098	Dec	22	15:23	113.4	126.4
2099	Nov	12	11:29	2099	Nov	12	11:29	107.8	140.8
2099	Dec	11	23:09	2099	Dec	11	23:08	113.1	-0.8
2100	Nov	2	3:14	2100	Nov	2	3:13	104.7	-105.7
2100	Dec	1	13:00	2100	Dec	1	13:00	111.8	139.0
2101	Oct	22	18:30	2101	Oct	22	18:29	101.2	14.7
2101	Nov	21	4:48	2101	Nov	21	4:48	109.9	-109.8
2102	Oct	12	4:51	2102	Oct	12	4:51	97.2	-149.8
2102	Nov	10	17:57	2102	Nov	10	17:56	107.2	41.1
2103	Oct	1	8:11	2103	Oct	1	8:10	93.0	150.9

2103	Oct	31	0:37	2103	Oct	31	0:37	103.8	-69.3
2104	Sep	19	8:14	2104	Sep	19	8:13	-88.6	141.2
2104	Oct	19	1:12	2104	Oct	19	1:12	100.0	-88.4
2105	Sep	8	12:22	2105	Sep	8	12:21	-84.4	70.2
2105	Oct	8	2:08	2105	Oct	8	2:07	95.8	-112.0
2106	Aug	28	23:47	2106	Aug	28	23:46	-80.5	-110.5
2106	Sep	27	9:37	2106	Sep	27	9:36	91.6	126.5
2107	Aug	18	16:03	2107	Aug	18	16:03	-76.9	-3.5
2107	Sep	16	23:46	2107	Sep	16	23:45	-87.4	-95.0
2108	Aug	7	8:43	2108	Aug	7	8:42	-73.7	97.5
2108	Sep	5	16:41	2108	Sep	5	16:40	-83.4	3.0
2109	Jul	27	21:13	2109	Jul	27	21:13	-70.9	-101.0
2109	Aug	26	7:43	2109	Aug	26	7:42	-79.6	129.1
2110	Jul	17	2:41	2110	Jul	17	2:40	-68.8	167.0
2110	Aug	15	16:44	2110	Aug	15	16:43	-76.0	-15.8
2111	Jul	6	3:29	2111	Jul	6	3:29	-67.3	142.9
2111	Aug	4	19:02	2111	Aug	4	19:02	-72.8	30.2
2112	Jun	24	6:50	2112	Jun	24	6:49	-66.6	80.9
2112	Jul	23	20:09	2112	Jul	23	20:09	-70.1	24.3
2113	Jun	13	17:15	2113	Jun	13	17:14	-66.8	-71.8
2113	Jul	13	2:54	2113	Jul	13	2:54	-68.2	-65.2

11.4 Centurial photometric Calendar

Table 11.4-1: First crescent visibility date, time (UTC) and location at start and end of Ramaḍan for Ccr=6.35°, 8 °, and 10 °

Centurial Ramaḍān Calendar																				
Ramaḍān (Ccr=6.35)							Ramaḍān (Ccr=8.0)							Ramaḍān (Ccr=10.0)						
Y	M	D	Time	Ccr	Lat	Long	Y	M	D	Time	Ccr	Lat.	Long	Y	M	D	Time	Ccr	Lat	Long
2015	Jun	16	21:43	6.35	-44.4	-81.5	2015	Jun	17	2:12	8.00	-33.4	-139.4	2015	Jun	17	6:55	10.00	-25.9	154.2
2015	Jul	16	10:25	6.35	-50.7	86.4	2015	Jul	16	15:00	8.00	-40.5	26.4	2015	Jul	16	19:55	10.00	-33.3	-42.8
2016	Jun	5	9:47	6.35	-41.4	101.1	2016	Jun	5	13:52	8.00	-29.7	47.5	2016	Jun	5	18:07	10.00	-21.8	-12.3
2016	Jul	4	19:41	6.35	-42.8	-47.7	2016	Jul	4	23:35	8.00	-33.5	-99.7	2016	Jul	5	3:52	10.00	-26.9	-159.2
2017	May	26	2:28	6.35	-37.2	-144.8	2017	May	26	6:22	8.00	-25.3	163.1	2017	May	26	10:25	10.00	-17.4	105.8
2017	Jun	24	11:11	6.35	-34.0	85.6	2017	Jun	24	14:35	8.00	-25.8	39.2	2017	Jun	24	18:24	10.00	-20.0	-15.1
2018	May	15	19:20	6.35	-32.0	-34.1	2018	May	15	23:20	8.00	-20.4	-89.2	2018	May	16	3:33	10.00	-12.5	-148.7
2018	Jun	14	5:07	6.35	-24.8	-178.3	2018	Jun	14	8:22	8.00	-17.8	136.3	2018	Jun	14	12:05	10.00	-12.7	82.8
2019	May	5	7:51	6.35	-25.6	143.1	2019	May	5	12:09	8.00	-14.8	82.0	2019	May	5	16:46	10.00	-7.4	14.9
2019	Jun	3	20:57	6.35	-15.1	-51.9	2019	Jun	4	0:21	8.00	-9.4	-99.5	2019	Jun	4	4:17	10.00	-5.2	-157.0
2020	Apr	23	13:12	6.35	-18.7	66.8	2020	Apr	23	17:42	8.00	-8.9	1.4	2020	Apr	23	22:38	10.00	-2.1	-71.4
2020	May	23	6:26	6.35	-5.2	170.6	2020	May	23	10:05	8.00	-0.8	117.4	2020	May	23	14:24	10.00	2.5	53.7
2021	Apr	12	14:02	6.35	-11.9	57.2	2021	Apr	12	18:21	8.00	-3.3	-6.2	2021	Apr	12	23:13	10.00	2.8	-78.5
2021	May	12	8:50	6.35	4.3	137.7	2021	May	12	12:36	8.00	7.4	82.2	2021	May	12	17:06	10.00	9.8	15.3
2022	Apr	1	17:34	6.35	-5.2	6.5	2022	Apr	1	21:20	8.00	2.1	-49.7	2022	Apr	2	1:42	10.00	7.4	-113.8
2022	May	1	9:50	6.35	13.3	125.1	2022	May	1	13:22	8.00	15.3	72.5	2022	May	1	17:38	10.00	16.8	8.8
2023	Mar	22	3:52	6.35	1.4	-146.3	2023	Mar	22	7:05	8.00	7.3	165.3	2023	Mar	22	10:53	10.00	11.7	108.3
2023	Apr	20	16:11	6.35	22.2	31.2	2023	Apr	20	19:19	8.00	22.9	-15.7	2023	Apr	20	23:07	10.00	23.4	-72.7
2024	Mar	10	19:19	6.35	7.6	-18.4	2024	Mar	10	22:14	8.00	12.0	-62.6	2024	Mar	11	1:43	10.00	15.4	-114.3
2024	Apr	9	5:08	6.35	30.6	-162.0	2024	Apr	9	7:58	8.00	30.0	155.2	2024	Apr	9	11:24	10.00	29.5	103.4
2025	Feb	28	11:46	6.35	12.9	94.5	2025	Feb	28	13:24	7.26	14.8	69.7	2025	Feb	28	18:15	10.00	18.3	-3.9
2025	Mar	29	21:23	6.35	38.1	-47.5	2025	Mar	30	0:13	8.00	36.1	-89.2	2025	Mar	30	3:37	10.00	34.6	-140.5
2026	Feb	18	0:28	6.35	17.4	-97.2	2026	Feb	18	3:42	8.00	19.1	-146.2	2026	Feb	18	7:36	10.00	20.4	154.9
2026	Mar	19	12:18	6.35	44.7	86.5	2026	Mar	19	15:24	8.00	41.3	40.0	2026	Mar	19	19:07	10.00	38.7	-15.9
2027	Feb	7	5:43	6.35	20.5	-178.3	2027	Feb	7	9:17	8.00	21.0	128.0	2027	Feb	7	13:35	10.00	21.4	63.1
2027	Mar	8	21:13	6.35	49.7	-51.9	2027	Mar	9	0:48	8.00	45.0	-104.0	2027	Mar	9	5:03	10.00	41.5	-167.2
2028	Jan	27	5:06	6.35	22.2	-171.4	2028	Jan	27	8:45	8.00	21.6	133.9	2028	Jan	27	13:10	10.00	21.2	67.7

Centurial Ramaḑān Calendar																				
Ramaḑān (Ccr=6.35)							Ramaḑān (Ccr=8.0)							Ramaḑān (Ccr=10.0)						
Y	M	D	Time	Ccr	Lat	Long	Y	M	D	Time	Ccr	Lat.	Long	Y	M	D	Time	Ccr	Lat	Long
2028	Feb	25	22:35	6.35	52.7	-78.3	2028	Feb	26	2:35	8.00	46.9	-135.1	2028	Feb	26	7:15	10.00	42.6	156.0
2029	Jan	15	5:57	6.35	23.3	173.5	2029	Jan	15	9:21	8.00	21.6	123.0	2029	Jan	15	13:29	10.00	20.3	61.5
2029	Feb	13	21:27	6.35	54.1	-67.7	2029	Feb	14	1:30	8.00	47.3	-123.4	2029	Feb	14	6:10	10.00	42.3	168.8
2030	Jan	4	13:36	6.35	24.2	55.9	2030	Jan	4	16:41	8.00	21.4	10.9	2030	Jan	4	20:24	10.00	19.3	-43.8
2030	Feb	3	1:17	6.35	54.3	-130.4	2030	Feb	3	5:06	8.00	46.6	178.5	2030	Feb	3	9:24	10.00	40.9	117.2
2030	Dec	25	3:10	6.35	25.0	-149.4	2030	Dec	25	6:07	8.00	21.0	168.6	2030	Dec	25	9:37	10.00	17.9	117.5
2031	Jan	23	12:14	6.35	53.6	60.5	2031	Jan	23	15:47	7.98	45.0	15.0	2031	Jan	23	19:47	10.00	38.7	-40.9
2031	Dec	14	18:30	6.35	25.6	-21.6	2031	Dec	14	21:37	8.00	20.4	-65.6	2031	Dec	15	1:15	10.00	16.5	-117.4
2032	Jan	13	3:12	6.35	51.9	-166.2	2032	Jan	13	6:50	8.00	42.4	148.1	2032	Jan	13	10:47	10.00	35.7	93.7
2032	Dec	3	6:43	6.35	26.4	154.7	2032	Dec	3	10:19	8.00	19.8	103.8	2032	Dec	3	14:27	10.00	15.0	43.9
2033	Jan	1	17:32	6.35	49.6	-22.6	2033	Jan	1	21:36	8.00	39.3	-74.2	2033	Jan	2	1:53	10.00	32.1	-132.6
2033	Nov	22	11:58	6.35	26.9	75.8	2033	Nov	22	16:10	8.00	18.9	16.1	2033	Nov	22	20:54	10.00	13.2	-52.6
2033	Dec	22	2:45	6.35	46.4	-158.9	2033	Dec	22	7:27	8.00	35.4	139.7	2033	Dec	22	12:20	10.00	27.8	71.1
2034	Nov	11	11:17	6.35	27.2	86.9	2034	Nov	11	15:50	8.00	18.0	22.0	2034	Nov	11	20:52	10.00	11.4	-51.5
2034	Dec	11	4:56	6.35	42.3	171.4	2034	Dec	11	10:04	8.00	30.8	102.3	2034	Dec	11	15:24	10.00	23.0	26.4
2035	Oct	31	11:45	6.35	28.0	81.6	2035	Oct	31	16:12	8.00	17.3	17.8	2035	Oct	31	21:02	10.00	10.0	-53.0
2035	Nov	30	4:21	6.35	37.7	-176.2	2035	Nov	30	9:22	8.00	26.0	115.0	2035	Nov	30	14:37	10.00	18.1	39.9
2036	Oct	19	19:12	6.35	28.9	-28.3	2036	Oct	19	23:19	8.00	16.8	-87.8	2036	Oct	20	3:42	10.00	8.8	-151.2
2036	Nov	18	8:28	6.35	32.5	126.0	2036	Nov	18	12:54	8.00	20.9	64.6	2036	Nov	18	17:35	10.00	13.1	-3.0
2037	Oct	9	9:11	6.35	29.1	125.1	2037	Oct	9	13:05	8.00	16.1	68.1	2037	Oct	9	17:10	10.00	7.6	7.7
2037	Nov	7	19:57	6.35	26.4	-42.6	2037	Nov	7	23:45	8.00	15.5	-96.1	2037	Nov	8	3:52	10.00	8.0	-154.7
2038	Sep	29	1:39	6.35	28.5	-118.6	2038	Sep	29	5:38	8.00	15.2	-177.9	2038	Sep	29	9:46	10.00	6.4	120.3
92038	Oct	28	12:05	6.35	19.6	79.4	2038	Oct	28	15:34	8.00	9.7	29.5	2038	Oct	28	19:24	10.00	2.8	-26.6
2039	Sep	18	16:01	6.35	26.9	28.6	2039	Sep	18	20:21	8.00	13.8	-37.2	2039	Sep	19	0:51	10.00	5.2	-104.2
2039	Oct	18	4:30	6.35	12.6	-163.5	2039	Oct	18	8:00	8.00	3.9	145.2	2039	Oct	18	11:57	10.00	-2.2	87.0
2040	Sep	7	0:19	6.35	24.4	-92.7	2040	Sep	7	5:02	8.00	12.1	-165.1	2040	Sep	7	10:00	10.00	3.9	119.3
2040	Oct	6	16:38	6.35	5.4	16.3	2040	Oct	6	20:22	8.00	-1.9	-39.3	2040	Oct	7	0:40	10.00	-7.2	-102.3
2041	Aug	27	2:29	6.35	21.5	-123.1	2041	Aug	27	7:15	8.00	10.5	163.2	2041	Aug	27	12:23	10.00	2.9	84.5
2041	Sep	25	21:34	6.35	-1.3	-56.5	2041	Sep	26	1:30	8.00	-7.2	-114.5	2041	Sep	26	5:59	9.95	-11.5	178.1
2042	Aug	16	4:24	6.35	18.8	-150.4	2042	Aug	16	8:43	8.00	9.1	142.1	2042	Aug	16	13:31	10.00	2.3	68.2
2042	Sep	14	22:09	6.35	-7.1	-64.8	2042	Sep	15	1:58	8.00	-11.7	-121.5	2042	Sep	15	6:30	10.00	-15.1	170.1
2043	Aug	5	12:20	6.35	15.7	91.0	2043	Aug	5	16:00	8.00	7.6	33.2	2043	Aug	5	20:10	10.00	1.7	-31.3

Centurial Ramaḍān Calendar

Ramaḍān (Ccr=6.35)							Ramaḍān (Ccr=8.0)							Ramaḍān (Ccr=10.0)						
Y	M	D	Time	Ccr	Lat	Long	Y	M	D	Time	Ccr	Lat.	Long	Y	M	D	Time	Ccr	Lat	Long
2043	Sep	4	1:45	6.35	-12.4	-118.1	2043	Sep	4	5:08	8.00	-15.6	-169.5	2043	Sep	4	9:14	10.00	-18.1	128.6
2044	Jul	25	2:52	6.35	12.5	-127.1	2044	Jul	25	6:03	8.00	5.9	-177.2	2044	Jul	25	9:44	10.00	1.1	125.7
2044	Aug	23	12:23	6.35	-17.4	80.9	2044	Aug	23	15:21	8.00	-19.1	35.9	2044	Aug	23	18:57	10.00	-20.5	-18.6
2045	Jul	14	20:37	6.35	9.5	-34.9	2045	Jul	14	23:38	8.00	4.5	-82.1	2045	Jul	15	3:10	10.00	0.7	-135.9
2045	Aug	13	4:20	6.35	-21.5	-159.9	2045	Aug	13	7:06	8.00	-22.0	158.4	2045	Aug	13	10:28	10.00	-22.3	107.6
2046	Jul	4	13:03	6.35	6.8	77.7	2046	Jul	4	16:12	8.00	3.2	28.9	2046	Jul	4	19:56	10.00	0.5	-28.4
2046	Aug	2	21:23	6.35	-24.9	-58.5	2046	Aug	3	0:15	8.00	-24.0	-100.3	2046	Aug	3	3:44	10.00	-23.3	-152.5
2047	Jun	23	23:40	6.35	4.5	-83.4	2047	Jun	24	3:06	8.00	2.2	-135.0	2047	Jun	24	7:13	10.00	0.5	162.2
2047	Jul	23	10:41	6.35	-27.5	100.0	2047	Jul	23	13:56	8.00	-25.3	52.2	2047	Jul	23	17:49	10.00	-23.7	-5.6
2048	Jun	12	2:53	6.35	2.8	-132.1	2048	Jun	12	6:31	8.00	1.7	172.6	2048	Jun	12	10:56	10.00	0.9	106.0
2048	Jul	11	16:41	6.35	-29.1	7.5	2048	Jul	11	20:22	8.00	-25.7	-46.0	2048	Jul	12	0:45	10.00	-23.1	-109.8
2049	Jun	1	3:31	6.35	1.2	-142.8	2049	Jun	1	7:03	8.00	1.3	164.0	2049	Jun	1	11:20	10.00	1.4	99.6
2049	Jun	30	17:06	6.35	-29.8	-0.3	2049	Jun	30	20:59	8.00	-25.3	-56.3	2049	Jun	1	1:35	10.00	-21.8	-122.6
2050	May	21	8:45	6.35	-1.0	137.2	2050	May	21	11:57	8.00	0.4	89.5	2050	May	21	15:50	10.00	1.5	31.5
2050	Jun	19	19:03	6.35	-30.5	-30.9	2050	Jun	19	22:49	8.00	-24.7	-84.2	2050	Jun	20	3:12	10.00	-20.3	-146.9
2051	May	10	20:52	6.35	-3.8	-45.9	2051	May	10	23:48	8.00	-1.0	-89.2	2051	May	11	3:20	10.00	1.1	-140.6
2051	Jun	9	3:48	6.35	-31.4	-162.4	2051	Jun	9	7:19	8.00	-24.1	148.8	2051	Jun	9	11:20	10.00	-18.8	91.0
2052	Apr	29	13:00	6.35	-7.0	71.9	2052	Apr	29	15:56	8.00	-2.7	29.0	2052	Apr	29	19:24	10.00	0.5	-22.5
2052	May	28	18:31	6.35	-32.1	-23.5	2052	May	28	21:57	8.00	-23.4	-70.7	2052	May	29	1:47	10.00	-17.1	-124.4
2053	Apr	19	4:35	6.35	-10.6	-161.4	2053	Apr	19	7:49	8.00	-4.8	151.3	2053	Apr	19	11:35	10.00	-0.5	95.3
2053	May	18	11:03	6.35	-32.6	89.7	2053	May	18	14:44	8.00	-22.4	39.0	2053	May	18	18:43	10.00	-15.3	-18.1
2054	Apr	8	14:51	6.35	-14.3	45.2	2054	Apr	8	18:38	8.00	-7.0	-10.9	2054	Apr	8	22:59	10.00	-1.7	-75.6
2054	May	8	0:40	6.35	-32.6	-112.3	2054	May	8	4:56	8.00	-21.0	-171.8	2054	May	8	9:25	10.00	-13.2	123.4
2055	Mar	28	17:32	6.35	-17.9	6.5	2055	Mar	28	21:51	8.00	-9.2	-58.0	2055	Mar	29	2:43	10.00	-3.0	-129.9
2055	Apr	27	7:37	6.35	-31.7	146.2	2055	Apr	27	12:30	8.00	-19.1	76.4	2055	Apr	27	17:35	10.00	-10.8	2.1
2056	Mar	16	16:37	6.35	-21.5	22.7	2056	Mar	16	21:06	8.00	-11.5	-45.0	2056	Mar	17	2:04	10.00	-4.5	-118.9
2056	Apr	15	8:32	6.35	-29.9	135.8	2056	Apr	15	13:41	8.00	-16.9	61.1	2056	Apr	15	19:02	10.00	-8.3	-17.8
2057	Mar	5	19:40	6.35	-25.4	-20.5	2057	Mar	5	23:56	8.00	-14.1	-85.9	2057	Mar	6	4:33	10.00	-6.4	-155.1
2057	Apr	4	9:53	6.35	-27.6	118.7	2057	Apr	4	14:43	8.00	-14.6	47.7	2057	Apr	4	19:46	10.00	-6.0	-27.2
2058	Feb	23	5:58	6.35	-28.8	-171.2	2058	Feb	23	9:56	8.00	-16.6	126.7	2058	Feb	23	13:43	9.79	-9.1	68.6
2058	Mar	24	17:41	6.35	-24.4	4.6	2058	Mar	24	21:52	8.00	-12.0	-58.0	2058	Mar	25	2:19	10.00	-3.7	-123.6
2059	Feb	12	21:01	6.35	-31.5	-34.2	2059	Feb	13	0:54	8.00	-18.8	-95.4	2059	Feb	13	4:56	10.00	-10.4	-158.4

Centurial Ramaḍān Calendar																				
Ramaḍān (Ccr=6.35)							Ramaḍān (Ccr=8.0)							Ramaḍān (Ccr=10.0)						
Y	M	D	Time	Ccr	Lat	Long	Y	M	D	Time	Ccr	Lat.	Long	Y	M	D	Time	Ccr	Lat	Long
2059	Mar	14	7:51	6.35	-20.5	155.1	2059	Mar	14	11:31	8.00	-9.2	99.2	2059	Mar	14	15:30	10.00	-1.4	39.1
2060	Feb	2	12:20	6.35	-33.5	99.2	2060	Feb	2	15:07	7.42	-24.4	53.8	2060	Feb	2	20:44	10.00	-12.4	-34.8
2060	Mar	3	0:36	6.35	-16.3	-94.3	2060	Mar	3	4:07	8.00	-6.4	-148.3	2060	Mar	3	7:59	10.00	0.6	152.6
2061	Jan	21	23:21	6.35	-34.6	-64.3	2061	Jan	22	3:56	8.00	-22.5	-138.0	2061	Jan	22	8:42	10.00	-14.4	147.2
2061	Feb	20	18:11	7.58	-5.5	0.7	2061	Feb	20	19:05	8.00	-3.8	-13.0	2061	Feb	20	23:11	10.00	2.3	-76.0
2062	Jan	11	3:16	6.35	-35.3	-120.9	2062	Jan	11	8:11	8.00	-24.1	159.1	2062	Jan	11	13:22	10.00	-16.5	77.6
2062	Feb	10	0:00	6.35	-8.7	-84.4	2062	Feb	10	3:55	8.00	-1.7	-145.2	2062	Feb	10	8:25	10.00	3.5	145.8
2062	Dec	31	2:56	6.35	-36.1	-115.5	2062	Dec	31	7:42	8.00	-25.9	166.6	2062	Dec	31	12:51	10.00	-18.8	85.7
2063	Jan	30	1:28	6.35	-6.2	-107.0	2063	Jan	30	5:29	8.00	-0.5	-169.1	2063	Jan	30	10:10	10.00	3.8	119.1
2063	Dec	20	5:45	6.35	-36.7	-158.5	2063	Dec	20	9:56	8.00	-27.5	132.9	2063	Dec	20	14:32	10.00	-21.0	60.1
2064	Jan	19	1:30	6.35	-4.4	-108.5	2064	Jan	19	5:13	8.00	0.0	-166.1	2064	Jan	19	9:38	10.00	3.4	126.4
2064	Dec	8	15:44	6.35	-36.5	49.5	2064	Dec	8	19:16	8.00	-28.5	-8.7	2064	Dec	8	23:16	10.00	-22.7	-71.8
2064	Nov	9	2:35	6.35	-61.7	-98.3	2064	Nov	9	6:32	8.00	-50.5	-170.6	2064	Nov	9	10:40	10.00	-42.6	121.9
2065	Nov	28	7:00	6.35	-35.7	178.0	2065	Nov	28	10:10	8.00	-28.9	126.4	2065	Nov	28	13:50	10.00	-23.8	68.9
2065	Dec	27	19:19	6.35	-1.3	-19.9	2065	Dec	27	22:11	8.00	0.7	-63.8	2065	Dec	28	1:39	10.00	2.2	-115.6
2066	Nov	17	23:22	6.35	-34.3	-71.6	2066	Nov	18	2:32	8.00	-28.6	-121.0	2066	Nov	18	6:14	10.00	-24.4	-178.4
2066	Dec	17	11:05	6.35	0.1	102.2	2066	Dec	17	13:53	8.00	0.9	59.8	2066	Dec	17	17:16	10.00	1.5	8.6
2067	Nov	7	12:11	6.35	-32.1	93.1	2067	Nov	7	15:35	8.00	-27.7	40.2	2067	Nov	7	19:37	10.00	-24.2	-21.6
2067	Dec	7	1:42	6.35	1.7	-118.5	2067	Dec	7	4:43	8.00	1.2	-163.7	2067	Dec	7	8:22	10.00	0.9	141.5
2068	Oct	26	17:51	6.35	-29.3	4.7	2068	Oct	26	21:32	8.00	-26.0	-51.6	2068	Oct	27	1:55	10.00	-23.5	-117.2
2068	Nov	25	10:29	6.35	3.2	107.7	2068	Nov	25	13:55	8.00	1.6	56.8	2068	Nov	25	18:02	10.00	0.4	-4.8
2069	Oct	15	18:02	6.35	-25.9	-0.5	2069	Oct	15	21:43	8.00	-23.8	-56.3	2069	Oct	16	2:10	10.00	-22.2	-122.6
2069	Nov	14	11:55	6.35	4.7	85.2	2069	Nov	14	15:39	8.00	1.9	30.0	2069	Nov	14	20:08	10.00	-0.2	-36.9
2070	Oct	4	19:56	6.35	-21.4	-31.1	2070	Oct	4	23:18	8.00	-20.5	-81.8	2070	Oct	5	3:24	10.00	-19.8	-142.5
2070	Nov	3	11:03	6.35	6.7	97.6	2070	Nov	3	14:46	8.00	2.7	42.8	2070	Nov	3	19:12	10.00	-0.4	-23.0
2071	Sep	24	4:44	6.35	-15.4	-163.2	2071	Sep	24	7:44	8.00	-15.9	151.8	2071	Sep	24	11:22	10.00	-16.3	97.0
2071	Oct	23	15:19	6.35	9.7	33.3	2071	Oct	23	18:47	8.00	4.2	-17.4	2071	Oct	23	22:51	10.00	0.1	-77.8
2072	Sep	12	19:30	6.35	-8.7	-25.2	2072	Sep	12	22:19	8.00	-10.6	-67.6	2072	Sep	13	1:43	10.00	-12.0	-117.7
2072	Oct	12	2:49	6.35	13.4	-137.8	2072	Oct	12	6:03	8.00	6.4	174.4	2072	Oct	12	9:48	10.00	1.3	118.8
2073	Sep	2	12:09	6.35	-1.5	86.6	2073	Sep	2	15:04	8.00	-4.8	42.3	2073	Sep	2	18:35	10.00	-7.3	-10.8
2073	Oct	1	18:28	6.35	17.4	-11.7	2073	Oct	1	21:46	8.00	8.9	-60.8	2073	Oct	2	1:29	10.00	2.7	-115.5
2074	Aug	23	1:50	6.35	6.1	-116.0	2074	Aug	23	5:10	8.00	1.5	-167.0	2074	Aug	23	9:06	10.00	-2.1	133.1

Centurial Ramaḍān Calendar

Ramaḍān (Ccr=6.35)							Ramaḍān (Ccr=8.0)							Ramaḍān (Ccr=10.0)						
Y	M	D	Time	Ccr	Lat	Long	Y	M	D	Time	Ccr	Lat.	Long	Y	M	D	Time	Ccr	Lat	Long
2074	Sep	21	9:35	6.35	21.5	124.1	2074	Sep	21	13:17	8.00	11.5	68.2	2074	Sep	21	17:22	10.00	4.5	6.7
2075	Aug	12	8:37	6.35	13.7	145.2	2075	Aug	12	12:28	8.00	7.7	85.6	2075	Aug	12	16:57	10.00	3.2	16.8
2075	Sep	10	19:35	6.35	25.4	-23.3	2075	Sep	10	23:58	8.00	14.0	-89.4	2075	Sep	11	4:39	10.00	6.3	-160.6
2076	Jul	31	9:14	6.35	21.0	139.7	2076	Jul	31	13:25	8.00	13.8	74.2	2076	Jul	31	18:13	10.00	8.5	0.0
2076	Aug	29	22:34	6.35	28.5	-64.7	2076	Aug	30	3:30	8.00	16.2	-140.3	2076	Aug	30	8:43	10.00	8.0	140.0
2077	Jul	20	10:25	6.35	28.3	126.2	2077	Jul	20	14:34	8.00	19.9	60.1	2077	Jul	20	19:13	10.00	13.9	-12.3
2077	Aug	18	22:45	6.35	31.0	-63.8	2077	Aug	19	3:46	8.00	18.3	-141.7	2077	Aug	19	8:59	10.00	9.8	137.6
2078	Jul	9	18:12	6.35	35.4	14.3	2078	Jul	9	22:05	8.00	25.9	-49.6	2078	Jul	10	2:21	10.00	19.1	-116.0
2078	Aug	8	3:30	6.35	32.9	-130.8	2078	Aug	8	8:05	8.00	20.2	155.6	2078	Aug	8	12:51	10.00	11.7	81.2
2079	Jun	29	8:29	6.35	41.5	165.4	2079	Jun	29	12:13	8.00	31.2	101.8	2079	Jun	29	16:13	10.00	23.9	37.6
2079	Jul	28	15:41	6.35	33.6	48.6	2079	Jul	28	19:43	8.00	21.4	-17.6	2079	Jul	28	23:57	10.00	13.3	-83.3
2080	Jun	18	1:24	6.35	46.7	-83.5	2080	Jun	18	5:17	8.00	35.8	-151.2	2080	Jun	18	9:22	10.00	28.2	142.7
2080	Jul	17	8:40	6.35	33.2	155.6	2080	Jul	17	12:24	8.00	22.1	93.8	2080	Jul	17	16:22	10.00	14.5	30.8
2081	Jun	7	16:18	6.35	50.9	55.7	2081	Jun	7	20:38	8.00	39.8	-20.3	2081	Jun	8	1:07	10.00	32.0	-92.0
2081	Jul	7	2:05	6.35	32.3	-105.1	2081	Jul	7	5:50	8.00	22.4	-166.8	2081	Jul	7	9:54	10.00	15.5	128.8
2082	May	28	1:10	6.35	53.9	-75.8	2082	May	28	6:00	8.00	42.9	-159.7	2082	May	28	11:01	10.00	35.2	119.5
2082	Jun	26	15:23	6.35	31.0	54.2	2082	Jun	26	19:21	8.00	22.5	-10.2	2082	Jun	26	23:46	10.00	16.4	-79.6
2083	May	17	3:34	6.35	56.1	-113.3	2083	May	17	8:34	8.00	45.4	160.9	2083	May	17	13:50	10.00	37.9	76.6
2083	Jun	15	21:32	6.35	29.7	-39.7	2083	Jun	16	1:41	8.00	22.5	-105.0	2083	Jun	16	6:24	10.00	17.3	-178.7
2084	May	5	5:03	6.35	57.5	-139.3	2084	May	5	9:40	8.00	47.2	142.2	2084	May	5	14:39	10.00	39.8	62.9
2084	Jun	3	22:52	6.35	28.6	-61.6	2084	Jun	4	2:53	8.00	22.7	-124.0	2084	Jun	4	7:33	10.00	18.2	163.6
2085	Apr	24	12:07	6.35	57.2	108.1	2085	Apr	24	16:05	8.00	47.6	42.2	2085	Apr	24	20:27	10.00	40.7	-26.6
2085	May	24	2:29	6.35	27.1	-117.3	2085	May	24	6:03	8.00	22.4	-172.9	2085	May	24	10:16	10.00	18.8	122.1
2086	Apr	14	1:50	6.35	54.9	-104.0	2086	Apr	14	5:14	8.00	46.3	-158.8	2086	Apr	14	9:05	10.00	40.1	141.4
2086	May	13	12:49	6.35	24.6	84.7	2086	May	13	15:53	8.00	21.2	37.3	2086	May	13	19:35	10.00	18.6	-19.1
2087	Apr	3	18:52	6.35	51.0	-6.5	2087	Apr	3	22:03	8.00	43.6	-55.9	2087	Apr	4	1:43	10.00	38.3	-110.9
2087	May	3	4:31	6.35	21.4	-152.8	2087	May	3	7:20	8.00	19.4	164.3	2087	May	3	10:44	10.00	17.7	112.6
2088	Mar	23	10:47	6.35	45.6	110.6	2088	Mar	23	14:04	8.00	39.7	60.9	2088	Mar	23	17:54	10.00	35.3	3.0
2088	Apr	21	21:30	6.35	17.7	-50.1	2088	Apr	22	0:21	8.00	16.9	-92.3	2088	Apr	22	3:49	10.00	16.3	-144.6
2089	Mar	12	21:00	6.35	39.1	-46.2	2089	Mar	13	0:34	8.00	34.6	-98.5	2089	Mar	13	4:47	10.00	31.2	-161.7
2089	Apr	11	10:58	6.35	13.2	107.0	2089	Apr	11	14:08	8.00	13.7	59.3	2089	Apr	11	17:58	10.00	14.2	1.8
2090	Mar	1	23:39	6.35	31.9	-87.5	2090	Mar	2	3:24	8.00	28.7	-142.4	2090	Mar	2	7:53	10.00	26.3	150.5

Centurial Ramaḍān Calendar																				
Ramaḍān (Ccr=6.35)							Ramaḍān (Ccr=8.0)							Ramaḍān (Ccr=10.0)						
Y	M	D	Time	Ccr	Lat	Long	Y	M	D	Time	Ccr	Lat.	Long	Y	M	D	Time	Ccr	Lat	Long
2090	Mar	31	17:04	6.35	8.3	14.6	2090	Mar	31	20:37	8.00	10.2	-38.8	2090	Mar	1	0:55	10.00	11.6	-102.4
2091	Feb	18	23:16	6.35	24.2	-82.1	2091	Feb	19	2:52	8.00	22.3	-134.8	2091	Feb	19	7:13	10.00	20.8	160.1
2091	Mar	20	16:57	6.35	3.2	16.5	2091	Mar	20	20:41	8.00	6.3	-39.7	2091	Mar	21	1:11	10.00	8.7	-106.5
2092	Feb	8	3:19	6.35	15.7	-141.0	2092	Feb	8	6:31	8.00	14.9	170.9	2092	Feb	8	10:25	10.00	14.4	112.4
2092	Mar	8	18:22	6.66	-1.6	-3.9	2092	Mar	8	21:16	8.00	1.8	-47.9	2092	Mar	9	1:32	10.00	5.2	-111.4
2093	Jan	27	14:18	6.35	6.6	55.5	2093	Jan	27	17:27	8.15	7.1	8.0	2093	Jan	27	20:40	10.00	7.5	-40.6
2093	Feb	26	1:00	6.35	-9.1	-101.0	2093	Feb	26	4:18	8.00	-3.4	-151.4	2093	Feb	26	8:11	10.00	1.0	149.6
2094	Jan	17	5:30	6.35	-2.6	-169.6	2094	Jan	17	8:18	8.00	-0.9	147.5	2094	Jan	17	11:41	10.00	0.5	96.1
2094	Feb	15	14:12	6.26	-16.3	63.2	2094	Feb	15	18:04	8.28	-7.8	3.0	2094	Feb	15	21:12	10.00	-3.5	-45.1
2095	Jan	6	20:21	6.35	-11.9	-30.1	2095	Jan	6	23:23	8.00	-9.0	-76.9	2095	Jan	7	3:01	10.00	-6.8	-131.5
2095	Feb	5	5:45	6.35	-22.4	-166.6	2095	Feb	5	9:07	8.00	-14.1	140.1	2095	Feb	5	12:55	10.00	-8.1	81.0
2095	Dec	27	6:03	6.35	-21.2	-171.5	2095	Dec	27	9:33	8.00	-17.1	133.9	2095	Dec	27	13:42	10.00	-14.0	70.0
2096	Jan	25	18:42	6.51	-27.8	1.8	2096	Jan	25	22:12	8.00	-19.5	-54.4	2096	Jan	26	2:30	10.00	-12.8	-120.6
2096	Dec	15	8:10	6.35	-30.1	160.1	2096	Dec	15	12:06	8.00	-25.0	98.0	2096	Dec	15	16:44	10.00	-21.0	26.4
2097	Jan	14	0:03	6.35	-34.9	-73.2	2097	Jan	14	4:37	8.00	-24.6	-147.4	2097	Jan	14	9:33	10.00	-17.5	135.1
2097	Dec	4	6:42	6.35	-38.7	-174.4	2097	Dec	4	10:45	8.00	-32.5	120.6	2097	Dec	4	15:27	10.00	-27.8	47.4
2098	Jan	2	23:33	6.35	-40.3	-62.7	2098	Jan	3	4:31	8.00	-29.5	-143.4	2098	Jan	3	9:47	10.00	-22.0	133.4
2098	Nov	23	9:23	6.35	-47.2	148.8	2098	Nov	23	13:12	8.00	-39.8	85.7	2098	Nov	23	17:35	10.00	-34.4	16.7
2098	Dec	22	23:35	6.35	-45.4	-59.4	2098	Dec	23	4:23	8.00	-34.3	-139.5	2098	Dec	23	9:25	10.00	-26.6	140.2
2099	Nov	12	19:27	6.35	-55.1	0.6	2099	Nov	12	23:02	8.00	-46.5	-61.0	2099	Nov	13	3:02	10.00	-40.3	-124.1
2099	Dec	12	6:21	6.35	-49.7	-157.4	2099	Dec	12	10:41	8.00	-38.5	127.1	2099	Dec	12	15:11	10.00	-30.8	54.3
2100	Nov	2	10:24	6.35	-62.0	138.9	2100	Nov	2	14:00	8.00	-52.3	74.9	2100	Nov	2	17:54	10.00	-45.2	11.8
2100	Dec	1	19:48	6.35	-52.9	1.1	2100	Dec	1	23:44	8.00	-41.9	-69.0	2100	Dec	2	3:50	10.00	-34.2	-134.9
2101	Oct	23	1:40	6.35	-68.0	-89.9	2101	Oct	23	5:38	8.00	-57.0	-161.1	2101	Oct	23	9:49	10.00	-49.2	131.3
2101	Nov	21	12:00	6.35	-55.2	116.8	2101	Nov	21	15:52	8.00	-44.5	48.2	2101	Nov	21	19:55	10.00	-36.9	-17.8
2102	Oct	12	12:40	6.35	-72.5	99.8	2102	Oct	12	17:14	8.00	-60.2	20.0	2102	Oct	12	22:00	10.00	-51.8	-55.1
2102	Nov	11	2:25	6.35	-56.3	-103.2	2102	Nov	11	6:29	8.00	-46.1	-173.2	2102	Nov	11	10:50	10.00	-38.9	117.0
2103	Oct	1	16:46	6.35	-74.6	26.4	2103	Oct	1	21:51	8.00	-61.5	-55.7	2103	Oct	2	3:06	10.00	-52.8	-135.3
2103	Oct	31	10:53	6.35	-56.1	123.5	2103	Oct	31	15:14	8.00	-46.7	51.7	2103	Oct	31	20:00	10.00	-39.8	-23.4
2104	Sep	19	17:02	6.35	-73.8	7.5	2104	Sep	19	22:06	8.00	-60.7	-66.5	2104	Sep	20	3:22	10.00	-52.1	-144.2
2104	Oct	19	12:46	6.35	-54.8	88.4	2104	Oct	19	17:10	8.00	-46.1	18.3	2104	Oct	19	22:06	10.00	-39.9	-58.1
2105	Sep	8	20:47	6.35	-70.2	-59.0	2105	Sep	9	1:18	8.00	-58.0	-119.5	2105	Sep	9	6:16	10.07	-49.5	168.5

Centurial Ramaḍān Calendar																				
Ramaḍān (Ccr=6.35)							Ramaḍān (Ccr=8.0)							Ramaḍān (Ccr=10.0)						
Y	M	D	Time	Ccr	Lat	Long	Y	M	D	Time	Ccr	Lat.	Long	Y	M	D	Time	Ccr	Lat	Long
2105	Oct	8	13:47	6.35	-51.8	66.8	2105	Oct	8	17:48	8.00	-44.2	4.5	2105	Oct	8	22:26	10.00	-38.7	-66.1
2106	Aug	29	7:51	6.35	-64.0	131.8	2106	Aug	29	11:44	8.00	-53.2	80.6	2106	Aug	29	15:55	10.00	-45.7	20.7
2106	Sep	27	20:34	6.35	-46.8	-40.3	2106	Sep	28	0:01	8.00	-40.6	-91.4	2106	Sep	28	4:03	10.00	-36.0	-152.3
2107	Aug	19	0:20	6.35	-56.3	-114.7	2107	Aug	19	3:48	8.00	-47.0	-160.9	2107	Aug	19	7:38	10.00	-40.4	144.5
2107	Sep	17	10:10	6.35	-40.2	113.2	2107	Sep	17	13:11	8.00	-35.5	68.3	2107	Sep	17	16:45	10.00	-31.9	14.8
2108	Aug	7	18:00	6.35	-47.7	-18.4	2108	Aug	7	21:24	8.00	-39.9	-65.0	2108	Aug	8	1:15	10.00	-34.2	-119.1
2108	Sep	6	3:18	6.35	-32.6	-144.6	2108	Sep	6	6:10	8.00	-29.3	172.8	2108	Sep	6	9:36	10.00	-26.8	121.4
2109	Jul	28	8:16	6.35	-38.2	131.1	2109	Jul	28	11:52	8.00	-31.8	80.4	2109	Jul	28	16:00	10.00	-27.1	20.3
2109	Aug	26	19:22	6.35	-24.2	-26.1	2109	Aug	26	22:24	8.00	-22.4	-71.3	2109	Aug	27	2:04	10.00	-21.0	-125.0
2110	Jul	17	15:32	6.35	-28.3	25.5	2110	Jul	17	19:21	8.00	-23.3	-29.6	2110	Jul	17	23:50	10.00	-19.6	-95.4
2110	Aug	16	5:49	6.35	-15.3	179.3	2110	Aug	16	9:11	8.00	-14.8	128.8	2110	Aug	16	13:15	10.00	-14.5	67.6
2111	Jul	6	17:04	6.35	-18.4	6.0	2111	Jul	6	20:52	8.00	-14.7	-49.5	2111	Jul	7	1:24	10.00	-11.9	-115.4
2111	Aug	5	8:53	6.35	-6.3	135.4	2111	Aug	5	12:31	8.00	-7.2	80.5	2111	Aug	5	16:55	10.00	-7.8	14.1
2112	Jun	24	19:42	6.35	-8.3	-29.9	2112	Jun	24	23:09	8.00	-5.8	-80.7	2112	Jun	25	3:20	10.00	-4.0	-141.6
2112	Jul	24	9:16	6.35	2.7	132.6	2112	Jul	24	12:52	8.00	0.6	77.8	2112	Jul	24	17:12	10.00	-0.9	12.0
2113	Jun	14	4:49	6.35	2.1	-162.0	2113	Jun	14	7:50	8.00	3.2	153.0	2113	Jun	14	11:31	10.00	4.1	97.9
2113	Jul	13	14:16	6.35	12.1	61.2	2113	Jul	13	17:34	8.00	8.7	10.1	2113	Jul	13	21:33	10.00	6.2	-50.8

11.5 Photometric anniversary of the Crucifixion

Table 11.5-1 *Crucifixion* Remembrance dates with Orthogonal Illumination

Date					Time		Sun Ecliptic		Moon Ecliptic		Sun Equatorial		Moon Equatorial		
Julian Prior to 1582 CE			Gregorian		TDT	ΔT (s)	Long (deg)	Lat (deg)	Long (deg)	Lat (deg)	RA (deg)	DEC (deg)	RA (deg)	DEC (deg)	Elonga-tion (deg)
33	Apr	3	Apr	1	17:37:53	10202	11.49	0.05	190.11	-0.53	10.55	4.59	189.08	-4.49	178.54
41	Apr	5	Apr	3	9:43:36	10123	13.18	0.06	191.91	-1.59	12.10	5.25	190.32	-6.17	178.01
52	Apr	3	Apr	1	0:51:56	10015	11.21	0.05	189.83	-1.14	10.29	4.48	188.58	-4.94	178.24
60	Apr	4	Apr	2	20:12:30	9937	13.02	0.06	191.84	-0.95	11.96	5.19	190.51	-5.56	178.52
79	Apr	5	Apr	3	3:17:40	9752	12.73	0.05	191.49	-0.35	11.69	5.08	190.43	-4.86	178.73
98	Apr	4	Apr	2	15:36:48	9569	12.65	0.05	191.23	0.29	11.61	5.05	190.44	-4.17	178.54
117	Apr	4	Apr	3	11:52:04	9387	12.89	0.05	191.27	0.96	11.83	5.14	190.74	-3.58	178.10
125	Apr	5	Apr	4	21:18:32	9311	14.30	0.06	193.00	0.99	13.14	5.69	192.34	-4.22	178.33
144	Apr	5	Apr	4	19:28:30	9130	14.61	0.06	193.15	0.32	13.43	5.82	192.22	-4.89	178.48
163	Apr	6	Apr	5	20:06:21	8949	15.03	0.06	193.53	-0.36	13.82	5.98	192.31	-5.67	178.48
182	Apr	6	Apr	5	17:39:41	8769	15.32	0.06	193.93	-1.05	14.09	6.09	192.41	-6.46	178.29
190	Apr	8	Apr	7	3:34:57	8693	16.75	0.07	195.47	-1.13	15.41	6.65	193.81	-7.13	178.33
209	Apr	7	Apr	7	23:02:53	8513	16.96	0.07	195.78	-0.44	15.61	6.73	194.37	-6.62	178.76
228	Apr	7	Apr	7	10:28:01	8332	16.84	0.07	195.75	0.21	15.50	6.68	194.59	-6.00	178.87
247	Apr	7	Apr	7	17:29:14	8152	16.55	0.06	195.42	0.82	15.23	6.56	194.52	-5.31	178.56
274	Apr	8	Apr	8	20:15:43	7893	18.08	0.07	197.10	0.45	16.65	7.16	195.93	-6.30	178.89
293	Apr	8	Apr	8	10:09:32	7711	18.07	0.07	196.90	-0.20	16.63	7.15	195.50	-6.82	178.83
312	Apr	8	Apr	9	7:38:25	7527	18.35	0.07	197.04	-0.87	16.90	7.26	195.37	-7.50	178.46
367	Apr	1	Apr	2	15:26:47	6991	11.59	0.04	190.61	1.17	10.64	4.62	190.21	-3.12	178.44
386	Apr	1	Apr	2	9:55:38	6804	11.76	0.04	190.86	0.49	10.80	4.69	190.18	-3.85	178.96
432	Apr	1	Apr	2	22:33:01	6349	13.12	0.05	192.43	-1.31	12.05	5.22	190.92	-6.12	178.56
451	Apr	2	Apr	3	6:33:15	6161	12.86	0.04	192.10	-0.71	11.81	5.12	190.85	-5.44	178.99
470	Apr	1	Apr	2	21:36:16	5974	12.89	0.04	191.99	0.05	11.84	5.13	191.01	-4.79	179.10
489	Apr	1	Apr	2	19:48:58	5787	13.21	0.04	192.20	0.62	12.14	5.26	191.46	-4.25	178.79
497	Apr	3	Apr	4	4:06:04	5709	14.57	0.05	193.82	1.34	13.40	5.79	193.24	-4.22	178.42
508	Apr	1	Apr	3	20:25:03	5601	13.62	0.04	192.58	1.28	12.52	5.42	192.07	-3.79	178.31
516	Apr	3	Apr	5	3:39:59	5523	14.95	0.05	194.10	0.67	13.74	5.94	193.24	-4.94	178.89
535	Apr	4	Apr	6	3:53:12	5337	15.35	0.05	194.50	0.01	14.12	6.09	193.35	-5.72	179.15
554	Apr	3	Apr	5	23:18:17	5153	15.56	0.05	194.76	-0.70	14.31	6.17	193.32	-6.46	178.97
562	Apr	5	Apr	7	10:51:17	5076	17.06	0.05	196.41	-1.45	15.70	6.75	194.56	-7.79	178.46
573	Apr	3	Apr	5	10:40:47	4970	15.44	0.05	194.67	-1.33	14.20	6.12	192.99	-7.01	178.51

Date					Time		Sun Ecliptic		Moon Ecliptic		Sun Equatorial		Moon Equatorial		
Julian Prior to 1582 CE			Gregorian		TDT	ΔT (s)	Long (deg)	Lat (deg)	Long (deg)	Lat (deg)	RA (deg)	DEC (deg)	RA (deg)	DEC (deg)	Elonga-tion (deg)
581	Apr	5	Apr	7	3:48:48	4894	17.17	0.05	196.57	-0.78	15.80	6.79	194.96	-7.24	179.05
600	Apr	4	Apr	7	12:58:45	4713	16.96	0.05	196.36	-0.15	15.61	6.71	195.02	-6.58	179.40
619	Apr	4	Apr	7	20:30:48	4534	16.68	0.05	196.03	0.46	15.35	6.60	194.95	-5.88	179.17
638	Apr	4	Apr	7	10:29:18	4356	16.67	0.05	195.91	1.11	15.34	6.60	195.08	-5.24	178.61
646	Apr	6	Apr	9	0:04:06	4282	18.25	0.06	197.76	0.83	16.81	7.21	196.69	-6.20	178.99
665	Apr	5	Apr	8	16:42:03	4107	18.34	0.06	197.73	0.16	16.90	7.24	196.41	-6.81	179.35
684	Apr	5	Apr	8	15:47:00	3934	18.69	0.06	197.99	-0.51	17.23	7.38	196.40	-7.53	179.16
692	Apr	6	Apr	9	23:36:07	3862	20.05	0.06	199.58	-1.63	18.49	7.89	197.45	-9.17	178.36
703	Apr	6	Apr	10	16:14:43	3764	19.11	0.06	198.39	-1.18	17.61	7.53	196.50	-8.30	178.66
711	Apr	7	Apr	11	23:43:02	3693	20.44	0.06	199.91	-0.98	18.86	8.04	198.01	-8.69	178.94
730	Apr	7	Apr	11	23:26:39	3525	20.82	0.06	200.29	-0.30	19.21	8.18	198.63	-8.21	179.42
749	Apr	7	Apr	11	17:26:53	3361	20.97	0.06	200.46	0.39	19.35	8.24	199.05	-7.64	179.33
768	Apr	7	Apr	11	3:26:13	3200	20.79	0.06	200.28	1.02	19.19	8.17	199.11	-6.98	178.81
776	Apr	8	Apr	12	21:26:01	3133	22.55	0.06	202.23	0.88	20.83	8.83	200.89	-7.84	179.00
814	Apr	8	Apr	12	13:52:16	2824	22.06	0.06	201.65	-0.35	20.37	8.65	199.88	-8.76	179.50
833	Apr	8	Apr	12	5:26:40	2675	22.11	0.06	201.61	-1.01	20.42	8.66	199.59	-9.36	178.93
907	Apr	1	Apr	6	11:10:21	2136	15.63	0.04	195.34	0.36	14.38	6.19	194.27	-5.71	179.51
926	Apr	1	Apr	6	4:06:53	2009	15.74	0.04	195.43	-0.32	14.48	6.23	194.09	-6.37	179.59
953	Apr	2	Apr	7	7:47:11	1837	17.32	0.04	197.18	-1.14	15.95	6.84	195.40	-7.80	178.89
972	Apr	1	Apr	6	15:26:56	1721	17.04	0.04	196.88	-0.53	15.70	6.73	195.35	-7.12	179.48
991	Apr	2	Apr	7	0:25:28	1611	16.83	0.04	196.60	0.09	15.49	6.65	195.33	-6.44	179.74
1010	Apr	1	Apr	7	17:04:25	1506	16.92	0.04	196.61	0.76	15.58	6.68	195.60	-5.83	179.15
1018	Apr	3	Apr	9	4:50:06	1464	18.43	0.04	198.34	1.21	16.98	7.26	197.38	-6.07	178.75
1029	Apr	1	Apr	7	16:09:21	1407	17.27	0.04	196.90	1.42	15.91	6.82	196.12	-5.33	178.50
1037	Apr	2	Apr	8	23:58:30	1366	18.62	0.04	198.47	0.54	17.16	7.33	197.24	-6.73	179.40
1056	Apr	3	Apr	9	0:05:48	1273	19.02	0.04	198.80	-0.13	17.53	7.49	197.30	-7.49	179.76
1075	Apr	3	Apr	9	23:48:09	1186	19.40	0.04	199.13	-0.81	17.89	7.63	197.35	-8.24	179.19
1083	Apr	5	Apr	11	7:59:31	1151	20.77	0.04	200.70	-1.32	19.17	8.15	198.62	-9.31	178.72
1094	Apr	3	Apr	9	17:48:02	1103	19.55	0.04	199.25	-1.47	18.03	7.68	197.20	-8.90	178.54
1102	Apr	5	Apr	12	6:24:14	1070	21.09	0.04	200.99	-0.65	19.47	8.27	199.14	-8.80	179.38
1121	Apr	4	Apr	11	21:48:29	994	21.14	0.04	201.00	0.02	19.51	8.28	199.41	-8.18	179.85
1140	Apr	4	Apr	11	5:59:42	924	20.89	0.04	200.70	0.63	19.28	8.19	199.36	-7.50	179.30
1148	Apr	6	Apr	13	0:59:23	895	22.69	0.04	202.70	1.28	20.97	8.87	201.48	-7.64	178.67

Date					Time		Sun Ecliptic		Moon Ecliptic		Sun Equatorial		Moon Equatorial		
Julian Prior to 1582 CE			Gregorian		TDT	ΔT (s)	Long (deg)	Lat (deg)	Long (deg)	Lat (deg)	RA (deg)	DEC (deg)	RA (deg)	DEC (deg)	Elonga-tion (deg)
1159	Apr	4	Apr	11	14:16:56	857	20.64	0.04	200.40	1.24	19.05	8.09	199.31	-6.82	178.70
1167	Apr	6	Apr	13	8:12:22	830	22.40	0.04	202.37	0.68	20.70	8.76	200.95	-8.07	179.27
1186	Apr	5	Apr	12	18:20:26	770	22.23	0.04	202.13	0.06	20.54	8.69	200.49	-8.57	179.86
1205	Apr	5	Apr	12	12:28:31	714	22.38	0.04	202.21	-0.62	20.68	8.75	200.30	-9.22	179.40
1213	Apr	6	Apr	13	23:17:18	691	23.86	0.04	203.90	-1.51	22.07	9.30	201.56	-10.67	178.53
1224	Apr	5	Apr	12	12:15:00	662	22.76	0.04	202.53	-1.28	21.04	8.89	200.36	-9.96	178.74
1232	Apr	6	Apr	13	19:42:43	641	24.10	0.04	204.08	-0.84	22.30	9.39	201.98	-10.12	179.20
1251	Apr	7	Apr	14	20:11:51	593	24.51	0.04	204.43	-0.17	22.69	9.54	202.56	-9.63	179.85
1270	Apr	7	Apr	14	19:13:13	549	24.86	0.04	204.72	0.51	23.02	9.66	203.09	-9.10	179.44
1289	Apr	7	Apr	14	11:43:12	508	24.95	0.04	204.76	1.17	23.10	9.69	203.37	-8.50	178.78
1316	Apr	8	Apr	16	15:14:51	455	26.51	0.04	206.47	0.14	24.58	10.27	204.60	-10.08	179.82
1335	Apr	8	Apr	16	22:45:02	420	26.24	0.04	206.14	-0.47	24.32	10.16	204.07	-10.53	179.56
1354	Apr	8	Apr	16	8:02:44	388	26.03	0.04	205.88	-1.09	24.12	10.09	203.59	-11.02	178.93
1447	Apr	1	Apr	10	6:47:26	255	19.63	0.02	199.56	-0.39	18.11	7.70	197.90	-8.02	179.62
1474	Apr	2	Apr	11	12:41:25	223	21.31	0.02	201.38	-1.03	19.68	8.33	199.36	-9.29	178.99
1493	Apr	2	Apr	11	1:28:34	203	21.24	0.02	201.25	-0.38	19.62	8.31	199.49	-8.65	179.64
1512	Apr	1	Apr	11	8:40:58	183	20.95	0.02	200.90	0.23	19.35	8.20	199.40	-7.95	179.75
1531	Apr	1	Apr	11	18:47:57	166	20.77	0.02	200.66	0.86	19.18	8.13	199.41	-7.28	179.12
1539	Apr	3	Apr	13	11:13:11	159	22.48	0.02	202.56	1.12	20.78	8.77	201.29	-7.74	178.85
1550	Apr	1	Apr	11	12:54:38	150	20.92	0.02	200.75	1.51	19.32	8.19	199.74	-6.70	178.46
1558	Apr	2	Apr	12	23:45:10	144	22.40	0.02	202.43	0.49	20.71	8.74	200.93	-8.28	179.49
1577	Apr	2	Apr	12	20:08:13	131	22.65	0.02	202.61	-0.19	20.94	8.83	200.84	-8.97	179.82
1586	Apr	3	Apr	3	19:49:08	125	13.65	0.01	193.74	-1.17	12.56	5.40	192.19	-6.50	178.84
1605	Apr	3	Apr	3	20:27:02	113	14.07	0.01	194.10	-0.50	12.94	5.56	192.78	-6.03	179.51
1624	Apr	3	Apr	3	18:12:08	88	14.37	0.01	194.33	0.17	13.22	5.67	193.26	-5.49	179.81
1643	Apr	4	Apr	4	8:33:34	59	14.36	0.01	194.27	0.83	13.22	5.67	193.46	-4.86	179.15
1651	Apr	5	Apr	5	23:45:28	47	16.02	0.01	196.13	1.18	14.76	6.32	195.32	-5.26	178.81
1670	Apr	5	Apr	5	11:31:33	23	15.92	0.01	195.97	0.54	14.66	6.27	194.92	-5.78	179.44
1689	Apr	4	Apr	4	18:36:37	9	15.62	0.01	195.62	0.07	14.38	6.16	194.36	-6.21	179.95
1708	Apr	5	Apr	5	5:37:46	9	15.48	0.01	195.40	-0.70	14.25	6.10	193.91	-6.71	179.31
1716	Apr	6	Apr	6	21:22:33	10	17.17	0.01	197.29	-1.34	15.82	6.75	195.42	-8.03	178.66
1727	Apr	6	Apr	6	0:41:54	10	15.67	0.01	195.53	-1.37	14.43	6.18	193.77	-7.37	178.64
1735	Apr	7	Apr	7	10:56:51	11	17.13	0.01	197.20	-0.70	15.79	6.74	195.58	-7.40	179.30

Date					Time		Sun Ecliptic		Moon Ecliptic		Sun Equatorial		Moon Equatorial		
Julian Prior to 1582 CE			Gregorian		TDT	ΔT (s)	Long (deg)	Lat (deg)	Long (deg)	Lat (deg)	RA (deg)	DEC (deg)	RA (deg)	DEC (deg)	Elonga-tion (deg)
1754	Apr	7	Apr	7	8:08:01	13	17.41	0.01	197.40	0.02	16.04	6.84	196.03	-6.85	179.99
1773	Apr	7	Apr	7	8:43:30	16	17.82	0.01	197.75	0.65	16.43	7.00	196.62	-6.37	179.34
1781	Apr	8	Apr	8	15:55:06	17	19.15	0.01	199.28	1.37	17.67	7.51	198.32	-6.28	178.61
1792	Apr	7	Apr	7	7:12:01	16	18.15	0.01	198.02	1.32	16.74	7.12	197.12	-5.85	178.67
1912	Apr	1	Apr	1	22:14:16	13	11.81	0.00	191.90	0.88	10.86	4.67	191.29	-3.90	179.11
1931	Apr	2	Apr	2	20:07:55	24	12.12	0.00	192.14	0.20	11.15	4.79	191.24	-4.61	179.79
1950	Apr	2	Apr	2	20:44:34	29	12.54	0.00	192.49	-0.47	11.53	4.96	191.30	-5.36	179.53
1958	Apr	4	Apr	4	4:00:15	32	13.89	0.00	194.03	-1.57	12.78	5.48	192.30	-6.98	178.43
1969	Apr	2	Apr	2	18:33:06	39	12.84	0.00	192.72	-1.14	11.81	5.07	191.26	-6.07	178.86
1977	Apr	4	Apr	4	4:19:04	48	14.29	0.00	194.36	-0.92	13.15	5.63	192.86	-6.51	179.08
1996	Apr	4	Apr	4	0:10:47	62	-14.51	0.00	194.53	-0.24	13.36	5.72	193.28	-5.95	179.76
2015	Apr	4	Apr	4	12:01:24	69	14.40	0.00	194.35	0.40	13.26	5.68	193.36	-5.29	179.59
2034	Apr	3	Apr	3	19:06:59	80	14.10	0.00	193.99	1.00	12.98	5.56	193.26	-4.59	178.99
2042	Apr	5	Apr	5	14:30:11	86	-15.94	0.00	196.02	1.00	14.68	6.27	195.14	-5.38	179.00
2061	Apr	4	Apr	4	21:54:05	116	15.65	0.00	195.67	0.39	14.41	6.16	194.58	-5.80	179.61
2080	Apr	4	Apr	4	11:23:38	157	15.61	0.00	195.56	-0.26	14.38	6.14	194.23	-6.37	179.73
2099	Apr	5	Apr	5	8:30:56	201	15.88	0.00	195.76	-0.93	14.63	6.25	194.16	-7.06	179.06
2107	Apr	7	Apr	7	17:30:11	220	17.30	0.00	197.37	-1.11	15.95	6.79	195.59	-7.85	178.88
2126	Apr	7	Apr	7	16:17:55	266	17.64	-0.01	197.65	-0.44	16.27	6.92	196.10	-7.33	179.55
2145	Apr	7	Apr	7	16:48:09	315	18.05	-0.01	197.98	0.23	16.65	7.08	196.67	-6.84	179.76
2164	Apr	7	Apr	7	13:34:34	359	18.31	-0.01	198.17	0.91	16.89	7.17	197.11	-6.29	179.09
2219	Apr	1	Apr	1	0:18:15	489	10.54	-0.01	190.59	-1.19	9.69	4.17	189.27	-5.29	178.80
2284	Apr	1	Apr	1	5:01:09	668	11.96	-0.01	191.98	1.32	11.00	4.72	191.53	-3.52	178.69
2303	Apr	3	Apr	3	4:19:40	726	12.33	-0.01	192.27	0.66	11.34	4.86	191.54	-4.24	179.35
2322	Apr	3	Apr	3	4:39:45	785	12.73	-0.01	192.60	0.01	11.72	5.02	191.59	-4.99	179.87
2341	Apr	3	Apr	3	0:36:59	848	12.96	-0.01	192.77	-0.69	11.92	5.11	191.48	-5.68	179.28
2349	Apr	4	Apr	4	11:49:55	874	14.47	-0.01	194.47	-1.34	13.33	5.69	192.79	-6.94	178.65
2360	Apr	2	Apr	2	12:35:50	912	12.85	-0.01	192.61	-1.32	11.83	5.07	191.08	-6.19	178.65
2368	Apr	4	Apr	4	5:25:31	940	14.60	-0.01	194.53	-0.68	13.45	5.74	193.12	-6.35	179.31
2387	Apr	4	Apr	4	15:07:51	1007	14.40	-0.01	-194.28	0.05	13.26	5.67	193.13	-5.68	179.87
2406	Apr	3	Apr	3	22:29:30	1077	14.11	-0.01	193.91	0.56	12.99	5.55	193.02	-4.97	179.42
2425	Apr	3	Apr	3	11:50:58	1149	14.06	-0.01	193.78	1.20	12.95	5.53	193.15	-4.33	178.78

2433	Apr	5	Apr	5	1:56:44	1181	15.69	-0.01	195.62	0.86	14.46	6.16	194.72	-5.35	179.15
2452	Apr	4	Apr	4	17:54:43	1256	15.75	-0.02	195.57	0.20	14.51	6.18	194.42	-5.94	179.74
2471	Apr	5	Apr	5	16:38:08	1334	16.09	-0.02	195.82	-0.47	14.83	6.31	194.39	-6.66	179.45
2479	Apr	7	Apr	7	0:42:37	1367	17.47	-0.02	197.41	-1.53	16.12	6.84	195.46	-8.25	178.45
2490	Apr	5	Apr	5	17:09:28	1414	16.50	-0.02	196.16	-1.13	15.21	6.47	194.45	-7.40	178.80
2498	Apr	7	Apr	7	0:35:16	1448	17.86	-0.02	197.71	-0.88	16.48	6.99	195.99	-7.76	179.09
2517	Apr	8	Apr	8	0:32:52	1532	18.25	-0.02	198.03	-0.21	16.84	7.14	196.55	-7.26	179.69
2536	Apr	7	Apr	7	19:13:18	1617	18.42	-0.02	198.15	0.47	17.00	7.20	196.92	-6.68	179.47
2555	Apr	8	Apr	8	5:53:58	1705	18.26	-0.02	197.94	1.10	16.85	7.14	196.97	-6.02	178.88
2667	Apr	1	Apr	1	4:59:45	2270	11.15	-0.02	190.69	0.95	10.25	4.39	190.20	-3.35	178.96
2675	Apr	2	Apr	2	12:37:33	2314	12.51	-0.02	192.24	1.13	11.52	4.93	191.69	-3.80	178.86
2694	Apr	2	Apr	2	12:12:44	2419	12.89	-0.02	192.56	0.46	11.86	5.07	191.73	-4.54	179.45
2713	Apr	3	Apr	3	5:57:08	2526	13.02	-0.02	192.65	-0.21	11.99	5.12	191.56	-5.20	179.56
2732	Apr	2	Apr	2	15:47:26	2635	12.83	-0.02	192.42	-0.84	11.81	5.05	191.09	-5.68	179.05
2740	Apr	4	Apr	4	9:58:17	2682	14.64	-0.02	194.40	-1.13	13.48	5.75	192.82	-6.71	178.83
2759	Apr	4	Apr	4	18:03:23	2795	14.37	-0.03	194.07	-0.52	13.24	5.64	192.75	-6.03	179.38
2778	Apr	4	Apr	4	2:31:48	2910	14.12	-0.03	193.73	0.09	13.00	5.54	192.67	-5.33	179.60
2797	Apr	3	Apr	3	18:20:53	3027	14.17	-0.03	193.64	0.75	13.05	5.56	192.85	-4.69	179.10
2805	Apr	5	Apr	5	6:45:50	3077	15.73	-0.03	195.44	1.34	14.50	6.17	194.73	-4.84	178.66
2816	Apr	3	Apr	3	16:58:39	3146	14.50	-0.03	193.86	1.41	13.36	5.69	193.30	-4.17	178.47
2824	Apr	5	Apr	5	1:07:10	3198	15.89	-0.03	195.46	0.68	14.65	6.23	194.50	-5.45	179.21
2843	Apr	6	Apr	6	0:55:44	3320	16.28	-0.03	195.73	0.02	15.01	6.37	194.49	-6.17	179.45
2862	Apr	6	Apr	6	0:56:58	3446	16.67	-0.03	196.07	-0.66	15.37	6.52	194.55	-6.93	179.09
2870	Apr	7	Apr	7	8:55:29	3499	18.06	-0.04	197.61	-1.32	16.67	7.05	195.73	-8.14	178.57
2881	Apr	5	Apr	5	19:48:04	3573	16.85	-0.03	196.23	-1.32	15.54	6.59	194.44	-7.60	178.51
2889	Apr	7	Apr	7	7:51:21	3627	18.40	-0.04	197.92	-0.66	16.99	7.18	196.28	-7.64	179.16
2908	Apr	8	Apr	8	0:13:40	3758	18.48	-0.04	197.98	0.01	17.06	7.21	196.59	-7.04	179.50
2927	Apr	8	Apr	8	9:00:57	3891	18.24	-0.04	197.69	0.63	16.84	7.12	196.56	-6.36	179.19
2946	Apr	7	Apr	7	16:52:13	4026	17.96	-0.04	197.32	1.23	16.58	7.01	196.44	-5.66	178.65
2973	Apr	8	Apr	8	20:41:47	4223	19.56	-0.04	199.01	0.23	18.07	7.62	197.63	-7.23	179.41
2992	Apr	8	Apr	8	13:56:10	4364	19.68	-0.04	198.96	-0.43	18.18	7.66	197.34	-7.82	179.15

Chapter 12 Conclusion

And there are those who talk and without knowledge or forethought,
Reveal a truth which they themselves do not understand.
And there are those who have the truth within themselves,
but they tell it not in words.

Khalil Gibran

Religious observances and social festivals depend on the lunar calendar. While the lunar cycle has long been characterized, as early as by Babylonian astronomers and Egyptians priests and later independently by the Mayans, the timing of when and where the new moon is first visible remained empirical and criteria-based until these days and age. The lack of attention to this subject has lingered for ages despite the pressing need for planning religious observances and festivals in all religions and cultures around the world.

Because of our human social nature, we seek togetherness in the celebration, which makes the topic of paramount importance. In the early human experience with religion, communication was slow, distances were substantial, and therefore, precise identification of the new moon was not required. Instant communication brought time and space to a universal realm. This, in turn, has brought a new challenge to the old tradition.

12.1 Celebration in the era of connectivity

Given today's connectivity, celebrations and observances are in need to be synchronized across time zones and geopolitical boundaries. Christianity adopted a rule-based approach to Easter celebration to avoid confusion. Muslims did not dear to change the divine dates. They elected to stick to the exact word from the scriptures. But lately, scholars, mainly from Muslim countries in Southeast Asia and trained in the west, were first to address the matter, realizing that discrepancies are untenable, they tried hard to bring the topic to the forefront of the Muslim world, only to fall on deaf ears. These scholars realized we have finally arrived at a time in which advancements in science and technology are leveraged to address the old problem of new moonlight visibility which plagued lunar calendars for ages. We are referring to the first light from the moon, which has long been the time-origin of the lunar month and not what is presented in the lunar ephemeris Tables. The time and location of a new moonlight first visibility proved to be hard to predict due to its monthly variation and has eluded astronomers of all cultures over the ages. The difficulties prevented the development of a useful photometric lunar calendar and brought the conjunction-based calendar instead, known as the astronomical lunar calendar. Even though Mesopotamian astronomer-priests were able to determine the lunar cycle with a remarkable degree of precision and have even established criteria for first visibility, they could not determine when the first day of the month might be -a task requiring global geographical consideration.

The characterization of the beginning of the lunar month through first light was dropped by modern astronomers. They choose to redefine the origin of the lunar calendar as the solar conjunction in the ecliptic plane, a time in which there is no light reflected off the moon. As a result, they discarded religious and social traditions. Instead of delivering a practical calendar that could be used by the public at large for day-to-day social, cultural and religious activities, they provided the Astronomical Lunar Calendar or Lunar Ephemeris. This calendar gives the four phases of the moon, an example of which is shown in Table 11.1-1. Today, one could see a designation of the phase of the moon in computer-generated solar calendars. But such information is not relied upon by civil societies, except perhaps for the few who are interested in fishing, hunting or tracking the lunar tide.

It is important to reflect on what hampered the development of a photometric lunar calendar, based on light visibility in the first place. For a start, the first dim light from the new moon is observed following sunset in a westerly horizon as it makes its hundreds of miles journey through a heated and turbulent section of the atmosphere. Accumulation of moisture, aerosols, and clouds prevent sky watchers from regular monthly sightings. Furthermore, common seasonal variations and overcast conditions prevent the collection of complete records, and even in dry regions, sandy and hazy conditions are common. Next, new moonlight observation is an event that is not repeatable from one cycle to another because the relative position of the moon-sun pair varies. As a result, the first light must be observed from different geographic latitudes and longitudes over several months and seasons. Even today, to carry a thorough technical investigation, observations need to take place from varying vantage points, including the middle of the oceans, and higher latitudes over many years. Not only one must hope for excellent weather conditions and the availability of a suitable site which is open to the west and characterized by a thin and dry atmosphere, but observations also need to be carried during a small of a few critical minutes following sunset. Finally, observations need to be over the course of several years to ensure the development of statistically relevant data.

Obviously, an investigation which meets such constraints would be a massive and abhorrently costly undertaken in these days and ages, let alone performing such work even in the recent past.

12.2 The proposal of The Witness organization

Despite all the above difficulties, we started by proposing a unique network of mainly portable observatories, distributed around the world and connected to a control center. The system should not only put an end to erroneous claims of sightings, such as those occurring frequently in the Muslim world but should also give an added capability to further bridge the gap between people of different faith and culture. Once implemented, the system is also expected to overcome sighting difficulties due to weather and other atmospheric adversities. The system, coined The *Witness,* provides a unique catalyst for the development of many key technology sectors. It is expected to gradually reinforce the culture of communication, science, and knowledge. *The Witness* was designed to have an immediate impact on scientific advancement in areas such as environmental monitoring, remote health care, and education. The system could jump-start a new scientific culture in optical engineering, atmospheric science, atmospheric physics, computer engineering, and telecommunication. *The Witness* system is the first project conceived by *The Witness* Organization, a non-profit, non-governmental entity based in Boulder, Colorado, USA. The top-level definitions and specification for the project have been drawn within by educators, lawyers, engineers, and scientists from about 20 countries. Members of the team would not wait for such a system to be in place and did not shy away from "chasing the new moonlight," gathering data from all over the world. They took measurements from the Death Valley in the US, the remote islands of the Pacific Ocean, the jungles of Africa, the high mountains of Chile, to name a few. They traveled to wherever the new moonlight is expected to manifest itself, even if there is a longshot at capturing the event. Over the course of this experience, which lasted about a decade, these observers faced several natural and other man-made hurdles. Some of the anecdotes included being chased by animals, refusal entry to countries following lengthy travel and confiscation of equipment at a border crossing, to name a few. On the bright side, the team had plenty of good luck. Hardware and software tools became accessible to support their work. They also had immediate communication, from satellite phone and Internet connectivity. Most importantly, they claim to have developed wonderful lifetime friendships.

When it became clear, that the first new moonlight is seen at a unique location on the globe and at a unique Universal Time, an immediate focus was placed on the physics of the problem and the law governing first sighting during twilight in the visible spectrum. Ultimately, first visibility based on the characteristics of scattering sunlight and the emergence of radiation from the crescent moon through the layers of the atmosphere was analytically established. The development is therefore independent of astronomical calculations. The principles of first visibility address twilight photometry when the brightness of the background sky decreases due to the reduction of scattered sunlight caused by the increase of earth shadow height. At the same time, a gradual reduction in the radiation intensity from the crescent moon takes place due to the increasing air mass traveled by its photons as it sets in the western horizon. The development evaluates the contrast between the brightness of the radiation from the new moon and the brightness of the background sky, as perceived by the standard human eye benchmark. To verify the physics of the problem, special equipment was built to monitor the brightness of the twilight sky. Measurements correlated well with theoretical work which predicted that the sighting of a new crescent moon is impossible if the contrast between its brightness against that of a background sky is immediately in decline following sunset. It was also found that when the moon elevation is higher than a certain level, the shadow of the earth encompasses the high-density layer of the atmosphere preventing scattering of sunlight, which represents the main obstacle to a sighting in a clear sky. These results were established for a standard atmosphere devoid of pollutants, humidity, and aerosols, making reliance on optical tools ultimately not necessary. The properties of a standard human eye were appended to the development to derive the relative sun-moon position at which first visibility occurs. It was found that this relative position is, in fact, a scattering angle limit. Finally, using a linearized form of the relative motions of the earth-sun-moon system from astronomy principles, we could determine the unique location and time of first new moon sighting.

In summary, the contrast in brightness between the new crescent and the background sky is adopted as criteria for first visibility and as an alternative to direct sighting requirements to initiate traditions and cultural events. This criterion is obviously useful when first visibility is hampered by difficult geography, air moisture, and aerosols, in an analogous way to the adoption of clock time as an alternative to monitoring the sun for certain social and religious events such as prayers, breaking fast, celebrating Easter, etc. We emphasize that this optical in nature and not based on astronomical computation. The astronomical calculation provides the time and location of first visibility.

The determination of the beginning of the lunar month is a major cultural and historical matter which requires long-overdue attention by entrusted officials. Even in these days and age, entire populations are unnecessarily invited to go out to monitor the new moon to start festivals and holidays. Some cultures even use the terminology of "Night of doubt," to refer to the night that follows conjunction. a symbol of lack of knowledge in the matter of lunar month chronology. People want to know when the lunar month starts so that they plan and celebrate collectively across artificial man-made boundaries. Future planning and foresight represent one of the most important traits of our collective human consciousness that defines and promotes our kind among the rest of earth's co-inhabitants.

12.3 Final word

We started this work by looking at celebrations and observances in the Muslim faith and ended up, not only addressing celebrations in other religions and cultures but also stumbled on what could be cross-faith celebrations. Could the holiest Islamic evening of the year be Christmas Eve? And, could a photometric lunar calendar unify celebrations in Christianity, Islam, and Judaism? Is the non-committal of Islam to a specific date for the Night of Destiny has something to do with the Messiah and Christianity?

We learned that a calendar needs consensus. The ultimate criteria for a photometric new lunar month must be the one we all agree upon. We depend on consensus to make the new reference for the beginning of the lunar month happen. Can we get together and adopt one? One would think it is an easy task. After all, it is the same moon and the same photometry. From month to month, the crescent moonlight starts shining over a different location. It does not make a choice for a country or a nation. It is an easy task if we simply think that this matter is just technical.

We hope that this contribution helps unify and synchronize religious and social traditions across the globe and promote peace and harmony.

This is the beginning of a movement to develop a photometric lunar calendar. If the reader is willing to contribute through his/her expertise in any of the fields covered by this topic, including religion, history, astronomy, photometry, communication, sociology, management, mathematics, physics, hands-on field detection, and marketing, please go the website and join Al-Shaahed -The Witness Organization at www.AlShaahed.Org

Bibliography

Al-Biruni, M. A. (1879). *The Chronology of Ancient Nations -An English Version of the Arabic Text of Al-Athar-Al-Bakiya of Al-Biruni by C. Edward Sachau.* London: William H Allen and CO.

Abell, G. (1973). *Exploration of the Universe -Updated Brief Edition.* Holt Rinehart and Winston.

Afzal, O. (1988). Lunar Islamic Calendar. *Proceedings of the IIIT Lunar Calendar Conference*, 1-23.

Ahmad, I.-A.-D. (1998). Afterword. *Proceedings of the IIIT Lunar Calendar Conference, 2nd. Edition* (pp. 93-96). Maryland: International Graphics.

Ahmed, M. (1999). Moon Calculator (Program documentation). In D. Ratledge, *Software and Data For Practical Astronomers.* Springer. Retrieved from http://extras.springer.com/1999/978-1-4471-0555-8/chap4

Al-`Alwani, T. J. (1988). Introductory Remarks. *Poceedings of the IIIT Lunar Calendar Conference*, xiii-xxvii.

Allen, C. W. (1976). *Astrophysical Quantities.* London: The Athlone Press Ltd.

Alpern, M. (1971). Rhodopsin Kinetics in the Human Eye. *J Physiol.* , 217(2): 447–471.

Ambraseys, N. C. (1994). *The Seismicity of Egypt, Arabia and the Red Sea: A Historical Review.* Cambrige: Cambrige University Press.

Aspenak, F., & Meeus, J. (2009). *Five Millennium Canon of Lunar Eclipses: –1999 to +3000 (2000 BCE to 3000 CE).* Greenbelt, Maryland: NASA Goddard Report, NASA/TP–2009–214172.

Beaumont, P. B. (1973). Border Cave – A Progress Report. S. Afr. *J. Science*, 41-46.

Blackwell, H. R. (1946). Contrast Thresholds of the Human Eye. *Journal of the Optical Society of America*, 624-643.

Bruin, F. (1977). The First visibility of the Lunar Crescent. *Vistas in Astronomy*, 21, 331-358.

Caldwell, A. J. (2001). *First Visibility of the Lunar Crescent.* African Skies, Cieux Africans, No.5 January 2001.

Clagett, M. (1976). *Greek Science in Antiquity: How human reason and ingenuity first ordered and mastered the experience of natural phenomena.* New York: McMillan Co. Inc.

Clagett, M. (1994). *Greek Science in Antiquity.* New York: Barnes and Noble, Inc.

Clark, R. N. (1990). *Visual Astronomy of the Deep Sky.* London: Cambridge University Press & Sky Publishing Corporation.

Cotterell, A. (1980). *The Encyclopedia of ANCIENT CIVILIZATIONS.* New York: The Rainbird Publishing Group Limited.

Crowley, D., & Heyer, P. (1991). *Communication in History -Technology,Culture, Society.* NewYork: Longman Publishing Group.

D'Erricoa, F., Backwel, L., & Villa, P. (2012). Early Evidence of San Material Culture Represented by Organic Artifacts from Border Cave, South Africa. *Proceedings of the National Academi of Sciences of the United States of America*, vol. 109 no. 33.

Danjon, A. (1936). *L'Astronomie, vol. 50, pp.57-65*, 57-65.

De Gaury, G. (1991). *The Rulers of Mecca.* New York: Dorset Press.

Dick, S. J. (1999). A History of the American Nautical Almanac Office. In A. D. Dick (Ed.), *Proceedings of the Nautical Almanac Office Sesquicentennial Symposium, U.S. Naval Observatory, March 3-4* (pp. 11-54). Washington, D.C.: U.S. Naval Observatory.

Doggett L.E. and Shaefer, B. (1994). Lunar Crescent Visibility. *Icarus*, 107, issue 2, 388-403.

Doggett, L. E., Seidelmann, P. K., & Schaefer, B. E. (1988). Moonwatch -July 14. *Sky &Telecope*, 76, 34-35.

Duffet-Smith, P. (1996). *Practical Astronomy with your Calculator* (Third ed.). Cambridge: Cambridge University press.

Duncan, D. E. (1999). *The Calendar, The 5000-Year Struggle to Align the Clock and the Heavens -and What Happened to the Missing 10 Days.* London: Fourth Estate Limited.

Engdah, E. R., & Villaseño, A. (2002). Global Seismicity: 1900–1999. In W. H. Lee, H. Kanamori, P. C. Jennings, & C. Kisslinger, *International Handbook of Earthquake and Engineering Seismology, Part A, Chapter 41* (pp. 665–690). Academic Press.

Fatoohi, L. J.-D. (1999). The Babylonian first visibility of the lunar crescent: Data and criterion. *Journal for the history of astronomy* , xxx, 51–72, p. 60.

Fotheringham, J. (1910). On the Smallest Visible Phase of the Moon. *Monthly Notices of the Royal Astronomical Society*, Vol. 70, p.527-532.

Garraty, J. A., & Gay, P. (1972). *The Columbia History of the World.* NewYok: Harper & Row Publishers.

Gautschy, R. (2014). On the Babylonian Sighting-Criterion for the Lunar Crescent and its Implications for Egyptian Lunar Data. *Journal for the History of Astronomy*, Vol. 45, 1, 79-90.

GREEN, R. M. (1985). *Spherical Astronomy.* Cambridge: Cambridge University Press.

Hadingham, E. (1984). *Early Man and the Cosmos.* Oklahoma: University of Oklahoma Press.

Hasanzadeh, A. (2012). Study of Danjon limit in moon crescent sighting. *Astrophys Space Sci* , 339:211–221 .

Hawkins, G. S., & White, J. B. (1965). *Stonehenge Decoded.* New York: Dell Publishing CO., Inc.

Hedin, A. E. (1991). Extension of the MSIS Thermospheric Model into the Middle and Lower Atmosphere. *J. Geophys. Res. 96* , 1159.

Hubert, A. A. (1998). *The Petroglyph Calendar- An Archeoastronomy Adventure.* Albuquerque: Hubert Allen and Associates, NM 87122.

Ibn-Khaldun. (c. 1350). *Kitāb al-'Ibar wa-Dīwān al-Mubtada' wa-l-Khabar fī Ta'rīkh al-'Arab wa-l-Barbar wa-Man 'Āṣarahum min Dhawī ash-Sha'n al-Akbār (Book of Lessons, Record of Beginnings and Events in the History of the Arabs and the Berbers and Their Powerful Contemporaries").* Retrieved from https://openlibrary.org/works/OL1597273W/Kitāb_al-'ibar

Ilyas, M. (1993). Lunar Crescent Visibility Criterion and Islamic Calendar. *Royal Astronomical Society*, 425-461.

Ilyas, M. (1994). Lunar Crescent Visibility Criterion and Islamic Calendar. *Quarterly Journal of the Royal Astronomical Society*, Vol. 35, NO. 4/DEC, P. 425-461.

Ioh, H. (2014). The Calendar in Pre-Islamic Mecca. *Arabica*, 471-513.

Johnson, R. A. (1985). *Matrix Analysis.* Cambrige University Press.

Karttunen, H., Kröger, P., Oja, H. P., & Donner, K. (1984). *Fundamental Astronomy.* Berlin: Springer Verlag.

Khan, A. R. (1988). A Review of the Principles of the Islamic Calendar. *Poceedings of the IIIT Lunar Calendar Conference* (pp. 24-38). Maryland: International Graphics.

Kieffe, H., & Stone, T. (2005). The Spectral Irradiance of the Moon. *The Astronomical Journal*, 129:2887-2901.

King, D. (1987). Some Early Islamic Tables for Determining Lunar Crescent Visibility. *From Deferent to Equant: A Volume of Studies in the History of Science in the Ancient and Medival Near East in Honor of E.S. Kennedy, King, D.A. and Saliba G. Annals of the NewYork Academy of Sciences, 500, New York.*

Krisciunas, K., & Schaefer, B. (1991). A Model of Brightness of Moonlight. *Astronomical Society of the Pacific*.

Labitzke, K., Barnett, J., & Edwards, B. (1985). *Handbook MAP 16, SCOSTEP.* University of Illinois.

Lane, A. P., & Irvine, W. M. (1973). Monochromatic phase curves and albedos for the lunar disk . *The astronomical journal*, Vol. 78, No.3.

Last, K. S., Hobbs, L., Berge, J., Brierley, A. S., & Cottier, F. (2016). Moonlight Drives Ocean-Scale Mass Vertical Migration of Zooplankton during the Arctic Winter. *Current Biology, 26*, Volume 26, Issue 2, Pages 244-251.

Lindberg, D. C. (1992). *The Beginnings of Wetern Science: The European Scientific Tradition in Philosophical, Religious and Institutional Context, 600 B.C. to A.D. 1450.* Chicago: The University of Chicago Press.

Lomgley, P. A., Goodchild, M. F., Maguire, D. W., & Rhind, D. W. (2001). *Geographic Information Systems and Science.* New York: John Wiley & Sons, LTD.

Marshack, A. (1991). *The Roots of Civilization: The Cognitive Beginnings of Man's First Art, Symbol and Notation.* NewYork: Moyer Bell Ltd.

Mason, S. F. (1979). *A History of the Sciences.* New York: Macmillan Publishing Co. Inc.

Mauder, E. (1911). *Journal of the British Astronomical Association*, Vol 21, 355.

Maunder, M. (1911). *Journal of the Britsh Astrnomical Association*, 21, 355-362.

McCartney, E. J. (1976). *Optics of the Atmosphere Scatering by Molecules and Particles.* New York: John Wiley & Sons Inc.

McPartlan, M. A. (1997). *The contribution of Qur'an and Hadith to Early Islamic Chronology, Ph.D. Thesis.* England: Depatment of Philosophy, University of Durham.

Middleton, W. (1958). *Vision Through The Atmosphere.* Toronto: University of Toronto Press.

Mommsen, A. (1883). *Chronologie: Untersuchungen über das Kalenderwesen der Griechen, Insonderheit der Athener.* Springer.

Monstein, C. (2001). The Moon's temperature at l=2.77 cm. *Research Collection, Radio Astronomy and Plasma Physics Group, ETH Zurich*.

Morrison, L., & Stephenson, F. R. (2004). Historical Values of the Earth's Clock Error ΔT and the Calculation of Eclipses. *J. Hist. Aston., Vol. 35 Part 3, 120*, 327-336.

Much, T. A. (1970). *Geology of the Moon: A Stratigraphic View.* Princeton University Press.

Neugebauer, O. (1955). Astronomical cuneiform texts: Babylonian ephemerides of the Seleucid period for the motion of the Sun, Moon, and the planets. i (London, 1955), 43.

Neugebauer, P. V. (1929). Astronomische Chronologie, ii, Table E 21. *Berlin and Leipzig.*

Odeh, M. (2004). New Criterion for Lunar Crescent Visibility. *Experimental Astronomy*.

Olmstead, A. (1966). *History of the Persian Empire.* Chicago: The University of Chicago Press.

Patat, F., Ugolnikov, O. S., & Postyliakov, O. V. (2006). UBVRI twilight at ESO-Paranal. *Astronomy & Astrophysics*, manuscript no. MS4992.

Penndorf, R. (1957). Tables of Refractive Index for Standard Air and the Rayleigh Coefficient.... *J. Opt. Soc., Am, 47*, 176-182.

Pilar Montanes-Rodriguez, E. P. (2007). Measurements of The Surface Brightness of the Earthshine with Applications to calibrate Lunar Flashes. *The Astronomical Journal*, 134:1145-1140.

Preston, D. (2009). *Cleopatra and Antony: Power, Love, and Politics in the Ancient World* . New York: Walker Publishing Company.

Rice, P. M. (2007). *Maya Calendar Origins: Monuments, Mythistory, and the Materialization of Time* . Austin, TX : University of Texas Press.

Ronan, C. A. (1982). *Science: Its History and development among the World's cultures.* New York: The Hamlyn Publishing Group Limited.

Rozenberg, G. V. (1966). *Twilight, a Study in Atmospheric Optics.* NewYork: Plenum Press.

Sasselov, D. (2012). *The Life of Super-Earths.* New York: Basic Books, Perseus Books Group.

Schaefer, B. (1988). Visibility of the Lunar Crescent. *Q. Jour. R.A.S. 29.*, 511-523.

Schaefer, B. (1996). Lunar Crescent Visibility. *Q.J.R. Astron. Soc.*, 37, 759-768.

Schaefer, B. A. (1993). Records for Young Moon Sightings. *QJRAS*, 34, 53-56.

Schafer, B. E. (1988). Calculating and Observing the Crescent Moon. *Proceedings of the IIIT Lunar Calendar Conference*, 79-88.

Schoch, K. (1928). Astronomical and calendrical tables, in S. Langdon and K. Fotheringham, The Venus Tablets of Ammizaduga: A solution of Babylonian chronology by means of the Venus observations of the First Dynasty. *Oxford.*

Seidelmann, B. E., Doggtt , L. E., & Kenneth, P. (1988). Calculating and Observing the Crescent Moon. *Proceedings of the IIIT Lunar Calendar Conference*, 66-71.

Seidelmann, P. K. (2012). *Explanatory Supplement to the Astronomical Almanac* . University Science Books, Mill Valley, California.

Stephenson, F. (1997). *Historical Eclipses and Earth's Rotation.* Cambridge: Cambridge Univ.Press.

Sultan, A. (2007). First Visibility of the Lunar Crescent: Beyond Danjon's Limit. *The Observatory*, 127, No 1196, 53-58.

Teeple, J. E. (1931). Maya Astronomy. *Contributions to American Archeology*, Vol 1, No. 104.

Thorpe, P. J. (1995). *Ancient Inventions.* NewYork: Ballantine Books, 1995.

Wertz, J. (1980). *Spacecraft Attitude Determination and Control.* D. Reidel Publishing Company.

White, J. (1970). *Ancient Egypt Its Culture and History.* New York: Dover Publications Inc.

Wilson, C. (2010). *The Hill-Brown Theory of the Moon's Motion -Its Coming-to-be and Short-lived Ascendancy.* NewYork: Springer.

Zombeck, M. V. (2006). *Handbook of Space Astronomy and Astrophysics.* London: Cambrige University Press.

Index

www.ingramcontent.com/pod-product-compliance
Lightning Source LLC
LaVergne TN
LVHW070933210926
848482LV00003B/231
* 9 7 8 1 7 3 3 1 6 1 4 0 4 *